Evolution's Clinical Guidebook

Evolution's Clinical Guidebook

Translating Ancient Genes into Precision Medicine

Jules J. Berman

ELSEVIER

ACADEMIC PRESS

An imprint of Elsevier

Academic Press is an imprint of Elsevier
125 London Wall, London EC2Y 5AS, United Kingdom
525 B Street, Suite 1650, San Diego, CA 92101, United States
50 Hampshire Street, 5th Floor, Cambridge, MA 02139, United States
The Boulevard, Langford Lane, Kidlington, Oxford OX5 1GB, United Kingdom

Notices
Knowledge and best practice in this field are constantly changing. As new research and experience
broaden our understanding, changes in research methods, professional practices, or medical
treatment may become necessary.

Practitioners and researchers must always rely on their own experience and knowledge in evaluating
and using any information, methods, compounds, or experiments described herein. In using such
information or methods they should be mindful of their own safety and the safety of others, including
parties for whom they have a professional responsibility.

To the fullest extent of the law, neither the Publisher nor the authors, contributors, or editors, assume
any liability for any injury and/or damage to persons or property as a matter of products liability,
negligence or otherwise, or from any use or operation of any methods, products, instructions, or ideas
contained in the material herein.

Library of Congress Cataloging-in-Publication Data
A catalog record for this book is available from the Library of Congress

British Library Cataloguing-in-Publication Data
A catalogue record for this book is available from the British Library

ISBN 978-0-12-817126-4

For information on all Academic Press publications
visit our website at https://www.elsevier.com/books-and-journals

Working together
to grow libraries in
developing countries

www.elsevier.com • www.bookaid.org

Publisher: Stacy Masucci
Acquisition Editor: Rafael Teixeira
Editorial Project Manager: Sandra Harron
Production Project Manager: Punithavathy Govindaradjane
Cover Designer: Christian Bilbow

Typeset by SPi Global, India

Other Books by Jules J. Berman

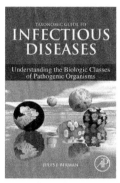

Taxonomic Guide to Infectious Diseases
Understanding the Biologic Classes of
Pathogenic Organisms (2012)
9780124158955

Principles of Big Data
Preparing, Sharing, and Analyzing
Complex Information (2013)
9780124045767

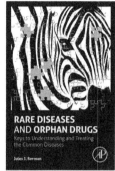

Rare Diseases and Orphan Drugs
Keys to Understanding and Treating
the Common Diseases (2014)
9780124199880

Repurposing Legacy Data
Innovative Case Studies (2015)
9780128028827

Data Simplification
Taming Information with Open
Source Tools (2016)
9780128037812

**Precision Medicine and The Reinvention of
Human Disease** (2018)
9780128143933

Principles and Practice of Big Data
Preparing, Sharing, and Analyzing
Complex Information, Second Edition (2018)
9780128156094

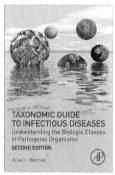

Taxonomic Guide to Infectious Diseases
Understanding the Biologic Classes of
Pathogenic Organisms, Second Edition (2019)
9780128175767

Contents

About the Author

Jules J. Berman received two baccalaureate degrees from MIT in Mathematics and in Earth and Planetary Sciences. He holds a PhD from Temple University, and an MD, from the University of Miami. His postdoctoral studies were completed at the US National Institutes of Health, and his residency was completed at the George Washington University Medical Center in Washington, DC. Dr. Berman served as Chief of Anatomic Pathology, Surgical Pathology, and Cytopathology at the Veterans Administration Medical Center in Baltimore, Maryland, where he held joint appointments at the University of Maryland Medical Center and at the Johns Hopkins Medical Institutions. In 1998, he was transferred to the US National Institutes of Health as a Medical Officer and as the Program Director for Pathology Informatics in the Cancer Diagnosis Program at the National Cancer Institute. Dr. Berman is a past president of the Association for Pathology Informatics and the 2011 recipient of the Association's Lifetime Achievement Award. He has first-authored more than 100 journal articles and has written numerous science books. His most recent titles, published by Elsevier, include:

Taxonomic Guide to Infectious Diseases: Understanding the Biologic Classes of Pathogenic Organisms, 1st edition (2012)

Principles of Big Data: Preparing, Sharing, and Analyzing Complex Information (2013)

Rare Diseases and Orphan Drugs: Keys to Understanding and Treating the Common Diseases (2014)

Repurposing Legacy Data: Innovative Case Studies (2015)

Data Simplification: Taming Information with Open Source Tools (2016)

Precision Medicine and the Reinvention of Human Disease (2018)

Principles and Practice of Big Data: Preparing, Sharing, and Analyzing Complex Information, 2nd edition (2018)

Taxonomic Guide to Infectious Diseases: Understanding the Biologic Classes of Pathogenic Organisms, 2nd edition (2019)

Preface

Everything has been said before, but since nobody listens we have to keep going back and beginning all over again.

Andre Gide

In 1973, Theodosius Dobzhansky, a distinguished geneticist and the awardee of the US National Medal of Science, delivered an influential essay entitled "Nothing in biology makes sense except in the light of evolution" [1]. In this essay, he argued that we need the theory of evolution to explain paleontology, zoology, and the biomedical sciences. The reason for this is that all of the observable phenomena of living (or once-living) organisms developed stepwise, over vast stretches of time. It is through the theory of evolution that scientists understand biological observations. Without evolution, biological observations are reduced to mere factoids.

Assuming that Dr. Dobzhansky was correct and that all biological observations must be interpreted within the framework of their evolutionary origins, then we would surmise that anyone interested in biology would be highly encouraged to study the field of evolution, as well as the complementary fields in which the steps of evolution are directly expressed in animals (i.e., embryology, histology, anatomy, microbiology, and physiology). These five fields, the former mainstays of medical education, have fallen behind the two relatively new fields of molecular biology and bioinformatics. Progress in the fields of molecular biology and bioinformatics, over the past three decades, has been so rapid and so exciting that we can hardly fault students who choose to concentrate their attention on these two new areas.

As a point of fact, advances in the genetics of human diseases have vastly outpaced our ability to analyze and understand our data. We have now identified thousands of gene variants that are the putative root causes of rare diseases, without achieving a deep understanding of the biological mechanism whereby the variant gene leads to the clinical expression of disease [2]. In the case of the common diseases of humans, particularly in the case of cancers, we have collected thousands of gene variants that are associated with subsets of affected individuals, but we seldom have a clear understanding of the biological roles these genes play in the development or progression of disease [3].

We now know that diseases develop through a sequence of biological steps that eventually lead to the appearance of a specific clinical phenotype (i.e., cellular pathology and consequent symptoms). A specific defect in a particular gene may lie at the root cause of the disease, but identifying a "causal" gene seldom tells us much about the subsequent steps that occur over days or months or years, leading to disease. In many cases, knowing those steps may be more important than simply identifying the root causal gene; the reason being that the most effective way to prevent, diagnose, or treat a disease may involve targeting those subsequent events and pathways. Furthermore, knowledge of the events and pathways that lead to the development of a specific disease may be directly applicable to other diseases, including subsets of common diseases [4].

At this point, you might be wondering how issues concerning disease development might relate to the topic of evolution. As it happens, the pathways leading to the development of disease are conserved cellular pathways, all of which came into existence through evolution, at particular moments in the development of ancestral species. If we want to understand the pathways that lead to disease, we might want to look at how and why these pathways evolved, the functions they serve at particular stages in an organism's development, and the particular cells in which those pathways are expressed.

Now we come to the premise of this book. We'll take Dobzhansky at his word, that "Nothing in biology makes sense except in the light of evolution," and we'll extend it to assert that "nothing in the field of Precision Medicine makes sense except in the light of evolution." *Evolution's Clinical Guidebook: Translating Ancient Genes into Precision Medicine* is an exploration of this assertion.

How to read this book

This book is primarily written for anyone who is searching for a clear, logical, and informed explanation of the relationship between evolutionary processes and the science of modern medicine. On a very simplistic level, evolutionary theory can be covered in a few paragraphs of a middle school textbook, leaving students with a credible accounting of how ancient animals may have given rise to all the species of animals that inhabit our planet today. If we want to understand exactly how particular diseases may have evolved and how we might use our understanding of evolutionary history to develop and test new and effective treatments for human diseases, then we must be prepared to think very deeply and to integrate the seemingly unrelated disciplines of embryology, biochemistry, paleontology, comparative anatomy, molecular biology, bioinformatics, and pathology.

Needless to say, it is impossible for anyone to absorb all the subjects covered in this book. Setting aside the many new concepts that will be discussed, the burden of mastering the terminologies of half a dozen scientific fields is too onerous to bear. Accordingly, the book is organized to eliminate the need for rote memorization, while permitting the reader to focus on the fundamental concepts explained in each section of each chapter. In some sense, the book is a collection of logical arguments. The facts in the book appear for the sole purpose of legitimizing the arguments. If you are a curious layman, the concepts that are developed throughout the book will satisfy your curiosity. If you are a medical researcher, then you can always return to the book and read it a second time for its factual content.

There are about 800 references included with the text and this should keep the serious scholar occupied for years to come. In addition, each chapter is accompanied by a glossary containing many of the discipline-specific terms appearing in the text. There are, in toto, over 300 glossary terms. Along with term definitions, most of the glossary items expand upon the concepts covered in the text, providing additional references and instructive examples. Readers who lack a strong background in the biological or medical sciences are encouraged to read the chapter glossaries, before proceeding to the subsequent chapters.

References

[1] Dobzhansky T. Nothing in biology makes sense except in the light of evolution. Am Biol Teach 1973;35125–9.

[2] Berman JJ. Rare diseases and orphan drugs: keys to understanding and treating common diseases. Cambridge, MA: Academic Press; 2014.

[3] Berman JJ. Neoplasms: principles of development and diversity. Sudbury: Jones & Bartlett; 2009.

[4] Berman JJ. Precision medicine, and the reinvention of human disease. Cambridge, MA: Academic Press; 2018.

1

Evolution, From the Beginning

OUTLINE

Section 1.1 In the Beginning

The beginnings and endings of all human undertakings are untidy.

John Galsworthy

Every story has a beginning. In the case of life on planet Earth, there is no one who can speak with any great authority on the subject. It happened too long ago, on a world that was very different from the world we live in. Nonetheless, the beginning of life on Earth is a topic that invites thoughtful speculation. The purpose of this chapter is to discuss how biological processes begin, in general terms. If nothing else, the chapter provides us an opportunity to explore fundamental concepts that will surface throughout the book: enzymes, pathways, natural selection, and evolution.

Abiogenesis

Abiogenesis is the creation of life from nonliving matter. For thousands of years, the mystery of the origin of living organisms has challenged philosophers, theologians, and scientists. In this section, we'll try to show that life is very simple to create, if we just accept the following three assumptions:

−1. All living things are composed of nonliving things.

Life on earth consists of cells, and cells are just bags of chemicals. It happens that these bags of chemicals can replicate themselves (i.e., self-replicate), accounting for why there are so many cells on the planet. Because all living matter is composed entirely of nonliving matter, we can guess that life may have gotten its start from some process involving nonlife.

Evolution's Clinical Guidebook. https://doi.org/10.1016/B978-0-12-817126-4.00001-1

   −2.   Natural selection operates equally well on living matter and nonliving matter.

It is a mistake to believe that natural selection is a force of nature, like magnetic fields or gravity, or heat. Natural selection is simply a convenient term describing a somewhat abstract truism. Basically, if conditions favor the persistence of one outcome over another, then you'll tend to see more instances of the favorable outcome.

For example, let's imagine that we are observing two mountains. One mountain is composed entirely of granite, and the other is composed entirely of sand. Over time, the granite mountain is likely to persist, while the mountain of sand is likely to flatten and vanish. Natural selection favors granite over sand. Too obvious? Let's extend the metaphor to include chemical reactions.

If one mountain is built from a chemical reaction that yields an insoluble product, and another mountain is built from a chemical reaction that yields a product that is slowly dissolved in water, then the insoluble mountain will persist. Moreover, if the chemical reaction that produces an insoluble product is sustained by readily available oceanic substrates, then we might expect to see new, insoluble mountains rising from the sea.

If there is a geographic location where the reaction occurs more efficiently than in other areas, then we might expect to see a greater number of mountains rising in this area. If there are risen mountains wherein a novel reaction generates the catalysts and substrates for the mountain-building reaction, then we might expect to see new mountains sprouting out from such mountains, replication of sorts.

We can carry this metaphor forward all day. The point is that natural selection applies to living and nonliving conditions. We shall see that the process of natural selection drives most, if not all, of the processes known as evolution.

   −3.   Catalysts are molecules that facilitate interactions between other molecules, sometimes leading to the synthesis of new types of molecules, including new types of catalysts.

In biology courses, we are taught that catalysts are specialized proteins, known as enzymes, and that these proteins are shaped in such a way that specific substrate molecules can be held and put into physical close proximity with one another, thus allowing them to interact chemically. The reaction product of such a catalyst-facilitated reaction, having no special affinity for the catalyst, is released, completing the reaction.

Catalysts need not be proteins [1–3]. My favorite example of a nonprotein catalyst is the dish upon which peanut butter and jelly sandwiches are prepared. It's all but impossible to spread peanut butter and jelly on a slice of bread without something to hold the bread in place while the ingredients are being spread. In a pinch, you can hold the bread in one hand, using the other hand to spread the ingredients with a knife, but it's very awkward and the bread always tears. The dish is the go-to item for anyone who wants to prepare a sandwich with the minimal expenditure of energy, in the least time, and with the least wastage of ingredients. The dish facilitates the reaction, but is not consumed therein. The dish is a catalyst, in every sense of the word.

Before there were proteins, reactions were driven exclusively by nonprotein catalysts, and we can guess that these early catalysts acted somewhat like the dishes that facilitate the construction of peanut butter and jelly sandwiches. They may have been hard, metallic surfaces, such as we see in rocks, holding substrates steady for the duration of a reaction. Today, cellular catalysts are proteins, but many of these catalysts are metal-protein complexes, such as hemoglobin, superoxide dismutase, the ferredoxins, and the cytochromes [4]. It seems as though we cannot escape our rocky start in life.

Let's look at hypothetical catalytic reaction. In the following case, X and Y are substrates. C1 is the first catalyst that we'll be examining, and Z is the product of the reaction.

```
X + (C1) -> X(C1)      The catalyst, C1, binds to the first substrate, X
Y + X(C1) -> XY(C1)    A second substrate, Y, binds with X and C1
XY(C1) -> (C1)Z        X and Y, bound to C1, yields the product Z
C1(Z) -> C1 + Z        Z dissociates from the catalyst
```

When basalt melts and aggregates, small bubbles, about the size of bacteria, form and interconnect with one another [5]. Examples of porous volcanic rocks include tufa, tuff, travertine, pumice, Stromboli basalt, and scoria. Any of these water-drenched rocks could have been crucibles for catalytic reactions, like the one shown above. It may well be that just as early humans were cave dwellers, so too were the early cells, though their caves were much smaller than ours (Fig. 1.1).

FIG. 1.1 Volcanic pumice is a highly porous rock. The lacunae in pumice, and other seabed rocks, may have served as the earliest substrate for the evolution of living cells, providing a hard catalytic substrate for metabolic reactions, and some level of compartmentalization providing sequestration of locally high levels of chemical substrates and products, while providing a method of egress for reactants and chemical messengers (e.g., nucleic acids, enzymes, and possibly viruses) to pass between lacunae. Source, Wikipedia, and entered into the public domain by deltalimatrieste.

Let us imagine that our first reaction (*vide supra*) occurred in a watery pocket inside a porous volcanic rock. The walls of rock lining the small internal pockets serve as sources of a catalyst (C1), and we suppose that X and Y are chemicals contained in the primordial water bathing the porous rocks. We'll allow Z, the product of the first reaction, to diffuse from one rock bubble to another, where a different catalyst, C2, is found and where Z and Y are abundant. Perhaps the following reaction ensues:

```
Z(C2) -> Z(C2)          Z binds to the catalyst C2
X + Z(C2) -> XZ(C2)     X binds to the Z and C2
XZ(C2) -> XXY(C2)       XY produced from Z and a new complex forms
XXY(C2) -> 2X + Y + C2  Y and X dissociate producing two molecules of X and one of Y
```

We can begin to imagine rocks containing millions and millions of tiny bubbles filled with substrates and catalysts and products, diffusing from one lacuna to another, and providing substrates for new reactions. We can imagine that if the product of the reaction participates as a substrate in some other reaction, then we might see a self-propagating system, producing a range of products. Under these circumstances, natural selection might intervene, increasing the likelihood that particular reactions may prevail over others, depending on the local concentrations of substrates, products, and catalysts. In time, the activity of certain catalysts may be modified (increased or decreased) based on chemical interactions with rock minerals. It all seems a little bit like a living process.

It is silly to think that we can ever determine the moment in time when life first emerged from lifeless chemicals. We can barely fathom a description of the earliest form of life, but let's try anyway. When we choose a definition of life that reflects the simplest organisms observed today, then we would say that the first form of life was a cell, enclosed by a lipid membrane, and capable of self-replication.

The earliest signs of cellular life, etched into ancient iron rocks, dates back to between 3.77 billion and possibly 4.28 billion years ago [6]. If we can accept this rough date, then the first life on earth appeared as early as a few hundred million years following the formation of planet earth (estimated at 4.5 billion years ago) [7].

Because rock bubbles are interconnected, early RNA virus-like molecules presumably may have moved from bubble to bubble, exchanging genetic materials along the way. Sometime later, DNA may have appeared. DNA is a much more stable molecule than RNA, less prone to replication error, and less susceptible to intrusion by RNA viruses that were freely commuting between rock bubbles. The evolution of DNA modifications (adherent proteins and base methylations, characterizing the early epigenome) may have developed as a defense against infection by RNA viruses. [Glossary DNA methylation, Epigenome, Genome]

What came next? After DNA appeared, as a template for RNA, it seems plausible that DNA viruses may have emerged. DNA viruses, being more stable than RNA viruses, could grow into large, complex entities, such as the megaviruses. At some point, cell membranes appeared. It is known that phospholipids spontaneously form lipid bilayers in agitated water. It seems possible that rock-dwelling organisms, endowed with an enclosing bilayer membrane assembled from phosphorylated small molecules, synthesized from an RNA template, would eventually float into the ocean, where they might encounter new sources of food. The late emergence of enclosing membranes, well after the initial development of membrane-less organisms dwelling within rocks, is supported by the profound structural differences in bacterial and archaean membranes [8, 9]. If the two classes of organisms had split off from a common, membrane-enclosed ancestor, you might expect them to have similar membranes. [Glossary LUCA]

Section 1.2 Bootstrapping Paradoxes

Before I speak, I have something important to say.

Groucho Marx

As a warning to the reader, this section on biological paradoxes is, without any doubt, the most difficult passage of the book. Paradoxes are always puzzling. Nonetheless, if we want to understand how the fundamental processes of living organisms got their start, we need to solve several "origin" problems. Readers are advised not to dwell too long on any of the examples included here. These same topics will pop up again in later chapters, along with additional background material that may provide additional insights. This section's discussion can always be revisited, if the need arises.

The term bootstrapping derives from an absurdist trope in which boys are instructed to pull themselves up by their bootstraps. Although it is certainly possible for a standing boy to pull his boots on with his bootstraps, it is impossible for a boy to gain a standing position by pulling the straps of a booted foot, no matter how hard he may pull. The term refers to a class of paradoxes in which some step in a process requires the completion of some earlier step, which itself requires the completion of a later step.

Bootstrapping paradoxes pop up everywhere in the natural sciences. In nearly all cases, these paradoxes are dismissed as being sophomoric (i.e., too silly to contemplate), or they are ignored for being philosophical (i.e., not a matter for serious scientific analysis). As it happens, bootstrapping paradoxes lie at the heart of some of the most powerful concepts in evolution and embryology, and the negative consequences of ignoring these issues has resulted, historically, in the delay of scientific advancement and the perpetuation of a host of superstitious ideations.

It is worth taking the time to explore the philosophical and the pragmatic aspects of bootstrapping. This is best accomplished by studying a few examples of bootstrapping paradoxes, and then formulating a general solution to the problem.

Which came first, the hardware or the software?

When you turn on your computer, the computer "boots up," the term being a shortened form of "bootstrapping." Every time you start your computer, a bootstrapping paradox must be solved. Here is the dilemma:

-1.   Computer hardware requires a software operating system for any kind of functionality. Without software, a computer is just a paperweight.

    –2.   Software operating systems cannot operate without functioning computer hardware. Without hardware, software is just a wish list.

    –3.   When you turn on the power to your computer, the hardware cannot begin to function until it receives software instructions, but the software instructions cannot be accessed by the hardware until the hardware begins to function.

Here is how computers solve their bootstrapping paradox.

At start-up, the operating system is nonfunctional. A few primitive instructions hard-wired into the computer's processors are sufficient to call forth a somewhat more complex process from memory, and this newly activated process calls forth other processes, until the operating system is eventually up and running. The cascading rebirth of active processes takes time, and explains why booting your computer may seem to be an unnecessarily slow process.

Basically, computer scientists solved the paradox by designing a tiny start-up kernel wherein both the hardware and the software are one and the same.

Which came first, the chicken or the egg?

The classic "chicken and egg" problem can be restated as a bootstrapping paradox:

    –1.   Eggs cannot come into existence without a laying hen, to produce the eggs.

    –2.   Hens cannot come into existence until they have been born, from eggs.

The chicken and egg paradox is known to everyone, but it is often posed as an example of an absurdist question. The glib response to the chicken and egg paradox is, "We have chickens, and we have eggs, so it really doesn't make any difference which one came first." As it happens, the chicken and egg paradox is one of the most enlightening of all biological questions, and students who seriously pursue its answer will be rewarded with a deeper understanding of the meaning of ancestry and of life.

First off, let's generalize the "chicken and egg" paradox, and reframe the paradox in terms that a cell biologist might appreciate. Adult animals contain two types of cells, distinguished by the method of division by which they were produced: mitotic cells and meiotic cells. The mitotic cells, also known as somatic cells, account for the vast bulk of the organism. The meiotic cells of the adult organism are the gametes: oocytes residing in the ovaries, and sperm cells inside the testes. The gamete cells are derived from primordial germ cells, and are produced early in the embryonic period, by mitotic cells of the endodermal layer. Backtracking, the mitotic cells of the embryo are the product of the fusion of two meiotic cells (i.e., a sperm fertilizing an egg). To summarize, we observe that meiotic cells produce mitotic cells. We also observe that mitotic cells produce meiotic cells. This poses a general question. How can a class of cells produce a second class of cells that produced the first class of cells [10]? [Glossary Endoderm, Gamete, Generalization, Germ cell line, Somatic]

Let's examine the meaning of ancestry, as it pertains to species. When we claim that a species derives from some preceding species, we are indicating that the cells of the child species were produced by the cells of the parent species (i.e., all cells come from cells). There are no interruptions or jumps from one species to another. Hence, the cells in the first human were cells from an earlier hominid, and so on through the lineage of living organisms, back to the earliest eukaryotic organism. We'll be examining this topic in detail in Section 4.2 "The Biological Process of Speciation." [Glossary Species]

The first eukaryotes were capable of meiosis and mitosis. We can say this because the enzymatic apparatus for both mitosis and meiosis share many of the same steps, and the genes for mitotic and meiotic division are found in every species of eukaryotic cell today, including single-cell eukaryotic organisms that are thought to reproduce asexually [11–15]. Hence, the first eukaryotes, much like the single-cell eukaryotes living today, were both egg and organism, at once. We humans are direct cellular descendants of those single-celled organisms, as are chickens. **The answer to the question, "Which came first, the chicken or the egg?" is that the chicken and the egg were the same cell back in the time when the chicken's direct cellular ancestor was a single-celled eukaryote.** [Glossary Meiosis, Mitosis]

We like to envision meiosis and mitosis as two forms of cell division. Let's radically change our perspective, and focus on the concept of haploid organisms and diploid organisms (and the fundamental roles played by meiosis and mitosis) [16].

All multicellular organisms that develop from an embryo (i.e., all members of Class Plantae and all members of Class Animalia) have haploid and diploid stages of existence. That is to say that every animal and plant can be envisioned as two different organisms each containing its own distinctive type of cells: a haploid organism containing the genes of one gender (i.e., male or female) and a caretaker diploid organism containing male and female genetic material. [Glossary Class]

The dichotomy between haploid organism and diploid organism is best understood by examining the life of plants, particularly the life cycle of the bryophytes, an ancient class of plant that flourished prior to the appearance of modern flowering plants (i.e., angiosperms). The bryophytes are nonvascular land plants and include the liverworts, hornworts, and mosses. The bryophyte life cycle consists of two phases, a haploid phase consisting of a male or a female gametophyte (i.e., plant grown from a gamete), and a diploid phase consisting of a sporophyte (i.e., plant grown from a diploid embryo). From the diploid sporophyte come the male and female gametes that grow as haploid gametophytes. From the haploid gametophytes come the sperm and eggs that fertilize one another to become a diploid embryo that grows as the sporophyte.

On first thought, we may believe that the bryophytes are unique among organisms, because they can be observed as growing plants that are either haploid (gametophytes) or diploid (sporophytes). If we think a bit more deeply, we can appreciate that all plants and all animals have life cycles that are equivalent to that of the bryophyte, alternating between haploid organisms (derived from gametes) and diploid organisms (derived from embryos). In the case of plants and animals, both organisms (haploid and diploid) may reside anatomically within the larger diploid organism (i.e., the haploid organisms of either gender are internalized in the diploid organism). In plants, the gametic organism is the ovule (in the fruit of the female) or the pollen (male). In animals, the gametic organism is the haploid oocyte (in ovary) or the haploid sperm (in testis).

We fool ourselves into thinking that eggs and sperm are just two examples of the 200+ cell-types in our bodies. Not so. **The egg and sperm are alternate organisms that happen to reside within us, following their own biological destinies**. A very long time ago, when our direct eukaryotic ancestors were single-celled organisms, the eukaryotic cell played both roles, dividing meiotically to produce haploid cells or mitotically to produce diploid cells. When single-celled eukaryotes evolved to become multicellular organisms, meiotic division was reserved for the gametes, and mitotic division was reserved for the somatic cells; but both types of cells retained the genetic information for either form of division. [Glossary Cell-type, Germ cell, Haploid, Haploid organisms]

Which came first, the enzyme or the enzyme-synthesizing machinery?

To build an enzyme, you need to have enzyme-synthesizing machinery, which is itself made of enzymes. The paradox here is that the product of the enzyme-synthesizing machinery (i.e., the enzyme) must exist prior to the existence of the enzyme-synthesizing machinery (which has created it).

To find the solution to this paradox, we must return to our discussion of catalysts, earlier in this chapter. Catalysts are facilitators of reactions, and they operate by holding the substrates of a reaction in close proximity, so that the reaction can proceed quickly. Proteins make excellent catalysts, because they can evolve as structures that fit just about any substrate. Other molecules (e.g., surfaces of rocks, minerals, RNA molecules) may also serve as catalysts, but only if they happen to have the required physicochemical features [1, 2, 17].

When we examine the process by which proteins are synthesized, we see that codons on a messenger RNA template molecule are sequentially processed by a ribosome to produce a protein composed of amino acids, wherein each amino acid corresponds to a specific triplet codon from the messenger RNA. The ribosome consists of proteins and specialized molecules of RNA. Among these specialized RNA molecules are the ribozyme, catalytic RNA molecules that facilitate the translation of messenger RNA into amino acid chains (i.e., proteins) [18]. Ribozymes are also capable of catalyzing their own synthesis. Under laboratory conditions, ribozymes can sequentially attach dozens of nucleotides to primer sequences, without the help of enzymatic proteins [19–21]. [Glossary Codon]

We can begin to see how messenger RNA and ribozymes could have catalyzed the synthesis of proteins in a world where life first existed, without the bacteria, archaea, and eukaryotes that populate our planet today [22, 23]. The answer to our paradox, "Which came first, the enzyme or the enzyme-synthesizing machinery," seems to be that the earliest enzyme-synthesizing machinery served as both the template (messenger RNA) and as the translational machinery (ribozymal RNA). Hence, no protein-based enzymes were involved in the creation of protein-based enzymes. [Glossary Ribozymes]

Which came first, RNA or DNA?

We know that RNA is synthesized from a DNA template. We also know that the DNA is replicated using a variety of proteins and nucleotide substrates, acting on a preexisting DNA molecule. If DNA synthesis requires proteins (translated from an RNA template) and if RNA synthesis requires DNA (serving as its template), then DNA could not have preceded RNA, and RNA could not have preceded DNA.

Biologists have long inferred that RNA preceded DNA (i.e., that there were RNA cells before there were cells with DNA). The logic for this inference is as follows:

-1. RNA is the template for proteins, and proteins are the molecules that provide the structure and the metabolic pathways for cellular life.
-2. RNA can be synthesized without DNA [3].
-3. RNA can also serve as the mechanism by which the RNA template is translated into specific proteins (i.e., the ribosome).
-4. RNA can serve as a template for the synthesis of DNA (via reverse transcriptase).
-5. Hence, cellular life can be sustained with RNA, amino acids, and various substrates that support enzymatic activity.
-6. DNA serves no purpose other than as a template for RNA.
-7. Because RNA cells can exist without DNA, while there is no mechanism to expect the first DNA templates to come into existence without the participation of an RNA template, we conclude that the first cells were RNA-based, and that the earliest DNA was synthesized from an RNA template.

Step 2 solves the paradox. It happens that RNA can be synthesized without the benefit of a DNA template. Montmorillonite, a soft phyllosilicate clay, can synthesize multimer RNA molecules from nucleotide substrates [3]. And that's not all. Montmorillonite also catalyzes the formation of lipid bubbles. Hence, montmorillonite clay, found near porous volcanic rocks containing water, nucleotides, amino acids, lipids, and a few other substrates, may have been the catalyst responsible for the first, lipid-delimited living cells. As an added tickler, montmorillonite deposits have been found on Mars, in abundance (Fig. 1.2).

So there we have it. RNA, created from nucleotides and a clay substrate, preceded DNA, and served as the template for the synthesis of DNA, the molecule that has become the template for RNA. Maybe. We really cannot rule out the possibility that early DNA molecules formed from mineral catalysis, or otherwise, without the participation of RNA. If this were the case, DNA may have preceded RNA. Furthermore, it is possible that both RNA and DNA may have arisen as polymers upon which spontaneously condensing amino acids (i.e., proteins) acted as templates for their own replication. Just such a scenario has been proposed [24]. In this case, the building block of a life on earth may have been proteins, and both RNA and DNA served as protein-encoded replication templates. There are many possibilities. [Glossary Amyloid world]

Which came first, the process of evolution or the product of evolution?
If evolution (i.e., the ability of a species to evolve) is a biological trait, like height and strength, then the ability to evolve must have been acquired as a new trait at some time in our ancestral past. It is through evolution that new traits are acquired. Therefore, evolution needs to exist in order for evolution to come into existence. This is a paradox.

This paradox is created from a false assumption: that evolution is a trait of living organisms that needs to be acquired. As discussed in Section 1.1, "In the Beginning," evolution is a fundamental condition of the universe, and the law of evolution by natural selection can

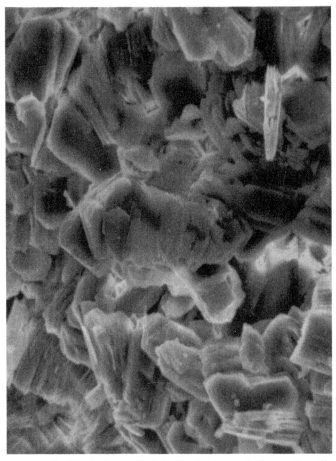

FIG. 1.2 Microscopic photograph of kaolinite, a soft phyllosilicate clay much like montmorillonite. The micropores permit a wide surface area for catalysis. Source, Wikipedia, from a public domain image prepared by the United States Geological Survey, an agency of the United States Department of the Interior.

be restated in terms of probability. Basically, anything that happens to have a greater chance of existence (e.g., by virtue of chemical stability, or strength, or mass) is more likely to persist, or increase in number, than things that have a lower chance of existence. Evolution does not create favorable mutations in organisms; it simply expresses the fact that animals with favorable mutations are more likely to persist and replicate than animals that lack such mutations. [Glossary Chance occurrence]

One can certainly argue that something other than evolution accounts for the world, as we find it today, but it would be very difficult to argue that evolution does not exist. It would be like saying that probability does not exist.

Which came first, the species or the class of animals into which the species is assigned?
Taxonomists create classifications of organisms based on their personal worldviews in which various organisms have preconceived relationships with other organisms.

Hence, the taxonomist's worldview contains a formed conception of the classification of things, which presupposes that the classification already exists as an abstraction. Essentially, the taxonomist cannot build a classification without first having the classification in her mind. This is a bootstrapping paradox, and this paradox provides the argument, accepted by some biologists, that classifications should not be built on perceived relationships among organisms, as such relationships are always biased by preconceived notions that might be false. [Glossary Classification]

However we choose to build a new classification, we need to start with some knowledge of species, and their relations to other species, and this information cannot come from a classification; because a classification is what we need to build.

In practice, the taxonomist begins with a root object that embodies the fundamental features of every class and every species within the classification. In the case of the taxonomy of living organisms, this root object might be "the living cell." Once the root object exists, the taxonomists can begin to create broad subclasses containing properties that are inclusive for the class and exclusive of other classes (i.e., Class Prokaryota, which lack a nucleus and Class Eukaryota, which have a nucleus). Then, based on observing properties of the prokaryotes, she might define additional classes that include some organisms and exclude others. This goes on until every organism has a class, and every class is a subclass of a parent class, in a lineage that extends backward to the root class. The root class, which contains every member of the classification, is itself the full embodiment of the classification (i.e., the first class of the classification contains the classification). [Glossary Nucleus]

Every thoughtful taxonomist will admit that a classification is, at its best, a self-correcting machine, not a factual representation of reality. We use the classification to create new hypotheses that can be used to see if the classification matches reality. The process of testing hypotheses may reveal that the classification is flawed, and that our early assumptions were incorrect. In this case, we revise our assumptions. With a little luck, we will find that the process of testing hypotheses will reassure us that our assumptions were consistent with new observations. By constantly testing the validity of the classification, we add to our understanding of the relations between the classes and instances within the classification. Thus, the process of creating a classification is very much a bootstrapping process.

A general solution for bootstrapping paradoxes

The generalized bootstrapping paradox can be broken down as follows:

```
Condition 1: A comes from B
Condition 2: B comes from A
Statement of Paradox: How can you create either A or B?
```

Because you can create either A if you had some B, and you can create B if you had some A, the bootstrap paradox is sometimes restated as "Which comes first: A or B?"

The solution often comes in the form that A and B were, at some past moment, one and the same.

- Instructions (i.e., software) hardwired into a chip (i.e., hardware)
- One organism with the basic pathways for both mitosis and meiosis
- One unicellular organism doubling as a somatic cell and as a germ cell
- A pathway composed of shorter pathways
- An enzyme that is also a template for enzyme synthesis.
- One primitive class constituting the entire classification

Whenever you encounter a bootstrapping paradox, try to rephrase it as a precedence paradox. After doing so, see if you can't find a solution based on a duality origin. If A and B were equivalent, at some point in their development, then the paradox disappears.

Section 1.3 Our Genes, for the Most Part, Come From Ancestral Species

The biochemist knows his molecules have ancestors, while the paleontologist can only hope that his fossils left descendants.

Vincent Sarich

The vast majority of the genes in extant mammalian species are ancient, arising hundreds of millions of years ago, sometimes billions of years ago. Here are a few observations that illustrate the point:

- -1.   There are over 500 core genes that are present in all extant metazoan (i.e., animal) species [25].
- -2.   Not only do we find homologs of human genes in single-celled eukaryotic organisms, but we also find homologies that extend all the way up our ancestral lineage and into the bacterial kingdom [as in the case of an actin homolog in the proteobacteria *Haliangium ochraceum* [26]]. [Glossary Homolog]
- -3.   Approximately 60% of the annotated protein coding genes in the mouse genome originate from prokaryotic and basal eukaryotic ancestors [27], and there is every reason to believe that the same can be said for all mammals, including humans.
- -4.   Nearly 75% of human disease-causing genes are believed to have a functional homolog in the fly [28].
- -5.   The human genome and the chimpanzee genome are between 97% and 92% alike, indicating that closely related species have nearly the same set of genes [29].
- -6.   Pax6 is a regulatory gene that controls eye development. This gene, like so many other basic regulatory genes, has been strictly conserved throughout animal evolution. Remarkably, Pax6 in mice is so similar to its homolog in insects that the corresponding genes from either species can be interchanged and function properly [30]. [Glossary Law of sequence conservation]
- -7.   Nearly identical gene families are found in many different classes of animals.

Most human genes evolved from duplications of other genes [31]. The duplicate gene, no longer serving an essential function (fulfilled by the original gene), can serve some new purpose after it has been modified by mutation and preserved in the species gene pool through natural selection. Because the new gene is a modification of the original gene, you might expect the two genes to have some functional similarity. When this is so, the two genes are said to be members of a gene family. When the members of a gene family are situated close to one another on a chromosome, they are referred to as gene clusters. Further duplications of members of a gene family may produce large gene families. As we might expect, large gene families evolve over long periods of time, and this tells us that large gene families are ancient and may be present (in part or in whole) in all the classes of organisms that descend from the gene family's founder organism. [Glossary Founder effect, Gene pool, Natural selection, Paralog]

Let us codify our discussion as a biological rule.

Rule: Gene families are seldom if ever species specific.

How can we justify this rule? We know that all gene families are ancient. However, individual species are, for the most part not ancient. Hence, gene families arise within lineages of organisms. Hence, we would expect multiple species arising through the lineage to have the same gene families.

An example of an ancient gene family are the globins, whose function in the red blood cells of many animals is the distribution of oxygen from an external source (e.g., air or water) to tissues throughout the body [32, 33]. The genes of the globin family are so ancient that they are present in bacteria. It is believed that the original globin molecule evolved in bacteria prior to the evolution of oxygenic photosynthesis (i.e., prior to 2.3 billion years ago) at a time when all organisms were anaerobic. Oxygen is toxic to anaerobic organisms. The purpose of the first globins in bacteria was to scavenge oxygen, thus protecting anaerobic cells from oxygen-induced toxicity. Following the advent of oxygenic photosynthesis, when the earliest aerobic organisms were evolving, the oxygen-scavenging properties of the early globins were modified to facilitate the extraction of oxygen from the air, and the delivery of oxygen to the intracellular enzymes participating in the pathways for oxidative phosphorylation. A variety of hemoglobinopathies affect humans, all involving the globin gene family [32].

One of the largest gene families is the olfactory receptor family, which employs about 800 genes in humans [34]. These genes code for chemical receptors that can detect and discriminate between tens of thousands of different odors [35]. Sometime in our ancestral past, the gene family diverged a bit, producing two major classes of receptors in humans, grouped by homology to our ancestors: the fish-like receptors, and the (later evolving) tetrapod-like receptors [36].

We will be discussing the elegant regulatory mechanisms of the olfactory receptors in Section 4.3, "The Diversity of Living Organisms." For now, let us just note that the olfactory receptor gene family and the retinal receptor gene family are related to one another, both producing rhodopsin-like proteins. Because olfactory receptor proteins are related to the rhodopsin molecules responsible for light reception in retinal cells, we might expect to find some inherited diseases that affect both vision and olfaction. One such example is Refsum disease, an inherited condition characterized by anosmia and retinitis pigmentosa. Retinitis pigmentosa is a progressive retinal disorder that may lead to blindness. The root cause of Refsum disease is a mutation that produces an accumulation of phytanic acid in cells. How the build up of phytanic acid produces anosmia and retinitis pigmentosa is unknown at this time, but it is intriguing to guess that the pathogenesis of this rare disease may be mediated by toxicity in rhodopsin-like gene-family proteins found in olfactory receptors and light receptors. [Glossary But-for, Pathogenesis, Rare disease, Root cause]

-8.   The human genome is packed with relic sequences from ancient ancestral organisms.

About 8% of our genome is derived from sequences with homology to known infectious retroviruses, and these sequences can usually be recognized by their subsequences that contain viral genes (e.g., gag, pol, and env genes) and by the presence of long terminal repeats. The viral sequences, remnants of ancient retroviral infections and the occasional nonretroviral infection, were branded into our DNA and subsequently amplified [37–39]. [Glossary Retrovirus]

This above list of findings indicates that species use the genetic tools that they inherit from their ancestors. We should not be surprised to learn that most of the evolution of extant organisms was played out long ago, within the genomes of our ancient ancestors. If this were not so, then every animal species would need to somehow create, for themselves, the genes for their metabolic pathways, their cell-types, and their body plans. If animals did not inherit their genes from ancestral organisms, then we would not have much need for evolution, and the concept of phylogenetic lineage would have very little meaning. Luckily for us, the theory of evolution holds up quite nicely.

Some key steps in eukaryotic development were stolen from other organisms.

All oxygenic photosynthesis in eukaryotes is accomplished with the use of chloroplasts, a symbiotic organelle formed from a captured cyanobacteria. As the story goes, an ancient member of Class Archaeplastida found that by engulfing a cyanobacteria, it too could photosynthesize. This indentured relationship between Archaeplastida and cyanobacteria became permanent, and every descendant of Class Archaeplastida, which includes all the green plants (i.e., Class Viridiplantae), have benefited from their ancestor's thievery (Fig. 1.3).

It is generally assumed that the acquisition of a cyanobacteria as a self-replicating organelle occurred only once in Earth's history. If this were true, then chloroplast-containing organisms other than those within Class Archaeplastida must have seized their chloroplasts from green algae or from other members of Class Archaeplastida (i.e., stolen from the original thieves). How do we know that our original assumption is true? We do so by counting membranes. Chloroplasts in the Archaeplastida are lined by two membrane layers, corresponding to the inner and outer membranes of the original, indentured cyanobacteria. The chloroplasts of non-Archaeplastida eukaryotes have three or four membrane layers, suggesting that a member of Archaeplastida was itself engulfed, and the membranes of its

FIG. 1.3 *Gloeotrichia echinulata*, a member of Class Cyanobacteria, the first and only terrestrial organism to evolve oxygenic photosynthesis. *Source: Wikipedia, from a public domain image commissioned by the US Geological Survey, and photographed by Barry Rosen.*

chloroplast, and of the Arhaeplastida containing the chloroplast, were entrapped permanently in the predatory cell. [Glossary Cyanobacteria, Chloroplast evolution]

When we discuss the origin of Class Eukaryota (Section 5.3, "Eukaryotes to Obazoans"), we will see that eukaryotic nuclei and mitochondria are held to be organelles appropriated from non-eukaryotic organisms, not the result of an evolutionary process occurring within the eukaryotic genome. In short, the evidence suggests that the root eukaryote, ancestor to all living eukaryotes, owed its existence to cellular abductions (Fig. 1.4). [Glossary Trait, Nonhomologous, Taxa, Taxon, Taxonomy]

Section 1.4 How do Metabolic Pathways Evolve?

The most savage controversies are those about matters as to which there is no good evidence either way.

Bertrand Russell

The mechanism by which enzymatic pathways evolve is yet another bootstrapping problem. What plausible explanation can explain a metabolic pathway involving 20 coordinated enzymes, when natural selection operates on individual genes, not groups of genes? [Glossary Pathway]

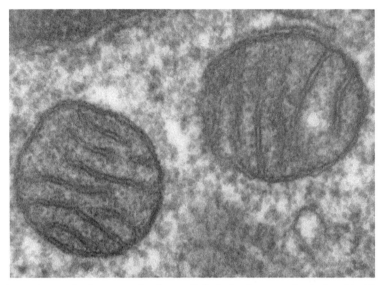

FIG. 1.4 An electron micrograph of a mammalian lung cell, featuring two round mitochondria. The two-layered parallel lines traversing the organelles are cristae. Cristae are infoldings of the mitochondrial inner membrane, providing a large surface on which cellular respiration may proceed. A single cell may contain hundreds of mitochondria. *Source: Wikipedia, from a work that was released into the public domain by its author, Louisa Howard.*

This is a question that has fascinated evolutionary biologists for years. One of the earliest proposed solutions was that metabolic pathways evolve backwards. The ancestral pathway begins with a single enzyme that catalyzes the transformation of the final substrate in the pathway to become the final product of the pathway. Once the last step has evolved, nature finds another enzyme that catalyzes the next-to-the-last substrate to produce a product that serves as the substrate last step of the pathway, and so on [40]. Thusly, pathways of arbitrary length can evolve backwards, enzyme by enzyme.

Perhaps so, but our current understanding of the biological diversity and resiliency of metabolic pathways suggests an alternate theory. The term "pathway" means something different today than it meant a few decades ago. According to traditional thinking, a pathway was a sequence of biochemical reactions, involving a specific set of enzymes and substrates that produced a chemical product. The classic pathway was the Krebs cycle. It was common for students to be required to calculate the output of the cycle (in moles of ATP) based on stoichiometric equations employing known amounts of substrate. Today, we know that pathways may involve the linked actions of receptors, activators, suppressors, and modifiers. The products of pathways are often cellular actions, rather than chemical products. These pathways do not always occur within an organelle or in an anatomically sequestered area of the cell. The pathways may interact with other pathways, and the direction of a pathway, and its biological consequences, may be complex and variable in different cells or in different physiological conditions within one cell. We can now think of pathways as opportunistic gangs of diverse chemicals; not as a disciplined platoon of enzymes. Thinking this way, we can imagine long pathways developing from partnerships with collections of smaller pathways. [Glossary Gene regulation, Epistasis]

For example, oxygenic photosynthesis required the independent evolution of two different pathways that combined to complete the full biosynthetic cycle [41]. The first pathway involves the capture of photons to hydrolyze water, yielding oxygen, NADPH, and ATP. The second, light-independent pathway, is the Calvin cycle, driven by high-energy molecules (NADPH and ATP) created by the first pathway to form simple sugars from carbon dioxide. The first and only organism to achieve oxygenic photosynthesis was a cyanobacterium, which mastered the pathway about 2.5 billion years ago. [Glossary Oxygen crisis]

One additional revelation has shaken our early notions of pathways, this being the observation that a single protein may have many different biological functions, and that a single function may be served by a multitude of different proteins. For example, actin is a major cytoskeleton protein found in many different types of cells and is a key participant in the contraction in myocytes. Actin is involved in numerous additional cellular functions, and is a regulator of gene expression.

In biological systems, pathway components change their functions depending on moment-to-moment circumstances. For example, cancer researchers discovered a protein that plays an important role in the development of cancer. This protein, p53, was at one point considered to be the primary cellular driver of human malignancy. It was thought that when p53 is mutated, cellular regulation is disrupted, and cells proceed down a slippery path leading to cancer. In the past few decades, as more information has been obtained, cancer researchers learned that p53 is one of many proteins that play a role in carcinogenesis, but the role changes depending on the species, tissue type, cellular microenvironment, genetic background of the cell, and many other factors. Under one set of circumstances, p53 may function in DNA repair; under another set of circumstances, p53 may cause cells to arrest the growth cycle [42, 43]. It is difficult to predict biological outcomes when pathways change their primary functionality based on cellular context. Various mutations in the TP53 gene have been linked to 11 clinically distinguishable cancer-related conditions, and there is little reason to assume that the same biological role is played in all of these 11 disorders [44]. [Glossary Ataxia telangiectasia, Carcinogenesis, Cancer progression, DNA repair, Druggable driver, Initiation, Li-Fraumeni syndrome, Mutation, Mutation types]

Likewise, the Pelger-Hüet anomaly and hydrops-ectopic calcification-"moth-eaten" (HEM) are both caused by mutations of the gene coding for the lamin B receptor. The Pelger-Hüet anomaly is a morphologic aberration of neutrophils wherein the normally multilobed nuclei become coffee bean-shaped, or bilobed, with abnormally clumped chromatin. The condition is called an anomaly, rather than a disease, because despite the physical abnormality, affected white cells seem to function adequately. HEM is a congenital chondrodystrophy that is characterized by hydrops fetalis (i.e., accumulations of fluid in the fetus) and skeletal abnormalities. It would be difficult to imagine any two diseases as dissimilar to one another as the Pelger-Hüet anomaly and HEM.

How could these disparate diseases be caused by a mutation involving the same gene? As it happens, the lamin B receptor has two separate functions: preserving the structure of chromatin, and serving as a sterol reductase in cholesterol synthesis [45]. These two different and biologically unrelated functions, in one gene product, account for two different and biologically unrelated diseases. [Glossary Neutrophils, Congenital disorder]

An example wherein different proteins have evolved to serve one equivalent function is found in the evolution of lens crystallins [46, 47]. In the following cases, the protein composing the principal lens protein, serving to produce a clear, refractive lens, varies among animals, on a phylogenetic basis, as shown:

- alphaA-Crystallin in vertebrates derives from gene coding for a small heat shock protein homolog
- delta1-Crystallin, found in birds and reptiles, derives from gene coding for argininosuccinate lyase homolog
- zeta-Crystallin, found in guinea pig, camel, llama, and llama, derives from gene coding for quinine oxidoreductase
  pi-Crystallin, found in gecko, derives from gene coding for glyceraldehyde-3-phosphate dehydrogenase
- nu-Crystallin, found in elephant shrew, derives from gene coding for retinaldehyde dehydrogenase
- omega/L-Crystallin, found in cephalopods and scallop, derives from gene coding for a homolog of aldehyde dehydrogenase
- J3-Crystallin, found in cubomedusan jellyfish, derives from gene coding for a homolog of saposin

In these cases, we see that the product of a gene may serve multiple unrelated functions (e.g., as zeta-Crystallin and as quinine oxidoreductase), and in which one function, lens refraction, is served in different species by the products of alternate genes. Of course, the evolution of this phenomenon, called gene sharing, has sparked lively debate, but we can take it as one more example of the molecular flexibility of metabolic pathways [48, 49].

If we accept that new pathways develop from preexisting pathways, we can draw a few tentative inferences.

Rule: All pathways are multifunctional, in whole or in part.

Rule: Pathways borrow proteins from other pathways.

Otherwise, there would be no mechanism to develop a new pathway through natural selection. These rules apply on the single molecule level and on the organized tissue level. Specialized proteins (like hemoglobin) must come from proteins that performed some other function (e.g., scavenging for oxygen). Likewise, specialized tissues (like feathers) must have come from tissues that performed some other function (e.g., keeping dinosaurs warm, providing a bit of lift when jumping over obstacles). The observations that new things (e.g., genes, cell types, tissues, species) always derive from modifications in old things is a property of evolved systems. If organisms were created in a single moment, as proponents of intelligent design would have us believe, then we would not observe this universal relationship between biological entities and their ancesters. [Glossary Exaptation]

Developing drugs that target pathways; not genes
Diseases with similar clinical phenotypes can often be grouped together according to shared pathways. Examples would include the channelopathies, ciliopathies, lysosomal storage diseases, and so on. At this point, our ability to sensibly assign diseases to pathways is limited because the effects of a mutation in a single gene may indirectly affect many different pathways, and those pathways may vary from cell type to cell type. We will see that syndromes involving multiple pathways and multiple tissues occur frequently when the mutation involves a regulatory element, such as a transcription factor [50]. Nonetheless, when one pathway plays a dominant role in the pathogenesis of a group of diseases, we can begin to ask how we might develop diagnostic tests and treatments that apply to every type of disease that is driven in whole or in part by such a pathway. In particular, we might ask how we might develop drugs that target the dominant disease-driving pathway for a set of related diseases. [Glossary Driver pathway, Lysosomal storage diseases, Pathway-driven disease, Phenotype, Syndrome, Transcription factor]

The notion that pathways can be targeted for treatment is relatively new and competes with the widely held belief that the best way to treat genetic diseases is by targeting the specific mutations that are thought to be the root cause of disease. For example, when we think about cancers, we tend to focus on the mutated oncogene that is thought to drive the tumor's malignant phenotype. Does it make sense to develop drugs that specifically inhibit the activity of the abnormal gene? You might think so, but there is a big problem with this approach. Specifically, a single type of cancer may result from one of many distinct mutations in any number of different genes. As one example, consider GISTs (GastroIntestinal Stromal Tumors). In GIST tumors, mutations may occur in either the c-KIT or the platelet-derived growth factor receptor alpha gene (PDGFRA) oncogenes. A drug that is active against the altered protein product of one mutated gene may not be active against all the variant mutant forms of the gene or of mutations in alternate genes associated with different examples of the same tumor. This means that even when an agent is targeted against a particular mutant oncogene, it may not be active against all the naturally occurring variants of oncogenes known to produce the tumor [51]. [Glossary Cause, Malignant, Malignant phenotype, Oncogene]

Furthermore, in those circumstances when a particular mutation is found in more than one cancer, we cannot assume that a gene-targeted therapy will be effective against all the cancers sharing the same root cause. For example, malignant melanomas having the BRAF V600E mutation respond well to treatment with vemurafenib, while colorectal carcinomas with the identical mutation do not respond to the same drug [52, 53]. [Glossary Melanoma, Nontoxic cancer chemotherapy]

Will the economics of drug development permit companies to develop drugs that target all the mutations that drive human diseases? Let's take a moment to examine the nitty-gritty realities of drug development. Some of the earliest and most successful new drugs have targeted specific mutations occurring in specific subsets of diseases. One such example is ivacaftor, which targets the G551D mutation present in about 4% of individuals with cystic fibrosis [54]. We shouldn't knock success, but it must be mentioned that the cost of developing a new drug is about $5 billion [55]. To provide some perspective, $5 billion dollars exceeds the total gross national product of many countries, including Sierra Leone, Swaziland, Suriname, Guyana, Liberia, and the Central African Republic. Many factors contribute to the cost of developing a new drug, but the most significant is the high failure rate of candidate drugs during clinical testing. About 95% of the experimental medicines that are studied in humans fail to be both effective and safe. The costs of drug development are reflected in the astronomical prices of new drugs. Americans should not pin their hopes on the belief that the Food and Drug Administration or the Center for Medicare and Medicaid will step in and put a stop to the price rises. The Food and Drug Administration can approve or reject drugs, but it does not regulate prices. Likewise, the Center for Medicare and Medicaid is not permitted to consider cost when it decides whether a treatment can be covered. Knowing this, some notable pharmaceutical companies have raised the prices of medications far beyond their manufacturing costs [56–58].

It is in the interests of society to develop drugs that have the widest possible user market [59]. Drugs that target a mutation that is specific for a few individuals with a rare disease, or a tiny subpopulation of individuals who have a common disease, are highly problematic. If we want to develop treatments that benefit the greatest number of individuals affected by a disease, it would be far more practical to find treatments that target the normal pathways that are dysregulated in disease. Here are just a few reasons that demonstrate the point:

– We know how to develop drugs that modulate the activity of pathways by targeting one or more of the normal enzymes of the pathways. Most of the drugs on the market exert their effect by targeting normal pathways.
– Drugs that target normal pathways may have generalized utility, alone or in combinations with other pathway-specific drugs, for classes of biologically related diseases.
– Pathway treatments will be effective against phenocopies of genetic diseases. Acquired phenocopies of genetic diseases often involve inhibitors of the same key pathways that drive their genetic counterparts. [Glossary Acquired disease, Phenocopy disease]
– Drugs that are active against a normal pathway will have an effect in every species of organism that inherits the pathway. Most of the cellular metabolic pathways are ancient, implying that candidate pathway drugs can be tested in any of the species that descend from the common ancestor in which the pathway first appeared. We will be returning to this point in Section 8.3, "New Animal Options."

We might hope that individuals in the early stages of common diseases, before multiple disease pathways converge to produce an intractable clinical phenotype, may be particularly amenable to treatments that interfere with the pathways that promote the ensuing steps in pathogenesis. [Glossary Complex disease].

Section 1.5 Cambrian Explosion

You can observe a lot by watching.

Yogi Berra

Studies of shale strata indicate that something very special happened, in the history of terrestrial animals, in a relatively short period, stretching from about 550 to 500 million years ago. In this 50 million year span, the body plans for every type of animal known today came into existence. We call this era the Cambrian explosion. The word "Cambrian" is a latinized form of the Welsh language word for Wales, where shale deposits of the Cambrian age were first studied. The word "explosion" tells us that paleontologists have come to think of a span of 50 million years as a blinding flash in the Earth's history (Fig. 1.5). [Glossary Phylum]

Of course, no sooner had evolutionary biologists agreed that the Cambrian period enclosed the origin of all extant animal phyla than a chorus of dissenters rose to argue

FIG. 1.5 A trilobite, exhibiting one of the body structures, making its first appearance during the Cambrian explosion. *Source: Wikipedia, entered into the public domain by its author, DanielCD.*

FIG. 1.6 A graphic display of all the geological periods of Earth's history, displayed as a spiral walkway. Look for the Cambrian period, with a few of its trilobite inhabitants. *Source: Wikipedia, a public domain work of the US Geological Survey.*

elsewise. Animals certainly preceded the Cambrian period, but such animals were soft and uncalcified and would be under-represented in the fossil record [60, 61]. Furthermore, our reliance on body plans, as the only measure of a phyla's emergence, is somewhat presumptuous. It may very well be that the defining expression of phyla may not have developed until well after the Cambrian explosion [62–64]. In particular, the bryozoans (tiny invertebrate aquatics that filter food particles from water) seem to have arisen sometime after the Cambrian period (Fig. 1.6).

Let's not quibble over a hundred million years plus or minus. There is general agreement that the Cambrian period, or thereabouts, was one of the most important periods in the evolution of animal life on this planet, raising a set of questions worthy of our immediate attention.

  – Question 1. Why then?

About 550 million years ago, planet Earth enjoyed a temperate climate. The supercontinent Rodinia had just broken into fragments, but there were no animals yet living on land. The seas were a soup of living organisms, and all of the organisms were small, many were microscopic. The waters of the relatively shallow oceans had, for the prior 1.5 billion years or so, been absorbing increasing amounts of oxygen produced largely by cyanobacteria. Beginning about 300 million years before the Cambrian explosion, and after the land and seas were saturated, the oxygen levels in the atmosphere had been steadily increasing.

The earliest metazoans (organisms developing from an embryo with a blastocyst) had evolved much earlier (about 760 mya), as had the eumatazoans (animals developing from two-layered embryos), and the first bilaterians (animals developing from a three-layered embryo) had appeared very recently (about 555 mya).

The availability of oxygen made possible the evolution of muscular animals (i.e., animals with cells having a high respiratory demand), and muscles permitted the evolution of predation (e.g., pursuing, grasping, tearing, and eating other animals) [64–66]. In turn, predation, rather than passive consumption of the decomposing debris of single-celled eukaryotes, provided an enhanced supply of metabolic energy that hastened the natural selection of bigger and more muscular carnivores.

The theory that increasing oxygen levels fueled the Cambrian explosion is strengthened by the recent observation that vampire squids, a cephalopod, lost their appetite for killing when they adapted to a low oxygen existence in pelagic waters. The vampire squid, despite its gruesome name, is a detrivore, living on cellular debris dropping to the seabed, in the fashion of organisms that lives in the low-oxygen environment of pre-Cambrian oceans [66]. We presume that carnivorous diets, and a consequent competitive predator/prey lifestyle, explain the fast tempo of evolution achieved in the Cambrian period. [Glossary Tempo and mode]

  – Question 2. How could it have happened so quickly?

Even if we accept that the emergence of the predator/prey lifestyle provided impetus for the rapid evolution of new, mobile body plans, we still need to have a biological mechanism whereby such evolution can be achieved. What would be the mechanism whereby new body plans could evolve in a short period of time?

Aside from the caveat that the evolution of new animal phyla may have extended to periods preceding and following the much-touted 50 million year Cambrian explosion, we can remark that the major evolutionary steps leading to the emergence of bilaterian animals (i.e., animals with a three-layer embryo and exhibiting bilateral symmetry about a central axis) occurred at about the same time that the Cambrian explosion began (555 million years ago). This evolutionary breakthrough was achieved, in no small measure, due to the appearance of genes that regulate the development of tissues and diverse cell types. In particular, the HOX family of regulatory genes has directed the development of body plans. HOX genes are conserved in all animals, a strong indicator of their evolutionary importance. [Glossary Homeobox]

Although exceedingly rare, 10 different HOX genes have been found to cause diseases in humans, and each affected gene may be associated with several different syndromes. In general, the reported cases of HOX gene diseases are syndromes of structural anomalies. For example, the Bosley-hand-foot-genital syndrome, associated with a mutation in HOXA13, is characterized by both limb and urogenital malformations. We observe that the human HOX diseases have very similar phenotypes to the mouse diseases resulting from mutations in homologous genes. This would suggest that within a class of animals, eutherians in this case, HOX genes have very similar functions [67]. This view is strengthened by the startling finding that Pax6 (a HOX-like gene that controls eye development) in insects can be swapped for its inserted human homolog, and still produce functioning eyes [30]. Furthermore, because the insect eye is structurally different from the human eye, this observation indicates that the genes that command eye development are different from the genes that implement the commands. [Glossary Malformation]

Because HOX genes control the embryologic development of body plans, it would seem that variations in HOX genes, occurring in the Cambrian period, may have accounted for the profusion of body plans [61, 62]. The HOX genes were particularly suited to support the fast evolution of body plans, because they were copied within the genome, thus providing a degree of functional redundancy and evolutionary "play."

- Question 3. Why were there few, if any, new body plans following the Cambrian period?

One of the most facile explanations has to do with competition [30]. In the beginning of the Cambrian period, there was a free market in body plans. The field of competitors was sparse at best. After the Cambrian period, the existing organisms ensured that no new competitors would enter the fray. This was accomplished by eating the new competitors or by consuming their supply of food and other necessary raw materials. This hypothesis is supported by the frequent observation that following a mass extinction of dominant predators (e.g., the dinosaur extinction occurring 66 million years ago), new classes of previously nondominant animals will soon emerge. When the dinosaurs died, hundreds of subclasses of small mammals, including many new carnivores, grew in number and ecologic influence, over the next 16 million years.

- Question 4. What's the big deal about body plans?

We've come to think of the Cambrian explosion as the time when all of our body plans suddenly appeared. Should we not ask ourselves whether all these body plans (basically skeletal arrangements and hard body parts) have any particular significance in the grand scheme of things [60]? For paleontologists, who are first and foremost devoted to studying hard body parts, the Cambrian explosion has obvious significance, but should the rest of us really care? In particular, should we be assigning the early animal phyla based on skeletal features, while glossing over the cellular and developmental innovations that were occurring in metazoan species, during the same time period?

The fundamental question regarding how we must organize the classes of living organisms is one of the most enduring problems in evolutionary biology. If we choose to ignore body plans, instead focusing attention in acquired metabolic pathways, or genes, or cell types, might we find some alternate basis for animal classification that displaces our traditional reliance on body structure [60]?

On one level, it seems self-evident that we need to define phyla based on class-specific traits that we can observe [62]. Of course, every new body plan occurs as the result of an evolutionary innovation in embryologic development. It is the embryo that creates the body plan from a recipe written in the genome. The skeleton is just what the finished product happens to look like. We don't give the cake credit for baking itself. Why should we give the body plan credit for creating its phylum?

In Section 3.1, "The Tight Relationship Between Evolution and Embryology" we will see that the great evolutionary advances, that gave rise to the classes of animals, are actually embryologic innovations. In fact, most of the ancestral classes that lead to the emergence of *Homo sapiens* can be defined by their characteristic embryologic patterns.

Section 1.6 After the Cambrian: Coexistence and Coevolution

It has become evident that the primary lesson of the study of evolution is that all evolution is coevolution: every organism is evolving in tandem with the organisms around it.

Kevin Kelly

The Cambrian marked the birth of the predatory world that has persisted to this day, wherein animals typically armed with five senses (sight, hearing, smell, taste, and touch) use their muscles and their sensitive nervous systems to pursue, grasp, kill, and devour weaker organisms [66]. Not surprisingly, post-Cambrian metazoan evolution features the acquisition of offensive structures (e.g., teeth), defensive structures (e.g., elaborate body armor), and mixed-function structures (e.g., brains). As previously observed, the apparent absence of new classes of metazoan organisms, coming after the Cambrian explosion, suggests that newly evolved animals could not compete with the established predators.

Had life on earth become nothing more than a gruesome endless battle among competing organisms? Thankfully, no. Just as organisms had evolved to attack other organisms, they had also evolved to rely on other types of organisms for all manner of necessary services, and the post-Cambrian period witnessed many different forms of interspecies mutualisms (e.g., organisms living inside other organisms, organisms providing metabolic aid to other organisms). In the meantime, predatory organisms specialized to pursue specific types of prey, thereby reducing the varieties of their prey. Over the following eons, a fluid truce was achieved, creating an earthly habitat in which millions of species could flourish. The two guiding principles of this truce can be summarized in two words: coexistence and coevolution. [Glossary Commensal]

Certainly, the sincerest form of coexistence occurs when one species lives inside another. Humans are a prime example of a species that has evolved to accommodate an enormous range of organisms. Indeed, every major kingdom of living organism takes residence somewhere in the human body, and this includes the following:

Viruses

A variety of potentially pathogenic viruses lie dormant in humans for long periods. Examples are the herpes viruses and the papilloma viruses. About 8% of the human genome is composed of inserted genomes of retroviruses [37–39]. In addition, nearly half of the human genome is filled with transposons; sequences of DNA such as LINE, SINE, and Alu that can be moved or are copied from one location in the genome to another [68]. As a group, the transposons behave very much like degenerate viruses that lack the ability to self-encapsulate within capsids. For the most part, the biological activities of transposons consist of copying themselves by hopping from one site to another within the human genome, and sometimes amplifying themselves [37]. Some transposons are active retroviruses, but most transposon sequences have degenerated over the eons to such an extent that they are no longer capable of retrotransposing. The high proportion of degenerate transposons in the human genome indicates that there has been little or no selective pressure to conserve their original sequences. [Glossary Dormancy]

Because much of the endogenous retroviral load in the human genome is due to amplification, and subsequent mutation, it is difficult to determine the number of retroviral species that established their niche in the human gene pool. Nonetheless, studies of these viral remnants would suggest that we harbor species from several dozen families of retroviruses, with an undetermined number of contributions from individual family members [69]. Based on comparisons of the viruses present in different species of primates, it would appear that the most recent acquisition of an endogenous retrovirus occurred in humans between 100,000 and 1 million years ago [70].

Aside from retroviruses inhabiting our genes, various other viruses infect humans and occasionally take life-long residence within our cells. Examples are several herpes viruses, dozens of human papillomaviruses, and cases of the hepatitis viruses B and C.

Archaea

Methanobrevibacter smithii, a methanogen, lives in the colon of over 95% of humans. At this time, we don't have a full account of all the different archaeans living in humans, but none to date have been associated with any disease [71] (Fig. 1.7).

Bacteria

The number of bacterial cells in the human body outnumber human cells by a factor of about 10 to 1.

Some of the bacteria that were once considered to be uniformly pathogenic have been shown to persist for extended periods, without producing any disease, suggesting that instances of a permanent carrier state are likely to apply to most, if not all, pathogens. Here are a few examples:

-1. We now know that *Bartonella* species can live in blood without causing disease, producing an asymptomatic bacteremia in a wide assortment of animals that includes humans. The mechanism of *Bartonella* transmission from animal to

FIG. 1.7 Some species of Class Archaea live intimately with members of Class Animalia. In a symbiotic threesome, single-celled eukaryotes living in the digestive tracts of termites digest cellulose, releasing hydrogen, which is metabolized by archaean prokaryotes to yield methane. *Source: Wikipedia, from a public domain work produced by the US Department of Agriculture.*

animal is not fully understood, but arthropod vectors (ticks, fleas, and lice) are suspected, as well as scratches and bites from infected animals (e.g., cats, rats) [72]. Generations of physicians and clinical microbiologists were trained to believe that healthy blood is a sterile medium, and that any organism found growing in a patient's blood is pathogenic (i.e., disease producing). We now know that this simply is not so. [Glossary Vector, Zoonosis]

−2. *Mycobacterium tuberculosis* can infect an individual, produce a limited pathologic reaction in the lung, and remain in the body in a quiescent state for the lifetime of the individual. In fact, it has been estimated that about one out of three individuals, worldwide, is infected with *Mycobacterium tuberculosis*, and will never suffer any consequences. Luckily, asymptomatic carriers of tuberculosis, in whom there is no active pulmonary disease, are non-infective to other individuals.

−3. *Staphylococcus aureus*, a bacterial pathogen that is known to produce abscesses, invades the host through tissues, and releases toxins. We find instances of *Staphylococcus aureus* circulating in the blood, without causing any symptoms, in a sizable portion of the human population [73]. [Glossary Invasion]

Single-celled eukaryotes that are not plants, animals, or fungi (informally known as protists)

Hundreds of these species, of varying levels of pathogenicity, have been identified in humans. Using Class Coccidia as one example, we have *Cryptosporidium parvum, Cyclospora cayetanensis, Cystoisospora belli (Isospora belli), Sarcocystis suihominis, Sarcocystis bovihominis,* and *Toxoplasma gondii.* About one-third of the human population has been infected by *Toxoplasma gondii* (the cause of toxoplasmosis). [Glossary Carrier]

As with the bacteria, we are finding that species of eukaryotes, once thought to be purely pathogenic, are often found living within apparently health humans. In particular, single-celled eukaryotes of Class Trypanosomatida, including the agents causing Chagas disease (*Toxoplasma cruzi*) and leishmaniases (*Leishmania donovani*), are commonly found living in apparently normal individuals. Likewise, the single-celled eukaryote *Toxoplasma cruzi* (Class Apicomplexa), known to cause toxoplasmosis, may persist in humans indefinitely without causing disease. [Glossary Protist versus protoctist versus protozoa]

Fungi

Many different fungal species reside in humans. Representatives of the genus *Candida* dwell in just about every crevice and cranny you can name (e.g., mouth, armpit, genitals).

Archaeplastida (Plants)

Archaeplastida is, broadly, the plant kingdom and consists of the green plants (Viridiplantae) plus red algae (Class Rhodophyta) and the glaucophyte algae (Class Glaucophyta). There seems to be only one class of organism, in the entire kingdom of plants, which is capable of causing an infectious disease in humans. These organisms are species in the genus Prototheca, a single-cell organism belonging to Class Viridiplantae. Both *Prototheca wickerhamii* and *Prototheca zopfi* have been isolated from human infections. *Prototheca wickerhamii,* and other single-celled descendants of Class Archaeplastida are ubiquitous in our environment, and are found in water and soil (Fig. 1.8). [Glossary Infectious disease]

Animals (Class Metazoa)

Yes, animals live in us. Aside from the many different animal parasites that infect humans, we have animals that reside in us, seldom causing disease. A good example is the mite *Demodex follicularis,* which lives in pilosebaceous units (hair follicles and sebaceous glands).

Along with raw coexistence, the post-Cambrian brought all manner of mutualisms (i.e., the condition in which organisms assist one another). Every high-school student is acquainted with the basic concepts of symbiosis and of coevolution. As an example, certain species of ants (specifically, the attine ants) began cultivating fungi in gardens, about 50 million years ago. For about 20 million years, various fungal species were cultivated, but about 30 million years ago, the attine ants switched to monoculture, using one species of fungus to the exclusion of all others. This approach worked to their benefit. Some attine nests have a volume equivalent to a bus, contain 5–10 million worker ants, and maintain up to 1000 fungal gardens.

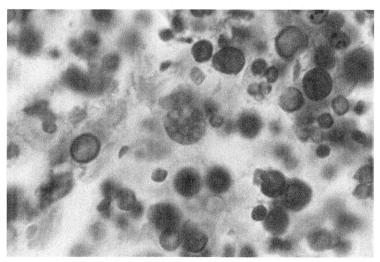

FIG. 1.8 Histologic image of *Prototheca wickerhamii* causing human infection (protothecosis). Prototheca is the only genus of Class Archaeplastida (formerly known as Class Plantae) known to produce human disease. *Source: CDC (US Center for Disease Control and Prevention), provided as a public domain work.*

The most widely known examples of mutualism come from members of Class Insecta (a subclass of arthropods that first appeared about 300 million years ago) and Class Angiosperma (flowering plants that may have appeared as early as 250 million years ago). The greatest proliferation of new insect species coincided with the greatest proliferation of new angiosperm species, when paired species of insect and plant cooperated as food donors (plants) and pollen distributors (insects). This mutual radiation of angiosperm and insect species occurred in a period roughly extending from 150 million years ago to 50 million years ago. The angiosperms seem to have gotten an additional species boost from the tail end of the eutherian radiation (occurring 66 million years ago to 50 million years ago), which provided angiosperms with a surplus of eutherian pollinators. In terms of numbers of species, insects are the most successful animals living today, and angiosperms vastly outnumber their sister class of seed plants, the gymnosperms. Nature seems to be telling us that it is better to make peace than war.

Glossary

Acquired disease Everything is acquired, even inherited diseases are acquired. In common parlance, an "acquired disease" results from conditions occurring in the environment or as the result of events, processes, or behaviors that happened after birth. Some so-called acquired diseases are genetic conditions that were not clinically expressed at birth (i.e., not congenital). For example, the condition known as focal epilepsy with speech disorder is referred to as an acquired disease because children reported to have this disease are typically born without clinical signs of the condition. Recent analysis indicates that the disease is caused by a germ-line mutation, and does not fit the definition of an acquired disease [74].

As an additional example, consider infectious diseases. It seems obvious that all infectious diseases are acquired. Nonetheless, there is great variation in susceptibility to infectious diseases, and genetics plays a role in determining who acquires an infection and who does not. The term "acquired" is applied throughout medical literature in such an inconsistent manner that it scarcely retains any biological meaning.

Amyloid world We tend to think of proteins as individual molecules; not as large aggregates. It seems intuitively obvious that any protein that serves as a catalyst (i.e., any enzyme) would be in the best position to bind to its intended substrate molecules if it were free-floating and untethered. All the introductory biochemistry books have graphic illustrations of enzymes that seem to float across the page, to participate in well-defined metabolic pathways.

As it happens, polypeptides and proteins (large polypeptides) can form large fibrillar aggregates, known as amyloid. These fibrils usually take the form of a large pleated protein sheet (the so-called beta-pleated sheet). Occasionally, amyloid is produced in large quantities, occupying extracellular spaces and causing pathologic consequences. The clinical presentation of amyloidosis will depend on the cellular source of the amyloid (e.g., immunocytes, neurons) and the molecular identity of the aggregate (immunoglobulin light chains, transthyretin, alpha-synuclein), and the quantity of amyloid produced.

It is easy to dismiss amyloid as a pathogenic material of no benefit to living organisms, but in the epoch preceding life as we know it, it is quite possible that amyloid protein conformations may have competed with non-amyloid proteins for a major role in terrestrial life forms [24]. Amyloid aggregates are chemically stable and resistant to most proteolytic metabolic pathways [75]. Amyloid proteins can serve as enzymatic catalysts, and they can act as templates in their own replication. If things had gone differently, ancient organisms may have evolved into an amyloid world.

Today, physicians think of amyloid in terms of its role in disease (i.e., amyloidosis). It is worth noting that amyloid participates in the life of organisms, playing a role in the formation of fimbriae in some types of bacteria, the epigenome of fungi, and in hormone release in humans [76].

Ataxia telangiectasia Also known as Louis-Bar syndrome and as Boder-Sedgwick syndrome, and caused by a mutation of the ATM gene, resulting in a defect in DNA repair. Cells of individuals with ataxia telangiectasia are highly vulnerable to radiation toxicity. The clinical phenotype of ataxia telangiectasia is cerebellar ataxia (i.e., a body movement disorder secondary to cerebellar impairment), telangiectases (i.e., small focal vascular malformations), immune deficits predisposing to ear, sinus, and lung infections, and a predisposition to malignancy (e.g., lung, gastric, lymphoid, and breast cancers).

"But-for" test From the field of law, the "but-for" test seeks to determine whether a sequence of actions leading to an event could have happened without the occurrence of a particular underlying action or condition. In the realm of death certification, the underlying cause of death satisfies the "but-for" test (i.e., but for the condition, the sequence of events leading to the individual's death would not have occurred).

Cancer progression The acquisition of additional properties of the malignant phenotype, in a tumor, over time. Progression is achieved through a variety of mechanisms [e.g., genetic instability [77], epigenetic instability, and aberrant cell death regulation] and results in the eventual emergence of subclones that have growth advantages over other cells in the same tumor. The presence of subclones of distinctive phenotype and genotype within a single tumor accounts for tumor heterogeneity [78].

Carcinogenesis The cellular events leading to cancer. Equivalent to the pathogenesis of cancer. Carcinogenesis in adults is a long process that involves the accumulation of genetic and epigenetic alterations that eventually produce growing clones of malignant cells. The conjectured sequence of events that comprise carcinogenesis begins with initiation, wherein a carcinogen damages the DNA of a cell, producing a mutant clonal founder cell that yields a group of cells that have one or more subtle (i.e., morphologically invisible) differences from the surrounding cells (e.g., less likely to senesce and die, more likely to divide, less genetically stable, better able to survive in an hypoxic environment). After a time, which could easily extend into years, subclones of the original clone emerge that have additional properties that confer growth or survival advantages (e.g., superior growth in hypoxic conditions). The process of subclonal selection continues, usually for a period of years, until a morphologically distinguishable group of cells appear: the precancer cells. Subclonal cells from the precancer eventually emerge, having the full malignant phenotype (i.e., the ability to invade surrounding tissues and metastasize to distant sites). The entire process may take decades.

Carrier In the field of genetics, a carrier is an individual who has a disease-causing gene that does not happen to cause disease in the individual. For example, individuals with one sickle cell gene are typically not affected by sickle cell disease, which usually requires homozygosity (i.e., both alleles having the sickle cell gene mutation) for disease expression. When two carriers mate, the offspring has a 25% chance of inheriting both alleles (i.e., being homozygous for the sickle cell mutation) and developing sickle cell disease.

In some cases, an individual may carry a low-penetrance disease gene, without any signs of disease. The carrier, in this instance, my pass the gene to his or her child, who may, in turn, develop the full clinical phenotype of the disease.

In the field of infectious diseases, a carrier is an individual who harbors an infectious organism, but who suffers no observable clinical consequences. If the carrier state is prolonged, and if infectious organisms cross to other individuals, a single carrier can cause an epidemic.

Cause The event or the condition that is responsible for a specific result. The argument is made herein that diseases develop in a sequence of biological steps, over time, involving multiple cellular pathways. Hence, there are many opportunities to contribute to the disease process, and there is no justifiable reason to think that a disease has a single cause.

Cell-type The number of different kinds of cells in an organism varies based on how you choose to categorize and count them, but most would agree that there are at least 200 different cell types in the adult body. The number of cell types that appear for a short period during in utero development, then disappear before birth, is not included in the count. It can be difficult to assign a cell type to a fetal cell whose precise function cannot be specified.

Chance occurrence Biomedical scientists speak glibly about biological phenomena that occur "by chance" or "at random," but there is no credible evidence to support the intuitively inviting hypothesis that events occur randomly in nature. As a point of fact, we have grown adept at identifying causes for many observed biological processes. For those biological processes for which a specific root cause is unknown (the so-called sporadically occurring diseases), the most we can say is that we do not know what is happening in such instances. We cannot say, with any certainty, that unexplained events occurred "by chance" or "at random."

It is worth noting that randomness is a mathematical abstraction. When we say that a biological process is random, or occurs by chance, what we are actually saying is that the outcome of a biological process can be reasonably modeled as a probabilistic construct. For example, when we flip a coin or throw dice, we can expect the outcomes to approximate a probabilistic event, even though we know that the outcomes are actually determined by physical forces (i.e., not probabilistic). Likewise, when an individual develops a disease due to a de novo germline mutation, it may seem that the disease occurred by chance, but we simply do not know the physical/chemical forces that yielded the specific mutation that served as the root genetic cause of the disease.

Chloroplast evolution Chloroplasts are the organelles (little membrane-wrapped replicating structures within cells) that produce glucose and oxygen, via a process that can be loosely described as:

```
carbon dioxide + water + light energy -> carbohydrate + oxygen.
```

With very few exceptions, chloroplasts are found in all members of Class Plantae. Aside from photosynthesis occurring in plants, we can also observe photosynthesis in cyanobacteria. Photosynthesis produced by cyanobacteria is thought to account for the conversion of our atmosphere from an anoxic environment to an oxygen-rich environment. Did photosynthesis, a complex chemical pathway, evolve twice in terrestrial history; once in cyanobacteria and once again in primitive plants?

Present thinking on the subject holds that the evolution of photosynthesis occurred only once, in the distant past, and all photosynthesis ever since, in cyanobacteria and in plants, arose from this one event. It is presumed that plants acquired photosynthesis when they engulfed photosynthesizing cyanobacteria that evolved into self-replicating chloroplasts. This conclusion is based on the observation that chloroplasts, unlike all other plant organelles, are wrapped by two membranous layers. One layer is believed to have been contributed by the captured cyanobacteria, and one layer presumably came from the capturing plant cell as it wrapped the cyanobacteria in its own cell membrane.

Class A taxonomic class is an abstract representation of a hypothesized group of subclasses that have the same ancestry and all of which share the same class attribute(s). A class is not a biological entity; it is a way of organizing related subclasses according to a set of rules or properties. When we indicate that humans are descended from a class (e.g., Class Mammalia), we are simply saying that we descended from one of the unidentified early species of Class Mammalia, and that we conform to the general definition of the Class (i.e., we meet the biological criteria for a mammalian species). Humans are also members of Class Eukaryota, because we descended from the eukaryotes and our cells have all the defining features of Class Eukaryota (i.e., a nucleus and one or more mitochondria). As a point of fact, humans are members of every class in the descendant lineage of classes that begins with Class Eukaryota and descends to Class Homo.

Taxonomists traditionally rank and split classes. For example, the class known as "Phylum" can be split into Superphylum, Phylum, Subphylum, Infraphylum, and Microphylum. The other ranks are likewise split. The ranks and subranks often have a legitimate scientific purpose. Nonetheless, current taxonomy order is simply too detailed for readers to memorize. Throughout this book, any and all ranks are referred to, simply, as "Class."

The word "class," lowercase, is used as a general term. The word "Class," uppercase, followed by an uppercase noun (e.g., Class Animalia), represents a specific class within a formal classification.

Classification A system in which every object in a knowledge domain is assigned to a class within a hierarchy of classes. The properties of the parent class are inherited by the child classes. Every class has one immediate parent class, although a parent class may have more than one immediate subclass (i.e., child class). Objects do not change their class assignment in a classification, unless there was a mistake in the assignment. For example, a rabbit is always a rabbit, and does not change into a tiger.

Classifications can be easily modeled with computational algorithms and are non-chaotic (i.e., calculations performed on the members and classes of a classification should yield the same output, each time the calculation is performed). A classification should be distinguished from an ontology. In an ontology, a class may have more than one parent class and an object may be a member of more than one class. A classification can be considered a simplified ontology wherein each class is limited to a single parent class and each object has membership in one and only one class [79].

Codon A codon is a sequence of three nucleotides that can be translated as an amino acid or as a stop instruction. Synonym: nucleotide triplet. There are 64 possible triplets, and only about 20 naturally occurring proteogenic amino acids. Hence, there is considerable redundancy in the code.

Commensal A symbiotic relationship between two organisms in which one of the organisms benefits and the other is unaffected, under normal conditions. A commensal may become an opportunistic pathogen when the host provides a physiologic opportunity for disease, such as malnutrition, advanced age, immunodeficiency, overgrowth of the organism (e.g., after antiobiotic usage), or some mechanical portal such as an indwelling catheter, or an intravenous line that introduces the organism to a part of the body that is particularly susceptible to the pathologic expression of the organism.

A commensal relationship between bacteria and an animal parasite may produce a pathogenic relationship in the parasite's host. For example, the bacterium *Wolbachia pipientis* happens to be an endosymbiont that infects most members of the filarial Class Onchocercidae [80]. *Onchocerca volvulus* is a parasitic filarial nematode in humans. The filaria migrate to the eyes, causing river blindness, the second most common infectious cause of blindness worldwide [81]. *Wolbachia pipientis* lives within *Onchochera volvulus*, and it is *Wolbachia* organism that is responsible for the inflammatory reaction that leads to blindness. Hence, *Wolbachia pipientis* is a commensal in *Onchocerca volvulus* and a pathogen in humans, simultaneously. Treatment for river blindness may involve a vermicide, to kill *Onchocerca volvulus* larvae, plus an antibiotic, to kill *Wolbachia pipientis*.

Complex disease A vague term often used to indicate that the pathogenesis of a disease cannot be understood. When the development of a disease involves numerous environmental factors, some known and others assumed, as well as multiple genetic and epigenetic influences, we have no way to fully understand how all of these factors interact with one another, and we have no way to fully describe the biological steps that lead to the clinical expression of disease. From the viewpoint of the clinician, it is often impossible to predict how different patients, all having the same complex disease, will respond to a given treatment. In this book, we use the term "common disease" interchangeably with "complex disease." The common diseases, other than those caused by infection, are always complex.

Congenital disorder Applies to congenital anomalies (i.e., structural or anatomic deformities) plus metabolic diseases, genetic or acquired, that are present at birth.

Cyanobacteria The most influential organisms on earth, cyanobacteria, were the first and only organisms to master the intricacies of oxygenic photosynthesis (>3 billion years ago). Oxygenic photosynthesis involves a photochemical reaction that uses carbon dioxide and water, and releases oxygen. All photosynthesizing life forms are either cyanobacteria, or they are eukaryotic cells (e.g., algae, plants) that have acquired chloroplasts. Chloroplasts, in turn, are organelles created in the distant past by endosymbiosis between a eukaryote and a cyanobacteria. Hence, all oxygenic photosynthesis that has ever occurred on the planet earth stems from a metabolic pathway that evolved in cyanobacteria. Before the emergence of oxygen-producing cyanobacteria, the earth's atmosphere had very little oxygen and organisms were anaerobic.

DNA methylation DNA methylation is a chemical modification of DNA that does not alter the sequence of nucleotide bases. It is currently believed that DNA methylation plays a major role in cellular differentiation, controlling which genes are turned on and which genes are turned off in a cell, hence determining a cell's "type" (e.g., hepatocyte, thyroid follicular cell, neuron). Because cells of a particular cell lineage divide to produce more cells of the same lineage, DNA methylation patterns must be preserved with each somatic cell generation. About 1% of DNA is methylated in human somatic DNA, and DNA methylation occurs preferentially on cytosine, most often at sites where cytosine is followed by guanine (designated as "CpG").

DNA repair With the exception of DNA, damaged cellular molecules need not be repaired. They are simply replaced with newly synthesized molecules. Because DNA is the template for its own replication, damaged DNA molecules cannot be replaced with new DNA molecules; the damage must be repaired in situ if the organism is to continue replicating faithfully. Repairing DNA is of such great importance to cellular life that mammals have at least seven different DNA repair pathways, each specializing in a particular type of damage repair (e.g., direct reversal repair, mismatch repair, nucleotide excision repair, homologous recombination, base excision repair, single strand break repair, nonhomologous end joining, and Fanconi Anemia DNA cross-link repair). Some types of DNA lesions are substrates for more than one pathway, and the ways in which these pathways interact are complex [82].

Dormancy The period from the time that a metastatic cell has seeded to a site that is non-adjacent to the primary tumor and the time that the seeded focus grows to a clinically detectable mass. Dormancy has a variable length, varying from days to decades. We know very little about the pathways that control dormancy. Most people who die of cancer succumb to multiple metastatic lesions; the primary cancer seldom kills. If we had a method that prolonged the dormancy of metastatic foci, it would bring enormous medical benefit to individuals with cancer.

Driver pathway A metabolic pathway that develops during the pathogenesis of disease, and that persists to play a necessary role in the clinical expression of the disease. A "driver pathway" is distinguished from a passenger pathway, the latter being a pathway that plays a role in the pathogenesis of the disease, but does not serve to maintain the clinical phenotype after the disease has fully developed.

The distinction between driver pathways and passenger pathways may have therapeutic importance. A driver pathway, even if it is present in a small portion of people affected with a disease, is likely to be a valid therapeutic target in the subset of patients who are shown to have the driver pathway. A passenger pathway would be the target for agents that prevent the development of the disease.

Druggable driver A driver pathway that serves as a molecular target for a therapeutic drug. The ideal druggable driver would have the following properties:
- 1. The pathway is necessary for the expression for disease, but is not necessary for the survival of normal cells (i.e., you can eliminate the pathway without killing normal cells).
- 2. There must be a pathway protein that is necessary for the activity of the pathway (i.e., if the protein is removed, the pathway cannot proceed).
- 3. The protein can be targeted by a drug. Among other properties, this condition informs us that the targeted protein must be chemically stable.
- 4. The protein target is itself not necessary for the survival of normal cells (i.e., targeting the protein must not kill the patient).

Endoderm There are three embryonic layers that eventually develop into the fully developed animal: endoderm, mesoderm, and ectoderm. The endoderm forms a tube extending from the embryonic mouth to the embryonic anus. The mucosa of the gastrointestinal tract, and the epithelial cells of the liver, pancreas, and lungs, all derive from endoderm.

Epigenome The epigenome, at its simplest, consists of cell-type specific chemical modifications to DNA that do not affect the sequence of nucleotides that comprise the genome. There are many kinds of non-sequence modifications that are included in the "epigenome." One of the best-studied epigenomic modifications is DNA methylation. The most common form of methylation in DNA occurs on cytosine nucleotides, most often at locations wherein cytosine is followed by guanine. These methylations are called CpG sites. CpG islands are concentrations of CpG sites. There are about 29,000–50,000 CpG islands [83]. The patterns of methylation are inherited among cells of the same type (e.g., neuron to neuron, liver cell to liver cell). Alterations in methylation patterns are referred to as epimutations. Epimutations may persist in those specialized cells that are descendants from an epimutated cell [84]. A simple way to think about the respective roles of genome and epigenome is as follows: The genome establishes the identity of an organism; the epigenome establishes the identity of the individual cell-types within the organism. Another major epigenome constituent consists of the histones, and non-histone nuclear proteins. Beyond this minimalist definition, there are expanded versions of the definition that would include any conformational changes in DNA that influence gene expression (e.g., heterochromatin), as well as protein interactions, RNA interactions, and any other physicochemical interactions that influence gene expression. As used in this book, the terms "epigenome" and "epigenetics" apply exclusively to non-sequence alterations in chromosomes that are heritable among somatic cell lineages [85, 86].

Epistasis A gene's role may be influenced by other genes, a phenomenon called epistasis. For example, a gene may be active only when a particular allele of one or more additional genes is also active. Because dependencies among genes can be expected in complex biological systems, the role of epistasis in the penetrance of disease genes and the pathogenesis of disease phenotypes is presumed to be profound. For example, there are at least 27 epistatic interactions claimed to be associated with the occurrence of Alzheimer's disease [87]. Epistatic interactions can be synergistic or antagonistic [88].

It should be kept in mind that genes do not directly interact with other genes; genes are just sequences of DNA in chromosomes. Interactions between genes must be mediated through other molecules, and there are a multitude of mechanisms whereby an epistatic effect may ensue. Examples might include coding for transcription factors that control the expression of other genes, modifications of regulatory systems that control gene expression, modifying the synthesis of proteins encoded by a gene, changing metabolic pathways that use the proteins encoded by genes, and so on. Epistatic interactions are complex, and attempts to predict the functional effect of single or multiple gene variations are typically futile [89, 90]. Because "epistasis" never tells us much about what is actually happening in cells, it is probably best to avoid using the term, whenever possible.

Exaptation Refers to a biological processes or properties that have been modified for some alternate purpose. For example, feathers, or something like feathers, existed before flight, when they may have served to preserve heat in small dinosaurs. Over time, feathered dinosaurs began to use their feathers for flight. It is through exaptation that traits evolve.

Founder effect Occurs when a specific mutation enters the population through the successful procreational activities of a founder and his or her offspring, who carry the founder's mutation. When all of the patients with a specific disease have an identical mutation, the disease may have been propagated through the population by a founder effect. This is particularly true when the disease is confined to a separable subpopulation, as appears to be the case for Navaho neurohepatopathy, in which the studied patients, all members of the Navaho community, have the same missense mutation. Not all diseases characterized by a single gene mutation arise as the result of a founder effect. In the case of cystic fibrosis, a dominant founder effect can be observed within a genetically heterogeneous disease population. One allele of the cystic fibrosis gene accounts for 67% of cystic fibrosis cases in Europe. Hundreds of other alleles of the same gene account for the remaining 33% of cystic fibrosis cases [91].

Perhaps the most notable "founder" is the so-called "mitochondrial Eve." Mitochondria are inherited whole from the cytoplasm of maternal oocytes. Hence, all of us can, in theory, trace our mitochondria back up to the founding woman of the human race. In fact, the mitochondrial lineage leads back to the very first eukaryotic cell. This is a corollary to the rule that "all cells come from cells," but in this case, "all mitochondria come from mitochondria."

Gamete A differentiated germ cell (i.e., an egg in the female or a sperm cell in the male).

Gene pool The aggregate collection of genetic material that is available to the species.

Gene regulation Gene expression is influenced by many different regulatory systems, including the epigenome (e.g., chromatin packing, histone modification, base methylation), transcription and post-transcription modifiers (e.g., transcription factors, DNA promoter sites, DNA enhancer sites, cis- and trans-acting factors, alternative RNA splicing, miRNA and competitive endogenous RNAs, additional forms of RNA silencing, RNA polyadenylation, mRNA stabilizers), translational modifiers and posttranslational protein modifications.

Disruptions of any of these regulatory processes produce disease in humans and other metazoans [92–101]. Moreover, anything that modifies any regulatory process (e.g., environmental toxins, substrate availability, epistatic genes) can influence gene regulation and hence, can produce a disease phenotype.

Generalization Generalization is the process of extending relationships from individual objects to classes of objects. For example, Darwin generalized his observations on barnacles, and other animals, to yield the theory of evolution by natural selection, thus explaining the development of all terrestrial organisms. Science would be of little value if observed relationships among objects could not be generalized to classes of objects.

Genome The collected assortment of an organism's hereditary information, encoded as DNA. For humans, this would mean the DNA found in chromosomes, plus the DNA from mitochondria.

Germ cell A gamete (i.e., an ova in females and a sperm cell in males), or a stem cell belonging to the cell lineage that produces gametes.

Germ cell line The germ cell line is a specialized lineage of cells that appears early in embryogenesis. The germ cell line produces gametes (ova in females and sperm in males). In addition, the earliest cells of the lineage have the ability to "erase" their epigenome, yielding an uncommitted, hence totipotent cell that can differentiate toward any lineage. Tumors that arise from the germ cell line are the tumors of the ova and sperm cell precursors (dysgerminomas and seminomas) and the tumors that arise from totipotent cells (e.g., teratomas and embryonal carcinomas). "Germ cell line" needs to be distinguished from "Germline," comprising the cells derived from the fertilized zygote.

Haploid From Greek haplous, "onefold, single, simple." Equivalent to the chromosome set of a gamete. In humans, this would be 23 chromosomes: one set un-paired autosomes (chromosomes 1–22) plus one sex chromosome (X or Y chromosome). The somatic cells of the body (i.e., all cells that are not gametes) are diploid (i.e., having twice the haploid set of chromosomes).

Haploid organisms All multicellular organisms (i.e., all members of Class Plantae, Class Metazoa, and Class Fungi) may have both haploid and diploid stages of existence. That is to say that every animal, plant, and fungus can be envisioned as two different organisms each containing its own distinctive type of cells (haploid or diploid). In humans, the haploid "organisms" are the gametes (oocytes or sperm).

Many fungi, algae, and bryophyte plants can grow as mature haploid or as diploid organisms, depending on their life cycle stage. The soma of most animals (i.e., the body of the animal, excluding its component of gametes) is diploid, but some male bees, wasps, and ants are haploid organisms because they develop from unfertilized, haploid eggs [102].

Homeobox Genes that code for transcription factors involved in anatomic development in animals, fungi, and plants. HOX genes are specialized homeobox genes found in metazoans that determine the axial relationship of organs. Mutations of homeobox genes are associated with remarkably specific, often isolated, anatomic alterations.

Examples are:
- MSX2 homeobox gene mutation, which produces enlarged parietal foramina.
- PITX1 homeobox gene mutation, which produces Rieger syndrome (hypodontia and malformation of the anterior chamber of the eye including microcornea and muscular dystrophy).
- PITX3 homeobox gene mutation, which produces anterior segment dysgenesis of the eye, moderate cataracts, and anterior segment mesenchymal dysgenesis.
- NKX2.5 homeobox gene, which produces atrial septal defect and atrioventricular conduction defects.
- SHOX homeobox (short stature homeobox) gene mutation causes Leri-Weill dyschondrosteosis (deformity of distal radius, ulna, and proximal carpal bones as well as mesomelic dwarfism).

The reason why homeobox mutations tend to produce diseases in isolated anatomic locations or involve some specific function, probably results from the coordinated regulatory activity of the individual homeobox genes. For example, one gene might regulate the synthesis of a group of proteins exclusively involved in growth of particular skull bones; another homeobox gene might regulate proteins involved in insulin production.

Homolog Genes from different organisms are considered homologous to one another if both descended from a gene in a common ancestral organism. Homologous genes tend to share similar sequences.

Infectious disease A disease caused by an organism that enters the human body. The term "infectious disease" is sometimes used in a way that excludes diseases caused by parasites. In this book, the parasitic diseases of humans are included among the infectious disease. The term "infectious disease" is often used interchangeably with "infection," but the two terms are quite different. It is quite possible to be infected with an organism, even a pathogenic organism, without developing a disease. As a point of fact, the typical human carries many, perhaps dozens, of endogenous pathogenic organisms that lie dormant under most circumstances. Examples are *Pneumocystis jiroveci* (the fungus that causes pneumonia in immunodeficient individuals), Varicella (the virus that may, when the opportunity arises, erupt as shingles), *Aspergillus* species (which not uncommonly colonize the respiratory tract, producing pneumonia in a minority of infected individuals), and *Candida* (a ubiquitous fungus that lives on skin and in mucosal linings and that produces diseases of varying severity in a minority of infected individuals).

Initiation In the field of cancer, the term "initiation" refers to the inferred changes in cells following exposure to a carcinogen, that may eventually lead to the emergence of a cancer in the cell's descendants. The process that begins with initiation and extends to the emergence of a cancer is called carcinogenesis. In molecular biology, the term "initiation" has a different meaning, referring instead to the necessary molecular events that allow a process (e.g., replication, transcription, or translation) to begin.

Invasion Invasion occurs when tumor cells move into and through normal tissues. It is a term that is often applied to cancerous growth, but which may also apply to the aggressive growth of microorganisms.

LUCA Abbreviation for Last Universal Common Ancestor, also known as the cenancestor, or (incorrectly) the progenote. Assuming that all organisms on Earth descend from a common ancestor, then LUCA is the most recent population of organisms from which all organisms now living on Earth have a common descent. LUCA is thought to have lived 3.5–3.8 billion years ago [103].

Law of sequence conservation If a sequence is conserved through evolution (i.e., if you can find a closely similar sequence that is present in all the animals within a phylogenetic class), then that sequence must perform a useful function for the organism. Furthermore, sequences that are highly conserved (i.e., with very little difference among class members) are likely to have a very important function. This law is so useful and so fundamental to genomics and to gene-related computational algorithms that it may as well be known as the First Law of Bioinformatics. The corollary to the law is that genomic sequences that degenerate over time, and for which there are large variations in the general population of a species, must not have a very important function in the organism.

Li-Fraumeni syndrome A rare inherited cancer syndrome associated with a mutation of the p53 tumor suppressor gene. Affected individuals are at risk of developing rhabdomyosarcoma, soft tissue sarcomas, breast cancer, brain tumors, osteosarcoma, leukemia, adrenocortical carcinoma, lymphoma, lung adenocarcinoma, melanoma, gonadal germ cell tumors, prostate carcinoma, and pancreatic carcinoma.

Lysosomal storage diseases As proteins, membranes, and other cellular constituents undergo degenerative changes, they are engulfed by lysosomes, and therein digested. When lysosomes lack enzymes required to digest cellular constituents, the engulfed molecules accumulate, causing cellular damage. The diseases caused by inherited lysosomal enzyme deficiencies are known collectively as lysosomal storage diseases. The tissues most affected are the tissues that produce the greatest amount of the accumulating cellular constituent. For example in Gaucher disease, a deficiency of glucocerebrosidase leads to the lysosomal accumulation of glucocerebroside in many tissues, but particularly so in macrophages and other white blood cells. In Tay Sachs disease, a deficiency of beta-hexosaminidase A leads to the accumulation of GM2 ganglioside, particularly in the nerve cells of the brain and the spinal cord.

Malformation A disorder of normal growth or development, often congenital. Technically, the term "malformation" includes growth abnormalities that arise at any time of life (e.g., angiomas appearing in adults). In common parlance, "malformation" is reserved for congenital conditions.

Malignant Any medical condition whose natural (i.e., untreated) biological course leads to the death of the patient. Most commonly, the term "malignant" is reserved for aggressive cancers.

Malignant phenotype The term "malignant phenotype" refers to the biological properties of sustained growth, invasiveness, and the ability to metastasize that characterize cancers. The term "malignant phenotype" may also apply to the morphological changes found in cancer cells that are associated with aggressive behavior of the tumor. These would include nuclear atypia (i.e., when the nucleus is large and misshapen), histologic disorganization, cellular crowding, loss of normal differentiation (i.e., loss of some of the cytologic features expected to be found in the type of cell from which the tumor arises), high rates of mitosis, and high numbers of degenerate or dead cells.

Meiosis A method of cell division wherein a diploid set of chromosomes recombines to produce two daughter cells, each with a double-haploid chromosome content (meiosis I); followed by a second round of division, in both daughter cells, to yield four haploid cells (meiosis II).

Melanoma Synonymous with malignant melanoma, a cancer of melanocytes, the pigment-producing cells of the skin. In the special case of melanoma, there is no lesion that takes the name "benign melanoma." Benign neoplasms of melanocytic origin are called nevi or moles.

Mitosis The phase in the cell cycle of somatic cells (i.e., not germ cells) wherein the replicated chromosomes condense and separate to form the nuclei of two daughter cells.

Mutation A term that should be confined to describing changes in the nucleotide sequences of genomic DNA. These would include point mutations (changes to a single nucleotide) and segment mutations (e.g., losses of a string of nucleotides) and structural variations that delete, amplify, or move segments of DNA.

Mutation types Some of the mutations that account for genetic diseases are: deletions (e.g., Duchenne muscular dystrophy), frame-shift mutations (e.g., factor VIII and IX deficiencies), fusions (e.g., chronic myelogenous leukemia, hemoglobin variants), initiation and termination codon mutations (e.g., a type of thalassemia), inversions (another type of thalassemia), nonsense mutations (familial hypercholesterolemia), point mutations (e.g., sickle cell disease, glucose-6-phosphate dehydrogenase deficiency), promoter mutations (e.g., yet another type of thalassemia), and RNA processing mutations, including splice mutations (e.g., phenylketonuria) [104].

Natural selection The tendency for favorable heritable traits to become more common over successive generations. The traits are selected from expressed genetic variations among individuals in the population. The genetic variations may take the form of genetic sequence variations (e.g., SNPs) or genetic structural variations.

Neutrophils Circulating white blood cells include lymphocytes, platelets, monocytes, and granulocytes (i.e., neutrophils, eosinophils, and basophils). The granulated cells can be distinguished by the dyes that can be absorbed by the different types of granules. Eosinophils retain a pink dye. Basophils retain a blue dye. And neutrophils do not retain dyes of either color.

Nonhomologous Two similar genes, protein, or organs are nonhomologous if they did not arise through descent from a common ancestor.

Nontoxic cancer chemotherapy Until the last two decades, all cancer chemotherapy was designed to be toxic to human cells. The underlying rationale for using toxic drugs is that tumor cells are more sensitive to certain types of toxins than are normal cells. A new generation of nontoxic drugs is aimed at inhibiting pathways in tumor cells that play a key role in the malignant phenotype. Because these new drugs are not intended to kill cells, they are often referred to as nontoxic chemotherapeutic agents. The term "nontoxic" is somewhat misleading because all medications have unintended toxic effects. Still, the new generation of nontoxic chemotherapeutic agents is vastly safer and more tolerable than the preceding generation of cytotoxic agents.

Nucleus The membrane-bound organelle that contains the genome and the apparatus necessary for transcribing DNA into RNA, for translating the RNA into protein molecules, and for replicating the DNA in preparation for cell division. Eukaryotes consist of all the organisms whose cells contain nuclei. The prokaryotes, bacteria plus archaea, are organisms that do not contain a nucleus. Every cell on earth is either a eukaryote or a prokaryote. The nucleus, though necessary for cell division, is not necessary for moment-to-moment cell survival. Mature red blood cells have no nucleus, but they manage to live for about 120 days.

Oncogene Normal genes or parts of genes that, when altered to become a more active form, or overexpressed, contribute toward a neoplastic phenotype in a particular range of cell types. The normal form of the oncogene is called the proto-oncogene. The altered, more active form of the gene, is called the activated oncogene. Activation usually involves mutation, or amplification (i.e., an increase in gene copy number), translocation, or fusion with an actively transcribed gene, or some sequence of events that increases the expression of the gene product. Some retroviruses contain activated oncogenes and can cause tumors by inserting their oncogene into the host genome.

Oxygen crisis Nobody knows with any certainty the history of terrestrial oxygenation, but a consensus of opinion seems to favor the following scenario:

— 1. About 3.5 billion years ago, a prokaryote evolved that could produce oxygen from water, through a rather inefficient pathway that did not involve photosynthesis. All life on earth at this time was prokaryotic, and virtually all living organisms were anaerobes for which oxygen was toxic. A relatively small amount of oxygen was produced, all of which was rapidly trapped by substances within the earth, particularly oxidizing minerals such as iron. As an incidental observation, virtually all of the iron found on earth is oxidized. Pure metallic oxygen is exceedingly rare and most of the pure elemental iron on this planet arrived recently in meteorites that had traveled billions of miles through airless space.

— 2. About 2.7 billion years ago, cyanobacteria evolved oxygenic photosynthesis, producing oxygen efficiently using photons. No organism since has independently evolved oxygenic photosynthesis. Early eukaryotes of Class Archaeplastida captured cyanobacteria and adapted them as organelles (chloroplasts) much later, probably no earlier than 2.1 billion years ago. At this point (i.e., about 2.7 billion years ago), a great deal of oxygen was being synthesized, but this oxygen was absorbed by the oceans and the emerging land masses, and very little oxygen made its way into the atmosphere [105].

— 3. About 2.5 billion years ago, the earth was saturated with oxygen, and the excess gas bubbled its way into the atmosphere.

— 4. About 2.3 billion years ago, atmospheric oxygen reached a level that was toxic to most of the existing anaerobic organisms, producing the so-called oxygen crisis or oxygen catastrophe.

— 5. The really large atmospheric oxygen concentrations, comparable to those we see today, did not come until about 1 billion years ago. By this time, there were many eukaryotes, all containing mitochondria capable of turning oxygen into metabolic energy. The most dramatic examples of organisms exploiting the high-energy metabolism provided by oxygen came about a half billion years later, in the Cambrian explosion [66].

Of course, the precise dates are vague, and readers are expected to make adjustments, based on local time zones.

Paralog A paralogous gene. Refers to genes found in different organisms that evolved from a common ancestor's gene through gene duplication. Paralogs permit the organism to get a new functionality from a gene without losing the functionality of the gene that has been duplicated. A paralog is a type of homolog. All homologs are either orthologs or paralogs, where orthologs arise through sequence changes that occur in one ancestral gene, as it undergoes variations in different descendant species.

Pathogenesis The biological steps preceding and leading to the development of a disease.

Pathway According to traditional thinking, a pathway is a sequence of biochemical reactions, involving a specific set of enzymes and substrates that produces a chemical product or that fulfills a particular function. The classic pathway was the Krebs cycle. It was common for students to be required to calculate the output of the cycle [in moles of adenosine triphosphate (ATP)] based on stoichiometric equations employing known amounts of substrate. As we have learned more and more about cellular biology, the term "pathway" has acquired a broader meaning. One pathway may intersect or subsume other pathways. Furthermore, a pathway may not be constrained to an anatomically sequestered area of the cell, and the activity of a pathway may change from cell type to cell type or may change within one cell depending on the cell's physiologic status. The individual enzymes that participate in a pathway may have different functions, in alternate pathways. Nevertheless, the term "pathway" is a convenient conceptual device to organize classes of molecules that interact with a generally defined set of partner molecules to produce a somewhat consistent range of biological actions.

Pathway-driven disease Refers to disorders whose clinical phenotype is largely the result of a single, identifiable pathway. Diseases with similar clinical phenotypes can often be grouped together as they share a common, disease-driving pathway. Examples would include the channelopathies (driven by malfunctions of pathways that involving the transport ions through membrane channels), ciliopathies (driven by malfunctions of cilia), and lipid receptor mutations (driven by any of the mutations involving lipid receptors). At this point, our ability to sensibly assign diseases to pathways is limited because the effects of a mutation in a single gene may indirectly effect many different pathways, and those pathways may vary from cell type to cell type.

Certain types of conditions do not fall easily into the "pathway-driven" paradigm. For example, it is difficult to speak of a class of diseases all driven by errors in transcription factor pathways. A single transcription factor may regulate pathways in a variety of cell types with differing functions and embryologic origins. Hence, the syndromes resulting from a mutation in a transcription factor may involve multiple pathways and multiple tissues and will not have any single, identifiable pathway that drives the clinical phenotype.

Phenocopy disease A disease that shares the same phenotype as a genetic disease, but without the genotypic root cause. Most of the phenocopy diseases are caused by a toxin or drug. We typically think of a phenocopy disease as the nongenetic equivalent of a genetic disease.

Here are examples of phenocopy diseases and their genetic counterparts:

– Acquired Conduction Defect—Inherited Conduction Defect.
– Acquired porphyria cutanea tarda—Inherited porphyria cutanea tarda.
– Acquired von Willebrand disease—inherited von Willebrand disease.
– Aminoglycoside-induced hearing loss—inherited mitochondriopathic deafness.
– Antabuse (disulfiram) treatment—inherited alcohol intolerance.
– Drug-induced methemoglobinemia—inherited methemoglobinemia.
– Fetal exposure to methotrexate—Miller syndrome [106].
– Methylmalonic acidemia caused by severe deficiency of vitamin B_{12}—Inherited methylmalonic acidemia.
– Osteolathyrism and scurvy—inherited collagenopathies. [Glossary Collagenopathy]
– Alcohol-induced sideroblastic anemia—Inherited sideroblastic anemia.
– Vitamin B_{12} deficiency—Inherited pernicious anemia.
– Cardiomyopathy due to alcohol abuse—inherited dilated cardiomyopathy [107].
– Lead-induced encephalopathy—inherited tau encephalopathy [108].
– Myopathy produced by nucleoside analog reverse transcriptase inhibitors (i.e., HIV drugs)—inherited mitochondrial myopathy.
– Pseudo-Pelger-Hüet Anomaly—Inherited Pelger-Hüetet Anomaly [109, 110].
– Thalidomide-induced phocomelia—Roberts syndrome and SC pseudothalidomide syndrome [111].
– Warfarin embryopathy—brachytelephalangic chondrodysplasia punctata [112].
– Drug-induced cerebellar ataxia [113] and hereditary spinocerebellar ataxia [114].
– Copper poisoning and Wilson disease.
– Chronic iron overload hemochromatosis and inherited hemochromatosis.
– Quinacrine-induced ochronosis and inherited ochronosis (i.e., mutation in the HGD gene for the enzyme homogentisate 1,2-dioxygenase) [115].
– Drug-induced Parkinsonism [116, 117] and autosomal-dominant inherited Parkinsonism [118].
– Acquired pulmonary hypertension due to hypoxia, thromboembolism, left-sided heart failure, or drugs and inherited pulmonary hypertension [119].
– Amphotericin B toxicity and renal tubular acidosis [120].
– Acquired and hereditary storage pool platelet disease [121].
– Acquired and inherited porphyrias [122, 123].
– Acquired iron overload and hemochromatosis and inherited hemochromatosis.
– Acquired cirrhosis and genetic cirrhosis (due to mutation in keratin 18) [124]. [Glossary Cirrhosis]
– Anticoagulant drugs that inhibit thrombus formation and inherited Factor X deficiency (a form of hemophilia).

– Phenocopy diseases provide important clues to the pathogenesis of rare and common diseases. The drug that produces a phenocopy disease is likely to share the same disease pathways observed in the genetic disease. Pharmacologic treatments for the phenocopy disease may be effective against the genetic form of the disease.

Phenotype The set of observable traits and features of a biological object. For example, we can describe a dog as best as we can, and that description would be our assessment of its phenotype. If we describe a lot of dogs of a certain breed, we might come up with some consensus on the phenotype of the breed. A "disease phenotype" is the medical community's consensus on the observed features that characterize a disease. For example, the term "cancer phenotype" refers to the properties of growth, persistence, invasion, and metastasis that characterize virtually every cancer.

Phylum A major class of organism. With respect to classes of animals, a phylum is a class wherein all the members have the same basic body type.

Protist vs protoctist vs protozoa For well over a century, biologists had a very simple way of organizing the eukaryotes [125]. Basically, the one-celled eukaryotes were called protists. The multicelled eukaryotes were assigned to the kingdom of plants or the kingdom of animals, often referred to a flora and fauna, respectively. The fungi were considered a subclass of plants. We now know that fungi are not plants and are not closely related to plants. Fungi are opisthokonts. As such, they are closely related to the metazoans (animals). Regardless, university-based mycologists (specialists in fungi) are typically employed in the Botany Department.

The protists, also known as protoctista, were considered the ancestral organisms for the multicellular organisms. Biologists applied the term "protozoa" to protist species that were the ancestral forms of animals (from "proto," the first and "zoa," animals). Modern taxonomists understand that the protists are a grab-bag of unrelated single-celled organisms. Furthermore, as the phylogenetic lineage of the various eukaryotes has improved, it becomes clear that the multicellular classes have a closer relationship to some of the single-celled organisms than those single-celled organisms had to other protists. Preserving the concept of Class Protoctista was making it difficult to appreciate the true phylogenetic relationships among the eukaryotes. Modern classifications of eukaryotic organisms simply dispense with Class Protoctista, assigning each individual class of eukaryotes to a hierarchical position determined by ancestry.

The term "protist" is reserved as an informal term to indicate any single-celled eukaryote. No longer does "protist" signify a formal taxonomic class [125, 126]. Strict taxonomists will refer to single-celled eukaryotic species by their proper taxa (e.g., Class Holozoa, Class Alveolata, Class Chromista, and Class Amoebozoa).

Rare disease As written in Public Law 107–280, the Rare Diseases Act of 2002, "Rare diseases and disorders are those which affect small patient populations, typically populations smaller than 200,000 individuals in the United States." [127]. This translates to a prevalence of <1 case for every 1570 persons. This is not too far from the definition recommended by the European Commission on Public Health, a prevalence <1 in 2000 people.

Retrovirus An RNA virus that replicates through a DNA intermediate. The DNA intermediate may become integrated into the host DNA, from which viral RNA is transcribed. When integration of the virus occurs in germ cells, the viral DNA can be inherited. Through this mechanism, the human genome carries a legacy of retroviral DNA. Ancient retroviruses account for about 8% of the human genome [69].

Ribozymes Ribozymes, a term conflated from "ribonucleic acid" and "enzymes" function as part of the large subunit ribosomal RNA, linking amino acids during protein synthesis. In addition, ribozymes act as catalysts in a variety reactions within the ribosome (e.g., including RNA splicing, transfer RNA synthesis, and viral replication) [18]. The finding that RNA molecules may act as genetic material and as enzymes has raised the possibility that ancient life forms relied on molecules much like ribozymes, which evolved to yield the various types of specialized RNA molecules that are essential to modern cells.

Root cause The earliest event or condition that is known to set in motion a chain of additional events that can result in some specified result. The term "root cause" is preferable to another term "underlying cause" that is often applied to the same concept. In this book, "underlying cause" is denigrated because any of the events that precede a result could be construed as underlying causes. The term "root cause" conveys the idea of a first or earliest event in a multi-event process. Of course, we can never be certain what the earliest event is in any process. For example, when a smoker dies of lung cancer, is the root cause of death "adenocarcinoma of lung" or "smoking" or "tobacco addiction" or "Unrestricted sales of cigarettes to minors" or "Invention of Cigarettes"? Where do we stop? In the case of inherited genetic diseases, the customary starting point (i.e., the root cause) is the introduction of a genetic error into the germ line of the affected individual. The idea here is that the pathogenesis of a genetic disease begins with the acquisition of the abnormal gene. If you think about it, we don't really know that this is the case. We know of many examples where an individual may carry a disease gene without developing the expected disease. In these cases, some condition required for disease development was not met. Was the condition some environmental factor that influenced the father's sperm or the mother's oocyte prior to fertilization? If so, might this condition, which preceded the acquisition of the genetic muta-tion, be the "root" cause? Again, even in the simplest of cases, it is difficult to assign a root cause with any certainty. We never know if we've looked backwards far enough. Still, we do the best that we can, and we apply the term "root cause" in this book with the understanding that we may need to modify our thinking, if evidence of an earlier event comes to light.

Somatic From the Greek, meaning body, refers to non-germ cells (i.e., not oocytes, not spermatocytes). The somatic cells, then, consist of the differentiated cells of the body and the stem cells in their lineage. Somatic cells do not divide by meiosis, and, under natural conditions, cannot pass acquired mutations to their progeny.

Species The modern definition of species can be expressed in three words: "evolving gene pool" [128]. This elegant definition is easy to comprehend and serves to explain how new species come into exis-tence [129, 130]. Because each member of a species (i.e., each organism) has a genome constructed from the species-specific gene pool, it is clear that membership within a species is immutable (e.g., a fish cannot become a cat and a cat cannot become a goat) inasmuch as their genomes are chosen from their respective gene pools.

   Species have a set of biological properties associated with any living entity: uniqueness, life, death, the issuance of progeny, and the benefit of evolution through natural selection. Hence, we should think of a species as a biological entity; not as an abstraction.

   It is estimated that 5–50 billion species have lived on earth, with >99% of them now extinct, leaving about 10 to 100 million living species [131].

Syndrome A syndrome is a constellation of physical findings that occur together. The symptoms of a syn-drome may or may not all have the same pathogenesis; and, hence, may or may not constitute a disease.

   The common cold is a disease, with a specific cause, that produces a syndromic pattern of headache, sniffles, cough, and malaise. All these symptoms arise after a viral infection at about the same time, and all the symptoms clear at about the same time. Multiple expressions of physical disorder, occurring in multiple body systems, are more likely to be syndromes than diseases, when the different conditions are inconsistently present or separated by intervals of time. Contrariwise, a localized pathological condition that is restricted to a particular set of cells is almost always a disease; not a syndrome.

   There are exceptions. It is possible to imagine some highly unlikely syndromes that are restricted to one cell type. *Clinorchis sinensis* is a species of trematode that causes biliary tract disease when the fluke take up permanent residence therein [132]. Localized reaction to the fluke infection results in chronic cholangitis (i.e., inflammation of the cells lining the bile ducts). In some cases, cholangiocar-cinoma (i.e., cancer of the bile duct) develops. In this case, cholangitis and cholangiocarcinoma arise through different biological pathways, despite having the same root cause (i.e., clinorchiasis). This being the case, it seems reasonable to count cholangitis and cholangiocarcinoma following *Clinorchis sinensis* infection as a syndrome, not as a disease; thus breaking the general rule that a localized pathologic condition is not a syndrome.

There are no hard or fast rules, but the distinction between a disease and a syndrome has therapeutic and diagnostic consequences. A true disease has one pathogenesis, despite its multiorgan manifestations and can be potentially prevented, diagnosed, or treated by targeting events and pathways involved in the development of the disease. A syndrome is a confluence of clinical findings that may or may not share the same pathogenesis.

Taxa Plural of taxon.

Taxon A taxon is a class. The common usage of "taxon" is somewhat inconsistent, as it sometimes refers to the class name, and at other times refers to the instances (i.e., members) of the class. In this book, the term "taxon" is abandoned in favor of "class," the term preferred by computer scientists.

Taxonomy When we write of "taxonomy" as an area of study, we refer to the methods and concepts related to the science of classification, derived from the ancient Greek taxis, "arrangement," and nomia, "method." When we write of "a taxonomy," as a construction within a classification, we are referring to the collection of named instances (class members) in the classification. To appreciate the difference between a taxonomy and a classification, it helps to think of taxonomy as the scientific field that determines how the different members within the classification are named. Classification is the scientific field that determines how related named members are assigned to classes, and how the different classes are related to one another. A taxonomy is similar to a nomenclature; the difference is that in a taxonomy, every named instance must have an assigned class.

Tempo and mode In the context of this book, "tempo and mode" refer to the evolutionary process. Tempo is the rate of evolution at any given epoch. Paleontologists have noticed periods of acceleration or deceleration in the rate of evolution of particular classes of species at particular periods in the history of evolution. Mode is the method by which evolutionary changes in classes or species of organisms are acquired. The tempo and mode of evolution have puzzled paleontologists and taxonomists for many decades [133]. In recent times, the same two terms, "tempo and mode" have been applied to molecular genetics, to address the rate of change in genes over time (i.e., the tempo of the molecular clock) and the mechanisms that drive those changes [134].

Trait Traits are phenotypic features that have a range of quantitative variance within a species. For example, height, weight, and skin color are traits that are present in every member of the species, and which vary considerably among members of a species. Molecular biologists have extended the concept of "trait" to include gene expression patterns, as these collections of cell-based data contain a range of genetic information expressed as a computed graphic [135].

Traits are polygenic and are inherited in a non-Mendelian fashion [136–138]. For example, at least 180 gene variants have been associated with variations of normal height. These 180 variants may represent only a fraction of the total number of gene variants that influence the height of individuals, as they account for only about 10% of the predicted spread [139].

Traits can evolve very quickly within a species. For example, it may take only a few dozen generations for breeders to produce a large dog, such as a mastiff, or a small dog, such as a chihuahua. Between species of the same class of organism, traits can have truly enormous variations. For example, *Wolffia globosa* (duckweed) has been described as the world's smallest flowering plant, at 0.1–0.2 mm in diameter. Compare this with the weight or size of another angiosperm species, the giant redwood. The difference can be upwards of a trillion-fold.

Less is known about the nonquantitative traits: features that you either have or you have not (e.g., blue eyes). The nonquantitative traits may have simple Mendelian inheritance patterns. For example, some horses have a natural ability to use a specialized gait known as the tolt. This gait is common in a certain bread of Icelandic horse and rare among all other breeds. Most horses, no matter how hard they try, cannot tolt. Recently, the tolt was found to come from a single gene mutation [140].

It should be noted that the term "trait" is used by geneticists to indicate that an individual carries a recessive disease mutation on a single allele. For example, sickle cell disease occurs when an individual carries two recessive alleles of the characteristic hemoglobin mutation. Individuals with only one allele affected, who may develop some tendency toward red cell sickling, but who do not develop the complete sickle cell disease phenotype, are said to carry the sickle cell trait.

As used herein, we confine our definition of "traits" to quantitative features present in every member of a species.

Transcription factor A protein that binds to specific DNA sequences to control the transcription of DNA to RNA. The human genome codes for several thousand different transcription factors [141].

Transcription factors regulate the gene expression in multiple organs, often with embryologic origins from more than one embryonic germ layer. Consequently, transcription factor diseases typically involve disparate tissues, and are thusly syndromic [142]. For example, Waardenburg syndrome types 1 and 2, whose root causes are mutations in the PAX3 and MITF transcription factor genes, are associated with lateral displacement of the inner canthus of each eye, pigmentary disturbance including frontal white blaze of hair, heterochromia iridis, white eyelashes, leukoderma, and cochlear deafness. Transcription factor diseases tend to have a dominant inheritance pattern, as insufficiency of one transcription factor allele is sufficient to produce a syndromic phenotype [50].

Vector An organism that transmits a disease-causing organism from one host to another. Diseases that are spread by vectors include: malaria, leishmaniasis, African trypanosomiasis, yellow fever, dengue fever, West Nile encephalitis, and chikungunya. Arthropods are the most common vectors of human diseases. All of the hundred or more arboviruses have arthropod vectors [143].

One vector can carry more than one type of infectious organism. For example, a single species of *Anopheles* mosquito can transmit *Dirofilaria immitis*, the O'nyong'nyong fever virus, *Wuchereria bancrofti*, and *Brugia malayi*. Obversely, one disease organism can be spread by more than one vector. For example, orbiviruses are spread by mosquitoes, midges, gnats, sandflies, and ticks [144].

Zoonosis A infectious disease of humans that it acquired from a nonhuman animal reservoir. The method of infection (e.g., vector) does not determine whether a disease is considered to be zoonotic. Malaria, passed by a mosquito vector, is not a zoonosis, because the reservoir for the organism that causes human malaria is, in most cases, some other infected human (i.e., not a nonhuman animal). The same is true for schistosomiasis, river blindness, and elephantiasis. Although these diseases are transmitted by nonhuman vectors, their typical reservoir is human. The term for an infection in humans that can be transmitted to nonhuman animals is anthroponosis.

References

[1] Wachtershauser G. Pyrite formation, the first energy source for life: a hypothesis. Syst Appl Microbiol 1988;10:207–10.

[2] Wachtershauser G. Before enzymes and templates: theory of metabolism. Microbiol Rev 1988;52:452–84.

[3] Huang W, Ferris JP. One-step, regioselective synthesis of up to 50-mers of RNA oligomers by montmorillonite catalysis. J Am Chem Soc 2006;128:8914–9.

[4] Cammack R, Rao KK, Hall DO. Metalloproteins in the evolution of photosynthesis. Biosystems 1981;14:57–80.

[5] Bai L, Baker DR, Rivers M. Experimental study of bubble growth in Stromboli basalt melts at 1 atm. Earth Planet Sci Lett 2008;267:533–54.

[6] Dodd MS, Papineau D, Grenne T, Slack JF, Rittner M, Pirajno F, et al. Evidence for early life in Earth's oldest hydrothermal vent precipitates. Nature 2017;543:60–4.

[7] Manhesa G, Allegre CJ, Duprea B, Hamelin B. Lead isotope study of basic-ultrabasic layered complexes: speculations about the age of the earth and primitive mantle characteristics. Earth Planet Sci Lett 1980;47:370–82.

[8] Koga Y, Kyuragi T, Nishihara M, Sone N. Archaeal and bacterial cells arise independently from noncellular precursors? A hypothesis stating that the advent of membrane phospholipid with enantiomeric glycerophosphate backbones caused the separation of the two lines of descent. J Mol Evol 1998;46:54–63.

[9] Valas RE, Bourne PE. The origin of a derived superkingdom: how a gram-positive bacterium crossed the desert to become an archaeon. Biol Direct 2011;6:16.

[10] Wilkins AS, Holliday R. The evolution of meiosis from mitosis. Genetics 2009;181:3–12.

[11] Dacks J, Roger AJ. The first sexual lineage and the relevance of facultative sex. J Mol Evol 1999;48:779–83.

[12] Ramesh MA, Malik SB, Logsdon JM. A phylogenomic inventory of meiotic genes; evidence for sex in Giardia and an early eukaryotic origin of meiosis. Curr Biol 2005;15:185–91.

[13] Malik SB, Pightling AW, Stefaniak LM, Schurko AM, Logsdon JM. An expanded inventory of conserved meiotic genes provides evidence for sex in *Trichomonas vaginalis*. PLoS ONE 2008;3:e2879.

[14] Lahr DJ, Parfrey LW, Mitchell EA, Katz LA, Lara E. The chastity of amoebae: re-evaluating evidence for sex in amoeboid organisms. Proc Biol Sci 2011;278:2081–90.

[15] Akopyants NS, Kimblin N, Secundino N, Patrick R, Peters N, Lawyer P, et al. Demonstration of genetic exchange during cyclical development of *Leishmania* in the sand fly vector. Science 2009;324:265–8.

[16] Margulis L, Sagan D. What is life? Oakland: University of California Press; 2000.

[17] Ronimus RS, Morgan HW. Distribution and phylogenies of enzymes of the Embden-Meyerhof-Parnas pathway from archaea and hyperthermophilic bacteria support a gluconeogenic origin of metabolism. Archaea 2002;1:199–221.

[18] Kruger K, Grabowski PJ, Zaug AJ, Sands J, Gottschling DE, Cech TR. Self-splicing RNA: autoexcision and autocyclization of the ribosomal RNA intervening sequence of tetrahymena. Cell 1982;31:147–57.

[19] Johnston WK, Unrau PJ, Lawrence MS, Glasner ME, Bartel DP. RNA-catalyzed RNA polymerization: accurate and general RNA-templated primer extension. Science 2001;292:1319–25.

[20] Zaher HS, Unrau PJ. Selection of an improved RNA polymerase ribozyme with superior extension and fidelity. RNA 2007;13:1017–26.

[21] Wochner A, Attwater J, Coulson A, Holliger P. Ribozyme-catalyzed transcription of an active ribozyme. Science 2011;332:209–12.

[22] Forterre P. The two ages of the RNA world, and the transition to the DNA world: a story of viruses and cells. Biochimie 2005;87:793–803.

[23] Gilbert W. Origin of life: the RNA world. Nature 1986;319:618.

[24] Greenwald J, Kwiatkowski W, Riek R. Peptide amyloids in the origin of life. J Mol Biol 2018;430:3735–50.

[25] Frederic MY, Lundin VF, Whiteside MD, Cueva JG, Tu DK, Kang SY, et al. Identification of 526 conserved metazoan genetic innovations exposes a new role for cofactor E-like in neuronal microtubule homeostasis. PLoS Genet 2013;9:e1003804.

[26] Wu D, Hugenholtz P, Mavromatis K, Pukall R, Dalin E, Ivanova NN, et al. A phylogeny-driven genomic encyclopaedia of Bacteria and Archaea. Nature 2009;462:1056–60.

[27] Neme R, Tautz D. Phylogenetic patterns of emergence of new genes support a model of frequent de novo evolution. BMC Genomics 2013;14:117.

[28] Pandey UB, Nichols CD. Human disease models in *Drosophila melanogaster* and the role of the fly in therapeutic drug discovery. Pharmacol Rev 2011;63:411–36.

[29] Wetterbom A, Sevov M, Cavelier L, Bergstrom TF. Comparative genomic analysis of human and chimpanzee indicates a key role for indels in primate evolution. J Mol Evol 2006;63:682–90.

[30] Erwin D, Valentine J, Jablonski D. The origin of animal body plans. American Scientist; 1997. March/April.

[31] Britten RJ. Almost all human genes resulted from ancient duplication. PNAS 2006;103:19027–32.

[32] Hardison RC. Evolution of hemoglobin and its genes. Cold Spring Harb Perspect Med 2012; 2. a011627.

[33] Storz JF. Gene duplication and evolutionary innovations in hemoglobin-oxygen transport. Physiology (Bethesda) 2016;31:223–32.

[34] Niimura Y. Evolutionary dynamics of olfactory receptor genes in chordates: interaction between environments and genomic contents. Hum Genomics 2009;4:107–18.

[35] Jessen JR, Jessen TN, Vogel SS, Lin S. Concurrent expression of recombination activating genes 1 and 2 in zebrafish olfactory sensory neurons. Genesis 2001;29:156–62.

[36] Glusman G, Yanai I, Rubin I, Lancet D. The complete human olfactory subgenome. Genome Res 2001;11:685–702.

[37] Griffiths DJ. Endogenous retroviruses in the human genome sequence. Genome Biol 2001;2: reviews1017.1—reviews1017.5.

[38] Horie M, Honda T, Suzuki Y, Kobayashi Y, Daito T, Oshida T, et al. Endogenous non-retroviral RNA virus elements in mammalian genomes. Nature 2010;463:84–7.

[39] Patel MR, Emerman M, Malik HS. Paleovirology: ghosts and gifts of viruses past. Curr Opin Virol 2011;1(4):304–9. 2011.

[40] Horowitz NH. On the evolution of biochemical syntheses. Proc Natl Acad Sci U S A 1945;31:153–7.

[41] Lane N. Life ascending: the ten great inventions of evolution. London: Profile Books; 2009.

[42] Madar S, Goldstein I, Rotter V. Did experimental biology die? Lessons from 30 years of p53 research. Cancer Res 2009;69:6378–80.

[43] Zilfou JT, Lowe SW. Tumor suppressive functions of p53. Cold Spring Harb Perspect Biol 2009;1: a001883.

[44] Vogelstein B, Lane D, Levine AJ. Surfing the p53 network. Nature 2000;408:307–10.

[45] Waterham HR, Koster J, Mooyer P, van Noort G, Kelley RI, Wilcox WR, et al. Autosomal recessive HEM/Greenberg skeletal dysplasia is caused by 3-beta-hydroxysterol delta(14)-reductase deficiency due to mutations in the lamin B receptor gene. Am J Hum Genet 2003;72:1013–7.

[46] Piatigorsky J. Gene sharing, lens crystallins and speculations on an eye/ear evolutionary relationship. Integr Comp Biol 2003;43:492–9.

[47] Piatigorsky J, Wistow GJ. Enzyme/crystallins: gene sharing as an evolutionary strategy. Cell 1989;57:197–9.

[48] Wistow G. Evolution of a protein superfamily: relationships between vertebrate lens crystallins and microorganism dormancy proteins. J Mol Evol 1990;30:140–5.

[49] Land MF, Fernald RD. The evolution of eyes. Annu Rev Neurosci 1992;15:1–29.

[50] Seidman JG, Seidman C. Transcription factor haploinsufficiency: when half a loaf is not enough. J Clin Invest 2002;109:451–5.

[51] Corless CL, Schroeder A, Griffith D, Town A, McGreevey L, Harrell P, et al. PDGFRA mutations in gastrointestinal stromal tumors: frequency, spectrum and in vitro sensitivity to imatinib. J Clin Oncol 2005;23:5357–64.

[52] Yang H, Higgins B, Kolinsky K, Packman K, Go Z, Iyer R, et al. RG7204 (PLX4032), a selective BRAFV600E inhibitor, displays potent antitumor activity in preclinical melanoma models. Cancer Res 2010;70:5518–27.

[53] Carlson RH. Precision medicine is more than genomic sequencing. Available at: http://www.medscape.com/viewarticle/870723_print; 2016 [viewed 11.03.17].

[54] Kotha K, Clancy JP. Ivacaftor treatment of cystic fibrosis patients with the G551D mutation: a review of the evidence. Ther Adv Respir Dis 2013;7:288–96.

[55] Herper M. The cost of creating a new drug now $5 billion, pushing big pharma to change. Forbes Magazine; 2013. August 11.

[56] Goldberg P. An old drug's 21st century makeover begins with 84-fold price increase. Cancer Lett 2005. May 13.

[57] Berenson AA. Cancer drug's big price rise is cause for concern. New York Times; 2006. March 12.

[58] Vanchieri C. When will the U.S. flinch at cancer drug prices? J Natl Cancer Inst 2005;97:624–6.

[59] Hurley D. Why are so few blockbuster drugs invented today? The New York Times; 2014. November 13.

[60] Erwin DH. The origin of bodyplans. Am Zool 1999;39:617–29.

[61] Valentine JW, Jablonski D, Erwin DH. Fossils, molecules and embryos: new perspectives on the Cambrian explosion. Development 1999;126:851–9.

[62] Bromham L. What can DNA tell us about the Cambrian explosion? Integr Comb Biol 2003;43:148–56.

[63] Budd GE, Jensen S. A critical reappraisal of the fossil record of the bilaterian phyla. Biol Rev Camb Philos Soc 2000;75:253–95.

[64] Love GD, Grosjean E, Stalvies C, Fike DA, Grotzinger JP, Bradley AS, et al. Fossil steroids record the appearance of Demospongiae during the Cryogenian period. Nature 2009;457:718–21.

[65] Brunet T, Fischer AHL, Steinmetz PRH, Lauri A, Bertucci P, Arendt D. The evolutionary origin of bilaterian smooth and striated myocytes. elife 2016;5:e19607.

[66] Sperling EA, Frieder CA, Raman AV, Girguis PR, Levin LA, Knoll AH. Oxygen, ecology, and the Cambrian radiation of animals. Proc Natl Acad Sci U S A 2013;110:13446–51.

[67] Quinonez S, Innis JW. Human HOX gene disorders. Mol Genet Metab 2014;111:4–15.

[68] Holmes I. Transcendent elements: whole-genome transposon screens and open evolutionary questions. Genome Res 2002;12:1152–5.

[69] Emerman M, Malik HS. Paleovirology: modern consequences of ancient viruses. PLoS Biol 2010;8: e1000301.

[70] Bannert N, Kurth R. The evolutionary dynamics of human endogenous retroviral families. Annu Rev Genomics Hum Genet 2006;7:149–73.

[71] Eckburg PB, Lepp PW, Relman DA. Archaea and their potential role in human disease. Infect Immun 2003;71:591–6.

[72] Jacomo V, Kelly PJ, Raoult D. Natural history of *Bartonella* infections (an exception to Koch's postulate). Clin Diagn Lab Immunol 2002;9:8–18.

[73] Banuls A, Thomas F, Renaud F. Of parasites and men. Infect Genet Evol 2013;20:61–70.

[74] Lesca G, Rudolf G, Bruneau N, Lozovaya N, Labalme A, Boutry-Kryza N, et al. GRIN2A mutations in acquired epileptic aphasia and related childhood focal epilepsies and encephalopathies with speech and language dysfunction. Nat Genet 2013;45:1061–6.

[75] Ciechanover A, Kwon YT. Degradation of misfolded proteins in neurodegenerative diseases: therapeutic targets and strategies. Exp Mol Med 2015;47:e147.

[76] Toyama BH, Weissman JS. Amyloid structure: conformational diversity and consequences. Annu Rev Biochem 2011;80:10.

[77] Benvenuti S, Arena S, Bardelli A. Identification of cancer genes by mutational profiling of tumor genomes. FEBS Lett 2005;579:1884–90.

[78] Swanton C. Intratumor heterogeneity: evolution through space and time. Cancer Res 2012;72:4875–82.

[79] Patil N, Berno AJ, Hinds DA, Barrett WA, Doshi JM, Hacker CR, et al. Blocks of limited haplotype diversity revealed by high-resolution scanning of human chromosome 21. Science 2001;294:1719–23.

[80] Slatko BE, Taylor MJ, Foster JM. The Wolbachia endosymbiont as an anti-filarial nematode target. Symbiosis 2010;51:55–65.

[81] Resnikoff S, Pascolini D, Etyaale D, Kocur I, Pararajasegaram R, Pokharel GP, et al. Global data on visual impairment in the year 2002. vol. 82. Bulletin of the World Health Organization; 2004. p. 844–851.

[82] Kothandapani A, Sawant A, Dangeti VS, Sobol RW, Patrick SM. Epistatic role of base excision repair and mismatch repair pathways in mediating cisplatin cytotoxicity. Nucleic Acids Res 2013;41:7332–43.

[83] Bogler O, Cavenee WK. Methylation and genomic damage in gliomas. In: Zhang W, Fuller GN, editors. Genomic and molecular neuro-oncology. Sudbury, MA: Jones and Bartlett; 2004. p. 3–16.

[84] Lancaster AK, Masel J. The evolution of reversible switches in the presence of irreversible mimics. Evolution 2009;63:2350–62.

[85] Berman JJ. Armchair science: no experiments, just deduction. Kindle edition. Amazon Digital Services, Inc.; 2014.

[86] Berman JJ. Rare diseases and orphan drugs: keys to understanding and treating common diseases. Cambridge, MD: Academic Press; 2014.

[87] Combarros O, Cortina-Borja M, Smith AD, Lehmann DJ. Epistasis in sporadic Alzheimer's disease. Neurobiol Aging 2009;30:1333–49.

[88] Lobo I. Epistasis: gene interaction and the phenotypic expression of complex diseases like Alzheimer's. Nature Edu 2008;1:1.

[89] Chi YI. Homeodomain revisited: a lesson from disease-causing mutations. Hum Genet 2005;116:433–44.

[90] Gerke J, Lorenz K, Ramnarine S, Cohen B. Gene environment interactions at nucleotide resolution. PLoS Genet 2010;6:e1001144.

[91] Estivill X, Bancells C, Ramos C. Geographic distribution and regional origin of 272 cystic fibrosis mutations in European populations. Hum Mutat 1997;10:135–54.

[92] Faustino NA, Cooper TA. Pre-mRNA splicing and human disease. Genes Dev 2003;17:419–37.

[93] Tanackovic G, Ransijn A, Thibault P, Abou Elela S, Klinck R, Berson EL, et al. PRPF mutations are associated with generalized defects in spliceosome formation and pre-mRNA splicing in patients with retinitis pigmentosa. Hum Mol Genet 2011;20:2116–30.

[94] Horike S, Cai S, Miyano M, Chen J, Kohwi-Shigematsu T. Loss of silent chromatin looping and impaired imprinting of DLX5 in Rett syndrome. Nat Genet 2005;32:31–40.

[95] Preuss P. Solving the mechanism of Rett syndrome: how the first identified epigenetic disease turns on the genes that produce its symptoms. Research News Berkeley Lab; 2004. December 20.

[96] Soejima H, Higashimoto K. Epigenetic and genetic alterations of the imprinting disorder Beckwith-Wiedemann syndrome and related disorders. J Hum Genet 2013;58:402–9.

[97] Agrelo R, Setien F, Espada J, Artiga MJ, Rodriguez M, Pérez-Rosado A, et al. Inactivation of the lamin A/C gene by CpG island promoter hypermethylation in hematologic malignancies, and its association with poor survival in nodal diffuse large B-cell lymphoma. J Clin Oncol 2005;23:3940–7.

[98] Bartholdi D, Krajewska-Walasek M, Ounap K, Gaspar H, Chrzanowska KH, Ilyana H, et al. Epigenetic mutations of the imprinted IGF2-H19 domain in Silver-Russell syndrome (SRS): results from a large cohort of patients with SRS and SRS-like phenotypes. J Med Genet 2009;46:192–7.

[99] Chen J, Odenike O, Rowley JD. Leukemogenesis: more than mutant genes. Nat Rev Cancer 2010;10:23–36.

[100] Martin DIK, Cropley JE, Suter CM. Epigenetics in disease: leader or follower? Epigenetics 2011;6:843–8.

[101] McKenna ES, Sansam CG, Cho YJ, Greulich H, Evans JA, Thom CS, et al. Loss of the epigenetic tumor suppressor SNF5 leads to cancer without genomic instability. Mol Cell Biol 2008;28:6223–33.

[102] Mable BK, Otto SP. The evolution of life cycles with haploid and diploid phases. BioEssays 1998;20:453–62.

[103] Glansdorff N, Xu Y, Labedan B. The last universal common ancestor: emergence, constitution and genetic legacy of an elusive forerunner. Biol Direct 2008;3:29.

[104] Weatherall DJ. Molecular pathology of single gene disorders. J Clin Pathol 1987;40:959–70.

[105] Holland HD. The oxygenation of the atmosphere and oceans. Philos Trans R Soc: Biol Sci 2006;361:903–15.

[106] Ng SB, Buckingham KJ, Lee C, Bigham AW, Tabor HK, Dent KM, et al. Exome sequencing identifies the cause of a mendelian disorder. Nat Genet 2010;42:30–5.

[107] Piano MR. Alcoholic cardiomyopathy: incidence, clinical characteristics, and pathophysiology. Chest 2002;121:1638–50.

[108] Zhu H-L, Meng S-R, Fan J-B, Chen J, Liang Y. Fibrillization of Human Tau is accelerated by exposure to lead via interaction with His-330 and His-362. PLoS ONE 2011;6:e25020.

[109] Wang E, Boswell E, Siddiqi I, Lu CM, Sebastian S, Rehder C, et al. Pseudo-Pelger-Huet anomaly induced by medications: a clinicopathologic study in comparison with myelodysplastic syndrome-related pseudo-Pelger-Hüet anomaly. Am J Clin Pathol 2011;135:291–303.

[110] Juneja SK, Matthews JP, Luzinat R, Fan Y, Michael M, Rischin D, et al. Association of acquired Pelger-Huet anomaly with taxoid therapy. Br J Haemat 1996;93:139–41.

[111] Schule B, Oviedo A, Johnston K, Pai S, Francke U. Inactivating mutations in ESCO2 cause SC phocomelia and Roberts syndrome: no phenotype-genotype correlation. Am J Hum Genet 2005;77:1117–28.

[112] Franco B, Meroni G, Parenti G, Levilliers J, Bernard L, Gebbia M, et al. A cluster of sulfatase genes on Xp22.3: mutations in chondrodysplasia punctata (CDPX) and implications for warfarin embryopathy. Cell 1995;81:1–20.

[113] Van Gaalen J, Kerstens FG, Maas RP, Harmark L, van de Warrenburg BP. Drug-induced cerebellar ataxia: a systematic review. CNS Drugs 2014;28:1139–53.

[114] Rossi M, Perez-Lloret S, Doldan L, Cerquetti D, Balej J, Millar Vernetti P, et al. Autosomal dominant cerebellar ataxias: a systematic review of clinical features. Eur J Neurol 2014;21:607–15.

[115] Penneys NS. Ochronosislike pigmentation from hydroquinone bleaching creams. Arch Dermatol 1985;121:1239–40.

[116] Langston JW, Ballard P, Tetrud JW, Irwin I. Chronic parkinsonism in humans due to a product of meperidine-analog synthesis. Science 1983;219:979–80.

[117] Priyadarshi A, Khuder SA, Schaub EA, Shrivastava S. A meta-analysis of Parkinson's disease and exposure to pesticides. Neurotoxicology 2000;21:435–40.

[118] Zimprich A, Biskup S, Leitner P, Lichtner P, Farrer M, Lincoln S, et al. Mutations in LRRK2 cause autosomal-dominant parkinsonism with pleomorphic pathology. Neuron 2004;44:601–7.

[119] Du L, Sullivan CC, Chu D, Cho AJ, Kido M, Wolf PL, et al. Signaling molecules in nonfamilial pulmonary hypertension. N Engl J Med 2003;348:500–9.

[120] DuBose Jr TD. Experimental models of distal renal tubular acidosis. Semin Nephrol 1990;10:174–80.

[121] Weiss HJ, Rosove MH, Lages BA, Kaplan KL. Acquired storage pool deficiency with increased platelet-associated IgG: report of five cases. Am J Med 1980;69:711–7.

[122] Bleiberg J, Wallen M, Brodkin R, Applebaum I. Industrially acquired porphyria. Arch Dermatol 1964;89:793–7.

[123] Cam C, Nigogosyan G. Acquired toxic porphyria cutanea tarda due to hexachlorobenzene. JAMA 1963;183:88–91.

[124] Ku NO, Wright TL, Terrault NA, Gish R, Omary MB. Mutation of human keratin 18 in association with cryptogenic cirrhosis. J Clin Invest 1997;99:19–23.

[125] Scamardella JM. Not plants or animals: a brief history of the origin of Kingdoms Protozoa, Protista and Protoctista. Int Microbiol 1999;2:207–16.

[126] Schlegel M, Hulsmann N. Protists: a textbook example for a paraphyletic taxon. Org Divers Evol 2007;7:166–72.

[127] Rare Diseases Act of 2002, Public Law 107-280, 107th U.S. Congress, November 6, 2002.

[128] DeQueiroz K. Species concepts and species delimitation. Syst Biol 2007;56:879–86.

[129] DeQueiroz K. Ernst Mayr and the modern concept of species. PNAS 2005;102(Suppl. 1):6600–7.

[130] Mayden RL. Consilience and a hierarchy of species concepts: advances toward closure on the species puzzle. J Nematol 1999;31:95–116.

[131] Raup DM. A kill curve for Phanerozoic marine species. Paleobiology 1991;17:37–48.

[132] Choi BI, Han JK, Hong ST, Lee KH. Clonorchiasis and cholangiocarcinoma: etiologic relationship and imaging diagnosis. Clin Microbiol Rev 2004;17:540–52.

[133] Simpson GG. The principles of classification and a classification of mammals. Bull Am Mus Nat Hist 1945;85:1–350.

[134] Fitch WM, Ayala FJ, editors. Tempo and mode in evolution. Washington, D.: National Academy Press; 1995.

[135] Gilad Y, Rifkin SA, Pritchard JK. Revealing the architecture of gene regulation: the promise of eQTL studies. Trends Genet 2008;24:408–15.

[136] Fisher RA. The correlation between relatives on the supposition of Mendelian inheritance. Trans R Soc Edinb 1918;52:399–433.

[137] Ward LD, Kellis M. Interpreting noncoding genetic variation in complex traits and human disease. Nat Biotechnol 2012;30:1095–106.

[138] Visscher PM, McEvoy B, Yang J. From Galton to GWAS: quantitative genetics of human height. Genet Res 2010;92:371–9.

[139] Zhang G, Karns R, Sun G, Indugula SR, Cheng H, Havas-Augustin D, et al. Finding missing heritability in less significant Loci and allelic heterogeneity: genetic variation in human height. PLoS ONE 2012;7:e51211.

[140] Goldberg R. Horses' ability to pace is written in DNA. The New York Times; 2012. September 11.

[141] Fulton DL, Sundararajan S, Badis G, Hughes TR, Wasserman WW, Roach JC, et al. TFCat: the curated catalog of mouse and human transcription factors. Genome Biol 2009;10:R29.

[142] Lee TI, Young RA. Transcriptional regulation and its misregulation in disease. Cell 2013;152:1237–51.

[143] Lemon SM, Sparling PF, Hamburg MA, Relman DA, Choffnes ER, Mack A. Vector-borne diseases: understanding the environmental, human health, and ecological connections, workshop summary. Institute of Medicine (US) Forum on microbial threats. Washington, DC: National Academies Press; 2008.

[144] Berman JJ. Taxonomic guide to infectious diseases: understanding the biologic classes of pathogenic organisms. Cambridge, MA: Academic Press; 2012.

2

Shaking Up the Genome

Section 2.1 Mutation Burden

During the time it takes to read this sentence, it can be estimated that the reader's DNA will incur on the order of 10 trillion DNA lesions.

Zachary D. Nagel, Isaac. A. Chaim, and Leona D. Samson [1]

I conclude that for a number of diseases the mutation rate increases with age and at a rate much faster than linear. This suggests that the greatest mutational health hazard in the human population at present is fertile old males.

James F. Crow [2]

Over the years, many estimates for the spontaneous mutation rate of DNA in cells have appeared in the literature. In humans, point mutations (i.e., mutations that occur in a single nucleotide base within the genome) seem to occur with a frequency of about 1 to 3×10^{-8} per base per division [3–5]. Cancer cells, which are generally characterized by genetic instability and multiple genetic abnormalities, seem to have even higher mutation rates [6]. When we extrapolate the mutation rate for the many cells that compose a human body, we might expect that trillions of point mutations occur every few seconds [1]. [Glossary Genetic instability]

There are many types of mutational alterations other than point mutations (e.g., genomic structural variations) [7]. Our knowledge of the likelihood of other types of mutation is limited. Furthermore, estimates of point mutation rates provide us with little information about the rate of mutation repair and estimates of rates of new mutations do not provide us with solid data on the accumulated burden of mutations, over our lifetimes. [Glossary Genomic structural variation, Next-generation sequencing]

Evolution's Clinical Guidebook. https://doi.org/10.1016/B978-0-12-817126-4.00002-3

With the advent of whole genome sequencing, scientists now have the tools to answer a great number of questions related to the true mutational burden of humans, such as:

- How many mutations are carried in our cells?

- What genes do these mutations affect?

- What diseases do these accumulated mutations produce?

- What steps might we take to ameliorate the process?

A direct way to answer these questions is to begin sequencing the genome of sampled cells, both gametes and somatic cells, obtained from newborns. Each individual in the group would be resampled at some planned interval (e.g., every 5 years) and their cells sequenced. In addition, whenever an individual develops a disease, samples of cells would be taken from affected tissues and from unaffected tissues. By doing so, for a population of individuals, we can get a better idea of the accumulating human mutation burden, over time, and we might also gain some insight into the role that acquired mutations play in various diseases. Of course, this experiment would take at least a century to complete, allowing time for subjects to enjoy their full life span before closing the books on the project. Obtaining a century's worth of guaranteed research funding can be such a challenge!

Is there an easier way? In a recent study wherein cell samples of chronically sun-exposed eyelid skin were sequenced and examined for oncogene mutations, researchers found that about a quarter of the sun-exposed skin cells contained activating mutations in oncogenes [8]. Furthermore, the oncogene mutations seemed to be under strong positive selection (i.e., skin cells with oncogene mutations were more likely to survive and divide than cells with other types of mutation, as judged by their increased frequency in clusters of sampled cells) [8]. This being the case, you might think that every inch of sun-exposed skin should be the site of multiple skin cancers. Of course, this is not the case. The reason that the occurrence of oncogene activation is many orders of magnitude less than the occurrence of cancer is that cancer develops as a sequence of events and none of these events are ordained. What these observations do tell us is that mutations accumulate in skin that has been damaged by chronic exposure to sunlight. Presumably, most of the mutagenic damage is due to ultraviolet radiation, which is known as a direct mutagen, and which is capable of penetrating just through the skin's epidermis. We might expect that whole genome sequencing on human tissues that have been exposed to suspected mutagens will provide us with a fairly good indicator of the genetic damage such agents inflict. [Glossary Exome sequencing, Mutagen, Proto-oncogene, Skin cancer]

It should be kept in mind that whole genome studies of human cells followed over time, or after periods of exposure to suspected mutagens, do not necessarily tell us much about the occurrences of cytotoxic mutations. A cell might suffer a lethal mutation and die. Such cells, and their lethal mutations, will not be included in whole genome sequencing of sampled tissues; such cells are dead and gone. Sampling is always limited to surviving cells; hence sampling is skewed to detect nonlethal mutations. This limitation not withstanding, we can use whole genome sequencing to compare mutagenic damage among different individuals and get a good idea of which individuals in a population will be particularly susceptible to the deleterious effects of environmental mutagens. A vast literature on the subject of inter-individual mutation rates would suggest that all the following factors will play significant roles:

–1. DNA repair efficiencies [1]

Relative efficiencies of any of the seven known DNA repair mechanisms available to human cells will influence an individual's mutation burden.

–2. Selective apoptosis

Apoptosis, also known as programmed cell death, is an active process that is initiated in cells that have been injured. This process serves to rid the organism of cells that have undergone an unacceptable degree of damage. Interindividual variations in the way that apoptosis is triggered and executed will influence the burden of mutations tolerated in a cell population over time.

–3. The generation time of cells

Dividing cells are more sensitive to mutagens than nondividing cells. The rate of cell division, which varies among the different cell types and which may vary among individuals in a population, will influence the mutation burden of cells.

–4. Metabolism of exogenous chemicals

Individuals may vary greatly in their metabolism of chemicals, favoring the activation of a chemical to a highly mutagenic form, or enhancing the deactivation of a potential mutagen. These differences among individuals can account for differences in mutation burden in affected cell populations.

–5. Selection pressures on "second allele" mutations

Observing the pathogenesis of genetic diseases can sometimes reveal a great deal about the mutation burden in humans. Individuals with neurofibromatosis type 1 are born with a mutation of the gene that codes for neurofibrin. Neurofibromas develop only in those individuals who acquire a second mutation, involving the second allele of neurofibrin in a somatic Schwann cell (the cell of origin of neurofibroma), after birth. Nearly everyone born with neurofibromatosis type 1 will develop at least one neurofibroma, indicating that a new (second) mutation occurred. We can count the number of neurofibromas growing in an affected individual to get a low-ball estimate of the number of new mutations occurring in one locus of one particular gene, in one minor cell type, over the lifetime of an individual. It is not uncommon for a single affected person to have hundreds of individual neurofibromas, each representing one such mutation. [Glossary Allele, Neurofibromatosis, Von Recklinghausen disease]

In the case of neurofibromatosis type 1, we have an opportunity to directly visualize the mutation burden for a particular gene's mutation. When we consider that there are many thousands of genes in the human genome and that we are only sampling one locus in one gene when counting neurofibromas, we can begin to understand that the somatic mutation burden in somatic cells is remarkably high. We should not be surprised to read that by the time we reach the age of 60 years, every nucleotide in our genome has been mutated in at least one cell of the body [9]. [Glossary Somatic mutation]

Diseases from mutations occurring randomly in somatic cells

Lethal mutations occurring in single somatic cells do not pose much of a health risk. A highly mutated single cell can sicken and die, without placing much of a burden on the organism. After all, it is estimated that the adult human body contains about 37 trillion cells [10]. Losing a few cells here or there is no great loss. The disease burden resulting from somatic mutations comes largely from proliferative clonal diseases, wherein a mutation eventually leads to the expansion of a clone of cells, with negative consequences for the organism. In general, such negative consequences take any one of three forms:

−1.   The expanding clone of mutation-carrying cells do not perform their intended purpose, and the organism suffers a dysfunction.

All of the fully differentiated cells in the body arise from less differentiated cells that are variously named reserve cells or stem cells. In the general scheme, each binary division of a stem cell produces one differentiated cell plus one replacement stem cell. In some cases, stem cells are morphologically well-characterized, and we can follow a succession of early stem cells that produce mid-level stem cells that in turn produce the final differentiated product. This is best studied in the hematopoietic (i.e., blood forming) system wherein we can identify blast cells that give rise to promyelocytic cells that, in turn, give rise to fully differentiated cells. If the number of intermediate stem cells is large, then the first stem cell in the lineage (i.e., the earliest stem cell) initiates a chain of differentiation events that may yield, at the end of the chain, many thousands of fully differentiated cells. Hence, reductions in a single stem cell may produce a large effect in the population of differentiated cells. [Glossary Differentiation, Stem cell]

Paroxysmal nocturnal hemoglobinuria provides the classic example of a maladaptation produced by the expansion of a mutated stem cell. In this condition, a hematopoietic stem cell that has acquired a mutation in the PIGA gene produces a clone of blood cells of every blood lineage (e.g., red blood cells, myelocytes, monocytes, and megakaryocytes) that lack an important cell surface protein [11]. Consequently, the affected red cells are prone to hemolysis, particularly at night. Though the disease produces a variety of clinical consequences, the occurrence of paroxysms of nocturnal hemolysis with hemoglobin spilling into urine is emblematic.

A PIGA gene mutation is just as likely to arise, by chance, in a liver cell as in a hematopoietic stem cell, but the PIGA mutation has no detrimental effect when it occurs in any cell other than the specific stem cell that produces a hematopoietic lineage.

A second example of clonal dysfunction is found in the acquired form of cyclic neutro-penia. Cyclic neutropenia is characterized by regular 21-day cyclic fluctuations in the number of blood cells. When the cycle is at low ebb, the associated neutropenia (low white blood cell count) can be associated with morbidity, including the risk of developing a severe infection. This condition, whose root cause is a mutation in the ELANE gene, can be inherited or can arise as an acquired disorder [12]. [Glossary Cytopenia]

The list of clonal hematologic disorders whose underlying cause is an acquired somatic mutation gets longer and longer, as we enhance our ability to detect new clonal mutations [13, 14].

–2.   The expanding clone of mutation-carrying cells eventually replaces much of the normal cell population, with an abnormally large population of mutated cells.

We see examples of this in hematopoietic populations that acquire through somatic muta-tion, a JAK2 mutation that produces a hyperproliferative population. In polycythemia vera, an acquired JAK2 mutation produces an overabundance of red blood cells [15]. Patients may be asymptomatic for years, but the increasing numbers of red blood cells eventually produce a constellation of symptoms that require medical attention. Acquired mutations of the JAK2 gene are involved in several myeloproliferative disorders, aside from polycythemia vera, including myelofibrosis, and at least one form of hereditary thrombocythemia [16, 17]. [Glossary Myeloproliferative disorder, Polycythemia]

–3.   The expanding clone produces an aggressive cancer.

Cancer is the most calamitous consequence of the acquired mutation burden in animals. Virtually every cancer contains acquired somatic mutations, and in many cases, the spe-cific mutation thought to be the root cause of the cancer can be identified [18–24].

Mutations affecting the human gene pool

Mutations in somatic cells are not passed onto the offspring. The only mutations that can be passed to offspring are mutations that arise in gametes or germ cells (i.e., the cells that produce gametes). Furthermore, mutations in gametes cannot be passed to the next generation if the mutation stops reproduction (i.e., produces a defective gamete that can-not successfully participate in the full reproductive cycle). In other words, a mutation in a gamete can only enter the population gene pool if it can be passed on to viable and fertile offspring [25].

Regarding the mutation burden within gametes, it has been estimated that the average human gamete acquires approximately 38 de novo base-substitution mutations plus three small insertion/deletions in complex sequences plus one splicing mutation plus some indeterminate number of insertions of transposons, microsatellite instabilities, segmental duplications and segmental deletions [9, 26]. Using genome-wide sequencing to identify single nucleotide polymorphisms (SNPs) in offspring that were not present in either par-ent's germline, it was estimated that there about 60 de novo SNPs in the genome in each of us, with 1–2 of these new SNPs in coding sequences [27]. Likewise, using genome sequencing and looking for new SNPs in haplotypes identified by their parental origins (i.e., maternal or parental), we can determine whether newly occurring SNPs are preferentially associated with either parent. It turns out that most de novo mutations in an offspring have a paternal (not maternal) origin. Furthermore, the older the father, the greater the germline mutation burden [28]. Other studies seem to confirm that the gametes of older fathers have more acquired mutations than the gametes of younger fathers [2, 26, 29]. [Glossary De novo mutation, Germline, Germline mutation, Haplotype, Microsatellite, Microsatellite instability, Single nucleotide polymorphism]

De novo disease mutations

Because de novo mutations represent an important obstacle for those among us who have visions of improving the human gene pool, we should pause here a moment to discuss these mutations, and how the circumstances of their occurrence will influence the ways we think about genetic diseases.

De novo is a term derived from Latin, meaning "new." Up until now, we have been discussing newly acquired mutations occurring in somatic cells. Although "de novo" means new and acquired mutations in adult organisms are undoubtedly new mutations, we traditionally reserve the term "de novo" to refer to new mutations that are present in the germline of developing organisms (i.e., in an early embryo).

There are basically two mechanisms by which a de novo mutation is acquired:

–1. From a new mutation that a parent acquires in his or her gamete, at some moment during their lifetime (i.e., not present in the germline of the parent).

In a sense, the parent does not have the mutation (only the parent's gamete has the mutation), and thus the mutation is not considered to be inherited from the parent. This accounts for Down syndrome being a noninherited disease. In most cases of Down syndrome, a nondisjunction results in an extra copy of chromosome 21 in either the father's sperm or the mother's oocyte. Consequently, the cells of the offspring's germline have three copies of chromosome 21 (i.e., trisomy 21). The child's Down syndrome clearly was produced by a mutation in the father or the mother, but Down syndrome is said to be de novo (i.e., noninherited) in this case, because neither the father nor the mother had trisomy 21 in their germline (i.e., neither the father or the mother had Down syndrome).

–2. From a new mutation that the offspring acquires very early in its zygote or in a very early embryonic cells (e.g., a cell of the morula or bastula).

When the mutation occurs in an embryonic cell (e.g., including the morula and the blastula, but not including zygote), then the resulting mutation will be expressed as a mosaic in the developed organisms (i.e., present in some somatic cells and not in others). The cells of the offspring that contain the mutation will be restricted to the clonal descendants of the original newly mutated embryonic cell.

If the mutation is present in the germline of either parent (i.e., present in all the cells of the parent), then the mutation is said to be inherited. A germline mutation in an offspring can be inherited or de novo, but not both. The implication here is that a de novo mutation can be passed from a father or a mother to the offspring, without being inherited from the parents. The notion that an individual can receive a noninherited disease from a mutation present in a parental cell is confusing, unless we fully understand the definitions of "de novo" and "germline." [Glossary Noninherited genetic disease]

Some of the most severe genetic diseases are de novo diseases. The reason being that a de novo mutation can occur anywhere in the genome, causing havoc to the organism. An inherited mutation has existed in the parent's germline without killing the parent, thus proving itself nonlethal, at least in the heterozygous state. For this reason, inherited disease mutations have taken residence in the human gene pool.

Examples of some rare diseases that are caused by de novo mutations would include:

- Baraitser-Winter syndrome, characterized by central nervous system and facial malformation.

- Borhing-Opitz syndrome, characterized by intellectual disability plus congenital malformations.

- CHARGE syndrome, an acronym for coloboma of the eye, heart defects, atresia of the nasal choanae, retardation of growth and/or development, genital or urinary abnormalities, ear abnormalities and deafness, and a leading cause of combined deafness and blindness in newborns.

- Kabuki syndrome, characterized by intellectual disability and congenital anomalies.

- KBG syndrome, characterized by a disease-typical facial dysmorphism, macrodontia of the upper central incisors, costovertebral anomalies, and developmental delay.

- Schinzel-Giedion syndrome, characterized by distinctive facial features, neurological problems, and organ and bone abnormalities.

De novo mutation diseases are always single allele diseases, in the sense that the likelihood that a de novo mutation will occur on two complementary alleles is essentially zero. Hence, when a de novo mutational disease produces the same mutation as is found in an inherited autosomal dominant disease, then the two diseases are clinically indistinguishable, although one form is inherited and the other is not. Consequently, only patients with the inherited form of the mutation will have a positive family history for the disease. Many autosomal dominant inherited diseases can occur either in inherited form or as de novo mutations. For example, about half of the cases of neurofibromatosis type 1 are inherited from an affected parent. The remaining half arise as de novo germline mutations. An individual with de novo neurofibromatosis type 1 can produce an offspring who inherits the mutation, thus producing a child with inherited neurofibromatosis. This is how de novo mutations eventually enter the human gene pool.

In general, de novo mutations are often suspected as the cause of diseases that occur in early childhood for which no other cause (e.g., no evidence of familial or parental inheritance, infectious etiology, or environmental influences) can be determined. Autism is a disease that is suspected to be a de novo mutational disease because it fits these aforementioned conditions [30]. [Glossary Etiology, Inheritance, Mendelian inheritance, Non-Mendelian inherited genetic disease]

Section 2.2 Gene Pools and Gene Conservation

The cell is basically an historical document, and gaining the capacity to read it (by the sequencing of genes) cannot but drastically alter the way we look at all of biology.

Carl Woese [31]

Life is a concept.

Patrick Forterre [32]

The gene pool is the repository for the collective genetic diversity of a species. Each individual organism of a species contains a unique genome composed of sequences that are floating somewhere in the community pool. Although the gene pool is an abstraction, it serves to clarify the way we think about natural selection, and it prepares us to better understand the biological entity known as "species" (to be discussed in greater depth in Section 4.1, A Species is a Biological Entity).

Natural selection isn't really very efficient at eliminating particular genes (or polymorphisms of genes). Basically, all natural selection does is weigh the odds in favor of one polymorphism versus another under one prevailing set of genomic/environmental circumstances versus another. [Glossary Polymorphism]

In point of fact, even if a particular polymorphism were entirely eliminated from the gene pool, you could safely bet that it will reappear in short order. The reason for this is that when you have a large number of organisms, and a high rate of mutation, then all possible (i.e., compatible with life) point mutations will eventually occur. Certainly, if the polymorphism arose in the past, it can arise again in the future.

It is tempting to think of certain genes as contaminants in the gene pool; particularly those genes that are the root causes of diseases. As we shall see, most of the genetic polymorphisms that are associated with disease may serve some useful, even vital, purpose under some sets of circumstances. [Glossary Association versus cause]

From the point of view of the species, the gene pool is the collection of genes that is available to the species as a whole. If the population of a species is large (let's pretend the population is infinite), then each existing gene polymorphism will occur in all of the possible different combinations with the other polymorphisms contained in the gene pool. For a given species with an infinite number of individuals, we can expect to see an enormous variety of genetically unique individuals, with no two individuals, with the exception of identical siblings, being genetically identical.

When there is a new stressor in the environment (e.g., a new toxin, or a climate change), it will be the gene pool, expressed as a large number of unique individuals, that provides the biological variation necessary for the survival of the species.

The Law of Sequence Conservation

The Law of Sequence Conservation is the guiding principle of bioinformatics. Put simply, if a sequence is conserved through evolution (i.e., if you can find a closely similar sequence that is present in a class of organisms, and their direct line of ancestors), then that sequence must perform a useful function for the organism. Furthermore, if a sequence is highly conserved (i.e., with very little sequence difference among class members and their ancestors), then that sequence is likely to have a vital function [33]. This law is so useful and so fundamental to the field of genomics and to the design of gene-related computational algorithms that it may as well be known as the First Law of Bioinformatics. [Glossary Bioinformatics]

One of the consequences of the Law of Sequence Conservation is that most monogenic diseases occur in conserved genes. Why is this so? Genes that are not conserved can tolerate mutations without causing disease (that is why they are not conserved). Genes that are highly conserved cannot tolerate mutation without producing disease (that is why they are conserved). This being the case, we can presume that just about every monogenic disease of humans (i.e., nearly all the rare inherited diseases and the rare de novo diseases) is also a monogenic disease of ancestors in the phylogenetic lineage (that is because conserved human genes are genes found in our ancestral lineage). [Glossary Monogenic disease, Single-gene disease]

Another consequence of the Law of Sequence Conservation is that we can infer that a gene has a vital function if we know that it is highly conserved. Regulatory genes are highly conserved in humans and in our ancestral species [33]. This tells us that regulatory genes have a very important function. Likewise, oncogenes are highly conserved, also telling us that oncogenes, despite their carcinogenic potential, serve an important purpose in the human genome and in the genomes of our ancestor.

How is gene conservation actually achieved? It's all done by natural selection. A new mutation in a gene is overwhelmingly likely to be deleterious. Hence, if a gene is vital to the function of an organism, and if a mutation occurs in this gene, then the organism that carries the mutation will be less likely to procreate and pass its genes to another generation, and the mutated form of the gene will never make much contribution to the species gene pool. Conversely, if a gene serves no function, then mutations in the gene will take place without reducing the fitness of organisms, and such mutations may greatly expand the size of the species gene pool. Hence, junk DNA gets progressively junkier as time passes.

There are three key conceptual points that we need to keep in mind when we invoke the Law of Sequence Conservation:

  –1. There is no biochemical method by which certain sequences in the genome are marked for conservation, while other sequences are marked for mutability.

It's all done through natural selection. If a sequence is important, then we might expect that mutations arising in the sequence will reduce the fitness of the affected organisms and that such mutations will become less frequent in the gene pool as the carrier organisms die [33].

  –2. Sequences that are not conserved and have no function are the ingredients for new genes.

Nonconserved sequences, that serve no function essential for survival, may mutate to produce new and useful sequences. In fact, it is only the nonconserved sequences from which new genes may evolve (because the conserved sequences are constrained not to contribute variants to the gene pool) [33].

  –3. Regulatory genes are highly conserved. Hence, genomic regulation is important to the species; perhaps as important as are the protein-coding genes.

Many of the most highly conserved sequences occur in noncoding regions (i.e., sequences that are not genes). We can infer that conserved noncoding sequences must have an important function, and the only known function of noncoding sequences is genomic regulation. In point of fact, it has been demonstrated that some of the ultra-conserved noncoding sequences are enhancers and are found to cluster in proximity to developmental genes [34].

  –4. If a mutation in a gene produces a disease in humans, we could reasonably expect that the same mutation might produce some deleterious effect in other species that occupy the human phylogenetic lineage.

When we look at conservation of a gene, we are usually studying a gene sequence found in multiple species of the same phylogenetic lineage, indicating that the gene is conserved ancestrally. This tells us that the gene is important up through the lineage. Hence, a mutation that produces a deleterious effect in one species is likely to produce a similar negative effect in other species of the same ancestral lineage.

Of course, there are exceptions. Lesch-Nyhan disease is a rare syndrome caused by a deficiency of hypoxanthine-guanine phosphporibosyl transferase (HGPRT), an enzyme involved in purine metabolism. In humans, HGPRT deficiency results in high levels of uric acid, with resultant renal disease and gout. Various severe neurological and psychological symptoms accompany the syndrome in humans, including self-mutilation. Neurological features tend to increase as the affected child ages. The same HGPRT deficiency of humans can be produced in mice. As far as anyone can tell, mice with HGPRT deficiency are totally normal [35]. How can this be? Once again, we are reminded that a gene mutation never causes a disease all by itself. The mouse evidently employs pathways that compensate for the deficiency in HGPRT. Likewise, mice with the most commonly mutated gene known to be the root cause of human nephronophthisis, a clinically serious inherited kidney condition, do not develop the disease [36].

Section 2.3 Recombination and Other Genetic Tricks

The evolution of sex is the hardest problem in evolutionary biology.

John M. Smith

The dream of every cell, to become two cells!

Francois Jacob

When we're taught evolution (IF we're taught evolution), the dual concepts of natural selection and mutation are locked together, as though natural selection waits for new mutations to confer a high level of fitness on a species. What you were taught is not strictly true. Although mutation is the basic fuel for evolution, its chief function is to produce natural variations in the gene pool of a species. The accumulation of natural variations in the gene pool occurs continuously, accumulating genetic variants since the first life on earth. The natural variants in the human population may have arisen recently, or remotely (i.e., in ancestral species). Natural selection, in turn, operates on those preexisting variations to produce unique individuals who may have some particular survival advantage; or not.

If it helps to use metaphors, evolution is much more like poker than it is like the lottery. We start with a deck of cards, and we shuffle it and deal each player a hand. Each of us makes the most of the hand we are dealt. The key mechanistic ingredients of poker are the deck (i.e., lots of cards with variants of any particular type of card, such as ace of hearts and ace of diamonds); and the shuffle. In the prior section, we focused on the deck (i.e., the gene pool). In this section, we'll focus on the shuffle (i.e., recombination events).

Basically, the human cell has three well-studied methods that shuffle the genome:

–1.   Multiple chromosomes

Pretend that we only had one pair of a single chromosome in each of our nuclei. Let's call this pair 1a and 1b. Then each offspring would have a pair of chromosomes wherein one of the pair came from the mother and one of the pair came from the father. There would be just four combinational possibilities: father-1-a/mother-1-a, father-1-b/mother-1-a, father-1-a/mother-1-b, and father-1-b/mother-1-b. This wouldn't allow for much variation in the offspring. If we ignored homologous recombination, there would be a lot of brothers and sisters and cousins with the identical gene sets. Adding chromosomes to the brew exponentially increases the number of possible outcomes. When we have dozens of chromosomes, we can virtually guarantee the absence of genetically identical offspring (except through the agency of twinning) anywhere in the species population.

−2.   Homologous recombination during meiosis

Homologous recombination is a shuffling trick that reassorts the alleles in a diploid chromosome. Each pair of chromosomes has one of the pair having a paternal origin and one of the pair having a maternal origin. This is most easily seen in the XY sex chromosome pair in males, wherein the Y chromosome is inherited from the father, and the X chromosome, in nearly all cases, is inherited from the mother. The same is true for all the other paired chromosomes. We say that the paired chromosomes are homologous to one another, a very poor choice of nomenclature that we seem to be stuck with. [Glossary Nomenclature, Reassortment, X-chromosome, X-linked inheritance, Y-chromosome]

During meiosis, each member of a chromosome pair moves adjacent to its homolog, whereupon a crossover of chromosomal fragments ensues. The points of crossover can occur at varying locations, changing from meiosis to meiosis. This shuffling process yields a set of chimeric chromosomes composed of complementary parts of the paired chromosomes. The net result of homologous recombination is that the offspring will contain no autosomal chromosomes (i.e., chromosomes other than the X or Y chromosome) that are identical to any of the autosomal chromosomes that were present in either parent [37]. This assures that the paired chromosomes of the offspring will have allele-allele combinations that were not present in either the father or the mother.

At this point, remembering the X and Y chromosome pair, you might be asking how it is that human offspring are not hermaphroditic, with a chimeric XY chromosome? It happens that the X and the Y chromosome are physically different from one another and lack the extensive homology observed in paired autosomal chromosomes. Without extensive homology between the X and Y chromosomes, we cannot expect much recombination and active suppression of crossover events reduces the possibility of recombination to a minimum. Any recombination between X and Y chromosomes is confined to a very small region known as the pseudoautosomal region that includes about 29 identified genes [38, 39]. Hence, for all practical purposes, males inherit an X chromosome of the mother plus the complete Y chromosome of the father.

−3.   Spliceosomes

In eukaryotes, DNA sequences are not transcribed directly into full-length RNA molecules, ready for translation into a final protein. There is a pre-translational process wherein transcribed sections of DNA, so-called introns, are spliced together, and a single gene can be assembled into alternative spliced products. Alternative splicing is one method whereby more than one protein form can be produced by a single gene [40, 41]. Cellular proteins that coordinate the splicing process are referred to, in aggregate, as the spliceosome. We can infer that the spliceosome plays an important role in eukaryotic life insofar as every animal, plant, fungus, and unicellular eukaryote conserves the spliceosomal machinery [42–44].

These three aforementioned shuffling mechanisms (multiple chromosomes, homologous recombination, and spliceosomes) produce a cascade of beneficial consequences:

```
Making effective use of the natural variation in the gene pool by ->
Creating new allele-allele gene combinations within individuals therefore ->
Yielding new cellular outcomes in the absence of new mutations, and ->
Creating a large variety of unique individuals, thus ->
Strengthening the species and providing the opportunity for further speciation
```

In a sense, humans do not evolve through the acquisition of new mutations; we evolve through the acquisition of new combinations of preexisting gene variants. One could argue that all of the preexisting gene variants arrived via mutations, so there really is no basic distinction to be made between the two modes of evolution. The point here is that the natural variants in the human gene pool are biologically tolerable; otherwise they would have been eliminated. A new mutation is likely to be deleterious, because it is rare to produce an improvement in a gene via random mutation. By relying on recombination of time-tested gene variants, we can be fairly confident that natural selection will operate on a relatively healthy gene pool.

Simple organisms that exist in very large numbers can evolve by mutation (not via meiotic recombination) because a mutation that causes a loss of an organism really doesn't hurt the species. Bacteria and viruses have high rates of mutation and do not have meiosis, evolving directly by mutation followed by selection. Animals, for the most part, have limited populations and cannot easily afford the losses from bad mutations. We animals depend on recombination events to produce unique and hardy individuals.

This sums up the story in humans, but there is large variation in the manner in which eukaryotic species shuffle their genomes. For example, nearly all species have multiple chromosomes, and humans have a chromosome number that puts us in the middle of the pack. The parasitic roundworm *Parascaris equorum* univalens has only one chromosome. As a rule, parasites have stripped-down genomes, deferring to their host organisms to do much of the physiologic heavy lifting (e.g., providing a sheltered place to live, supplying food, and maintaining a steady body temperature). No such excuse is available to the jack jumper ant (*Myrmecia pilosula*), which is not parasitic and which also has but a single pair of chromosomes. The males of the species are haploid, and thus have just one chromosome; if they had any fewer, they would disappear. Other insects luxuriate in their chromosomes. For example the atlas blue butterfly (*Polyommatus atlantica*), has about 450 chromosomes. Plants tend to have many more chromosomes than do humans; but not always. The black mulberry (*Monus nigra*) has 308 chromosomes, while common wheat (*Triticum aestivum*) has only 42. This is just one more example of the genomic free-for-all that will be discussed in detail in the next section of this chapter.

Regarding the benefits of the meiotic lifestyle, not every animal avails itself of the process. The bdelloid rotifers, a tiny animal that is best observed under a microscope, famously decline the benefits of sex and meiosis, preferring to reproduce exclusively by mitotic parthenogenesis [45]. The all-female league of bdelloids has managed to diversify over 500 species, thus defying generations of evolutionary biologists who would tell them that prolific speciation is impossible absent the gene shuffling methods described earlier in this section. Apparently, the bdelloids are able to acquire new genes directly from other organisms, a process known as horizontal gene transfer. The bdelloids bypassed the messiness of sex, preferring to simply take the genes they need (about 10% of their genome) from other organisms [45]. As discussed in Section 1.2, "Bootstrapping Paradoxes," all eukaryotes, including the bdelloids, apparently have the genetic machinery for meiosis. We have no way to exclude the possibility that the bdelloid rotifers may indulge in sexual reproduction, under some circumstances, but we can at least say that the bdelloid rotifers have valued their chastity for a very long time [45]. [Glossary Horizontal gene transfer, Parthenogenesis]

The bdelloids are not the only animals that acquire genes through horizontal gene transfer. Tardigrades, also known as water bears, are tiny animals that have obtained about one-sixth of their genome from many types of living organism, including bacteria, archaeans, plants, and fungi; but possibly without the indulgence of other animal species [46]. Their mode of gene acquisition seems to be working quite well for the tardigrades, as they are widely considered the hardiest animal on earth. Aside from enduring a wide range of environmental conditions here on earth, the tardigrades are held to be capable of surviving the extreme cold, and the blistering cosmic radiations of outer space.

What are the conditions that might sway a species to follow a life of crime, perpetually stealing genes from other species, rather than earning their genes through the hard work of evolution? The answer is simple. Evolution works through the selection of naturally occurring variants that exist within the gene pool. Evolution is not a very efficient way to achieve metabolic breakthroughs that involve two or more independently evolved pathways that work together to perform a new function. If two pathways are required to achieve one metabolic process, then each pathway must evolve without the benefit of selective pressure for the complete pathway (which requires two intact pathways to yield a survival benefit). Achieving a complex, combined pathway through natural selection is possible, but it is highly improbable. When such pathways are achieved in one species of organism, they tend to take many millions of years to evolve. There's no guarantee that the feat will be repeated by other organisms any time soon. In this case, the mechanism through which the pathway will be attained is most likely going to be through genetic acquisition of a complete pathway, from an endowed organism (e.g., horizontal gene transfer); not through natural selection of preexisting gene variants.

Section 2.4 Genomic Architecture: An Evolutionary Free-for-All

How is it that you keep mutating and can still be the same virus?
Chuck Palahniuk, in his novel, Invisible Monsters

When we use the word "genome," we prefer to think that we know what we're talking about. Understanding life would be easy if the differences between one organism and another were accounted for entirely by differences in the sequence of nucleotides that compose the genome (i.e., the totality of DNA within each cell). Our understanding of living organisms seemed to get even simpler when we learned, in the past few decades, that most of the genes in eukaryotes, including humans, are more or less equivalent between species (see Section 1.3, "Our Genes, for the Most Part, Come from Ancestral Species"). Gene-wise, there really is very little difference between a human and a dog.

Contrary to our newfound appreciation of the genetic similarities among diverse organisms, we are finding enormous variation in the manner in which all these homologous genes are actually regulated. In point of fact, there seems to be a dearth of fundamental principles of genomic organization and regulation that universally apply to living organisms. What follows here is a listing of some of the diverse ways in which DNA is organized, modified, packaged, and regulated.

Variations in the proteins that bind to DNA

Histones are the major proteins in chromatin (i.e., the material composing chromosomes). When histones are deacetylated (via deacetylases), the chromatin tightens, thus reducing transcription in the affected area. Histones are present in eukaryotes, and in Archaean prokaryotes [47]. You might think that histones were an essential feature of every living cell; but there is at least one notable exception to the rule.

Sperm cells lack histones. Instead, they contain protamines, a family of arginine-rich proteins synthesized by sperm cells in metazoans and spermatids in plants [48, 49]. The substitution of protamine for histone results in a highly condensed and dehydrated spermatocyte. In the first hour after a sperm fertilizes an oocyte, protamine molecules are replaced by histones derived from the egg's cytoplasm.

Variations in epigenomic methylation

DNA and RNA are the repositories of genetic information in all terrestrial organisms. The epigenome is something of an add-on, and different organisms have developed different ways of regulating the genome and of producing their various committed cell types. In particular, DNA methylation, the major modifier of the mammalian epigenome, is absent or rare in Drosophila (an insect genus) and in *Caenorhabditis elegans* (a nematode) [50].

Variations in intron content

Introns are noncoding regions of DNA that interrupt genes (i.e., the nucleotide sequences that code for proteins), creating a genomic landscape of discontinuous fragments of coding genes (i.e., exons). The transcribed fragments must be spliced together before they can be further processed. This extra splicing step achieves a level of modularity and permits the translation of more than one protein product from a single gene and is thought to promote the rapid evolution of new proteins.

Spliceosomes are conserved in all eukaryotic organisms. Insofar as the purpose of spliceosomes is to cope with the presence of introns, we would guess that introns would be present in every eukaryotic cell. This is not the case. As a general rule, the intron is associated with large genomes. Organisms that have small genomes (e.g., many viruses, prokaryotes, and some unicellular eukaryotes) are virtually back-to-back protein-coding and RNA-coding genes and are intron deficient [42, 51]. Conversely, introns have been found in organisms where we would not expect to find them, including some viruses and prokaryotes [52].

Tolerated variations in chromosome number

Humans have a diploid set of 46 chromosomes. We think of this number as inviolate, a species-specific constant wherein changes always indicate pathologic conditions (as seen in the trisomies and mononsomies). In fact, within a mammalian species, there may be different total chromosome counts. For example, the last remaining wild horses have 66 chromosomes, while domesticated horses have 64 chromosomes. Despite this difference, wild horses can be mated with domestic horses to produce fertile offspring [53]. [Glossary Karyotype]

Gene size not correlated with much of anything

We have no objective way of measuring the complexity of different organisms, but we like to hold onto a few beliefs, based on "common sense," even when there is no supporting evidence. These include:

–1. Humans are fairly complex organisms. In particular, our brains are more complex than the brains of any other living organism.

–2. The more complex the organism, the more genes it needs to function. Thus, the most complex organisms would have the largest genome.

Neither of these statements have much merit. The holometabolous insects (i.e., insect with distinctive metamorphic stages of maturation) produce multiple functioning forms of life, with a single genome. Furthermore, compared to the paltry one-body lifestyle of humans, insects do it all in miniature. As for the human brain, there really is very little evidence to suggest that humans have more complex brains than dolphins, pigs, octopi, or even birds. As for the second assertion, pretending for a moment that we had a way of measuring the complexity of an organism, there seems to be absolutely no correlation between genome size and any organismal trait.

Humans have about 3 billion nucleotides in their genome. The unicellular eukaryote *Polychaos dubium* has 670 billion nucleotide bases and the flowering plant Paris japonica has 150 billion [54]. The marbled lungfish has the largest known genome of any vertebrate, with 133 billion base pairs.

The onion, *Allium cepa*, is a diploid ($2n=16$) plant with a haploid genome mass of about 17 pg. For comparison, humans have a haploid genome mass of about 3.5 pg. Among onions there is a wide range of genome sizes. Why is it that the onion *Allium ursinum* has five times the genome size as the onion *Allium altyncolicum*?

Lest you protest that the human genome contains more genes than does the larger genome of plants, keep in mind that the human genome contains about 20–25 thousand genes. Rice has 46–56 thousand genes [55].

Finally, we note that there are seemingly complex organisms, developing through an embryo and having multiple tissues in the mature organism, that have remarkably simple genomes. For example, the bladderwort flower (*Utricularia vulgaris*) lacks noncoding regions [51]. Its genome, despite the apparent complexity of the bladderwort flower and its membership in Class Angiospermae (flowering plants), does not avail itself of the epigenomic regulatory mechanisms conferred on genomes by noncoding regions. For all practical purposes, the bladderwort has simply omitted the genomic architectural structures that evolved for nearly every other eukaryotic class and species. Scientists do not know the reason for this fundamental deviation from the norms of eukaryotic existence, but one statement seems to be obviously true. If an organism that lacks noncoding regions, then it is not susceptible to any of the diseases that arise from mutations in noncoding regions, or from dysfunctions of the regulatory roles played by noncoding regions (see Section 3.5, "Pathologic Conditions of the Genomic Regulatory Systems") (Fig. 2.1). [Glossary Noncoding mutational diseases]

Polytene chromosomes

Eukaryotes may contain polytene chromosomes characterized by puffs of highly replicated strands of DNA. These puffs of DNA that are found in some organisms, and not in others, are restricted to particular cell types within the organism. Polyteny provides intense levels of genetic expression and massively enlarges the size of the genome in affected cells (Fig. 2.2).

Polytene chromosomes are found in dipteran flies, springtails (Class Collembola), single-celled eukaryotes, and plants. In flies, polytene chromosomes are found in

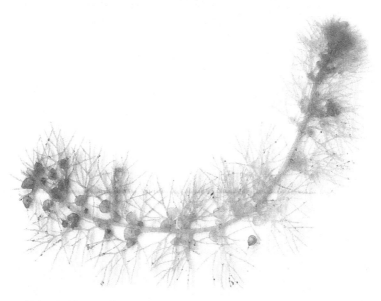

FIG. 2.1 A fragment of the bladderwort flower, *Utricularia vulgaris*, a complex organism lacking noncoding segments of DNA. *Source: Wikipedia, entered into the public domain by its author, Veledan.*

FIG. 2.2 Polytene chromosome from fly salivary gland. The green puffs attached to the chromosome are expansion sights where strands of DNA have been highly replicated. *Source: Wikipedia, image entered into the public domain by LPLT.*

nondividing salivary gland cells. Polyteny is also found in mammals, including humans, minks, and foxes, where it seems to be restricted to trophoblast cells [56].

In basic biology classes we are taught that every nucleated cell in an organism has a diploid (somatic cells), haploid (oocytes and spermatocytes), or $2n$-ploid (multinucleate giant cells such as megakaryocytes and syncytiotrophoblasts) genome. The polytenic genome shatters this myth.

Structural genomic variants in healthy animals

We are taught that all members of a species have the same sized genome. Genomic differences that account for phenotypic differences among members of a species presumably come from small variations, such as SNPs and epigenomic modifiers.

Genomic structural abnormalities involving chunks of DNA often have negative developmental consequences; but not always. GSVs (genomic structural variants) include alterations in karyotype or cytogenetic alterations observable with special techniques, as well as changes too small to see with a microscope, such as microdeletions, large insertions, inversions, and translocations. GSVs also include duplications and other copy-number alterations as well as gene conversions [57]. In humans, we can occasionally encounter apparently healthy individuals with inverted chromosomes [58], as well as individuals who are missing pieces of chromosomes, without clinical sequelae. GSVs among different individuals in the human population occur frequently and may account for more phenotypic variations in the human population than do SNPs [59]. [Glossary Microdeletion, Translocation, Copy number, Genetic surplus disorder, Trinucleotide repeat disorder]

As it happens, the genome can occasionally tolerate considerable genetic attrition without ill effect. In one study, a 2356 kb stretch of DNA was deleted from the mouse genome. Mice homozygous for the deletions were, as far as anyone could determine, absolutely normal; and could even reproduce [60]. We might imagine that mice are simple organisms compared to humans. Surely, the human genome could not tolerate any such genetic foolery. Wrong. Sequences of several megabases in length have been found to be present in apparently normal individuals but absent in the human reference genome. One group has identified hundreds of microsequences present in at least 1% of the human population, but absent in the reference genome [61]. [Glossary Homozygosity]

An exceptional gene variation in humans involves pseudoautosomal region 3. The pseudoautosomal regions are several homologous sequences in X and Y chromosomes. These are the only areas of the X and Y chromosomes wherein meiotic recombination occurs. Pseudoautosomal region 3 is found in 2% of the human population; the remaining 98% of humans manage to carry on, blissfully unaware of the missing piece in their genomes [39].

Permissible aneuploidy

Aneuploidy is the presence of an abnormal number of chromosomes (for the species) in a cell. In our minds, we associate aneuploidy with cancer, insofar as virtually every clinically advanced cancer occurring in humans is characterized by severe aneuploidy. In highly aggressive cancers, we might see some cells with over one hundred chromosomes, and other cells with fewer than the diploid number of chromosomes (i.e., fewer than 46 chromosomes). We can guess that the aneuploidy observed in cancer cells is tolerated because cancer cells have simplified physiologies (growth and invasion and little else). Moreover, cancer is a somatic disease, meaning that aneuploidy is restricted to the tumor cells and does not tell us much about the potential effect of aneuploidy on the whole organism (Fig. 2.3).

Considering all the mechanisms in place to preserve the integrity of DNA (i.e., at least seven DNA repair mechanisms in human cells) and of chromosomes (i.e., the precision of mitosis and meiosis [62]), we might assume that small deviations from the normal

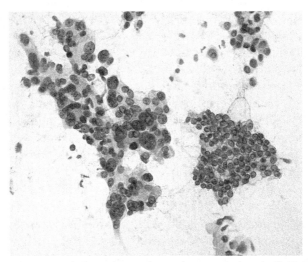

FIG. 2.3 A cytologic specimen composed of two clumps of normal pancreatic ductal cells (right lower corner of figure) and malignant pancreatic adenocarcinoma cells (large clump occupying entire left half of figure, plus scattered small clumps and single cells of the upper right corner of the figure). The nuclei of the normal cells are uniformly sized, evenly spaced, round, with a uniform *blue color*—gray in print version. The nuclei of the malignant cells have large nuclei, of varied size and shape, and nonuniform texture. These are typical signs of cytologic atypia and highly suggestive of the presence of aneuploidy. *Source: Wikipedia, released into the public domain by its author, Ed Uthman.*

number and configuration of chromosomes would be lethal to cells. In fact, minor imbalances in the number of chromosomes are associated with fetal death, or with nonviable newborns, when they involve the large autosomes (i.e., trisomies and monosomies of the smaller chromosomes can be compatible with life) [62, 63]. [Glossary Genomic disorder]

As it happens, a rare disease teaches us that generalized aneuploidy can be tolerated. Mosaic variegated aneuploidy syndrome is caused by a mutation in the BUB1B gene, which encodes a key protein in the mitotic spindle check point, or with the gene encoding the centrosomal protein CEP57 [64]. This disease is characterized by widespread aneuploidy in more than 25% of the cells of the body. Mosaic variegated aneuploidy syndrome is associated with some occurrences of growth deficiencies, microcephaly, and developmental delay. Individuals with mosaic variegated aneuploidy syndrome may have an increased risk of developing cancer in childhood [65]. Nonetheless, this aneuploidy syndrome is compatible with human life and is another indicator of the permissible range of chromosomal quantity in metazoans.

Nonuniversal genetic code

The genetic code is the mapping between RNA triplets and their respective amino acid translations. We have all seen charts of this coding system, which was, from its inception, recognized as universal. Aside from the obvious point that we know nothing about nonterrestrial genetics, and have absolutely no basis to assert that the code applies outside our planet, we have learned that the genetic code does not apply uniformly here on earth.

In particular, a "stop" triplet in one organism's code may serve some other purpose in another organism. A triplet that codes for a particular amino acid in one organism may code for some other amino acid in another organism. In fact, a triplet coding for one amino acid in the nucleus of an organism may code for a different amino acid in the same organism's mitochondria.

Such exceptions to the "universal" code are admittedly rare, but their occurrence indicates that natural selection operates in a world of limitless possibility.

Silly ciliates

The next time you yield to the implacable urge to watch a cheesy science fiction movie about strange aliens, pay special attention to the obligatory scene wherein the crew's medical officer examines a cell scraping taken from a sample of alien tissue, or blood, or goo left at the scene of some particularly gruesome extraterrestrial mayhem. The scientist will inevitably peer into a microscope and exclaim, "These cells are regenerating at an impossible rate!" If the movie scene provides you, the avid viewer, with a peek at what the doctor sees, it will be two general issue cells, with normal mitotic architecture, dividing somewhat rapidly. Really now, couldn't the writers come up with something a little more imaginative? Here on earth, we have eukaryotic organisms whose methods of cell division and genetic reproduction far exceed the wildest notions of Hollywood movie makers.

The ciliates (members of Class Ciliophora) are single-celled eukaryotes that are physically distinguished by the profusion of cilia visible on their surfaces (Fig. 2.4).

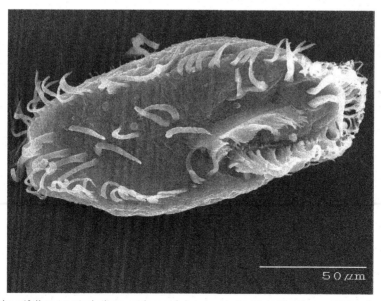

FIG. 2.4 *Oxytricha trifallax*, a typical ciliate, with ovoid shape and plentiful cilia. *Source: Wikipedia, from a US government public domain work.*

Each ciliate organism is endowed with two nuclei, with each nucleus having its own peculiar genomic architecture and cellular purpose. One nucleus is large and polyploid (the macronucleus), and it serves the general physiologic needs of the cell (e.g., producing RNA and maintaining the cell's phenotype). The other nucleus (the micronucelus) is tiny and does not express its genes. It serves as the genetic repository preserving the germline of the organism. The micronucleus also produces the macronucleus, through a process that requires the amplification of its genome, followed by some severe editing. The macronucleus divides by a kinky process known as amitosis, but each amitotic division reduces its viability. After about 200 such divisions, the micronucleus must generate a new macronucleus. On such occasions, two ciliates may couple, and their micronuclei exchange DNA via a conjugal mechanism that involves the meiotic production and exchange of haploid chromosomes. If movie producers want to wow their audiences with a glimpse of other-worldly cellular behavior, they need to go no further than the ciliates. There are an estimated 30,000 species of these organisms, and they can be found swimming in any pond or river, just waiting for their big chance to make it into Hollywood (Fig. 2.5). [Glossary Polyploidy]

There is only one species of ciliates known to cause human disease: Balantidium coli. Balantidiasis occurs in locations where drinking water is contaminated with pig or human feces.

What have we learned in this section? We know that all living organisms on earth have the same genetic material (DNA, or RNA in the case of some viruses), using the same or a

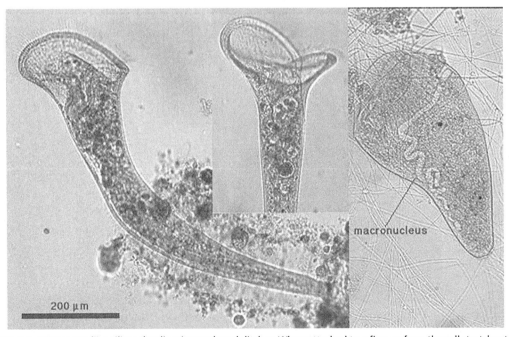

FIG. 2.5 *Stentor roselii*, a ciliate that lives in ponds and ditches. When attached to a firm surface, the cell stretches to form a trumpet shape that can reach up to 1.2 mm in length. Its cilia serve to brush passing microorganisms into its wide mouth. *Source: Wikipedia, from a public domain image.*

closely similar genetic code. We know that DNA replicates and serves as a template for RNA (or sometimes the reverse). RNA, in turn, serves as a template for proteins. Everything else about our genomic lifestyles is up for grabs. Different organisms have their own ways of regulating and exploiting the basic processes of life. Hence, physiologic functions such as mitosis, meiosis, reproduction, tissue repair, and aging may occur quite differently in different organisms; and may not occur at all. [Glossary Aging]

Section 2.5 Rummaging Through the DNA Junkyard

I would be quite proud to have served on the committee that designed the E. coli *genome. There is, however, no way that I would admit to serving on a committee that designed the human genome.*

David Penny [66]

A corollary to the Law of Sequence Conservation is that sequences that degenerate over time, and for which there are large variations in the general population of a species, must not have a very important function in the organism. This being the case, then why do cells tolerate their high level of junk DNA. Perhaps, if junk DNA accumulated with no cost to the species, then we might say that maintaining the extra sequences might be easier than trying to slim the genome. We know, however, that every cell pays a price to haul its junk DNA. At every cell division energy is expended to replicate the genome and the larger the genome, the more energy must be expended. Why doesn't our genome simply eject our surplus DNA? Stripping down the genome to its base essentials is a common strategy employed by a variety of obligate intracellular parasites. [Glossary Facultative intracellular organism, Obligate intracellular organism, Parasite]

Is there a reason to save our junk DNA? Before answering this question, we need to think about the difference between function and purpose. Function relates to organismal activity (e.g., participation in a metabolic pathway, serving as a structural component) [66]. Purpose relates to destiny [67]. A gene's purpose may only be to expand the gene pool, with sequences that may be useful, in some future set of circumstances. The genes-in-waiting serve no function whatsoever, but they have an important purpose.

The poet John Milton wrote, in "On His Blindness" that "They also serve who only stand and wait." In this section, we will see that junk DNA, though it serves no useful function for individual organisms, has a very important role to play in the fitness and evolvability of every species. One of the recurring themes in this book is that evolution is something that applies to species, not to individual organisms. When we focus on the properties of species, we will find that some of the enduring mysteries of biology will simply evaporate.

As it happens, junk DNA fulfills several important purposes:

Purpose 1. Fueling evolution

Every evolving organism must contain some nonfunctional (i.e., junk) DNA [67]. Lets look at the line of reasoning that supports this inference.

1. Our proof begins by imagining a species that has a completely functional genome (i.e., a species in which every gene codes for a useful protein).

2. If this were the case, then every gene in the organism would be conserved, through natural selection. Otherwise, the organism would lose the functionality of one or more of its useful genes, thus reducing the fitness of the species.

3. If every gene in the organism were conserved, then the species could not evolve. We can infer this because evolution involves the introduction of variant genes, and there can be no variant genes if every gene has been conserved.

4. But we know that every species evolves. Remember, a species is defined as an evolving gene pool.

5. Hence, there can be no species whose genome is completely functional [67].

A somewhat poetic way of expressing the same inference is: "A perfect being cannot evolve." If every gene is perfectly adapted to create a perfect being, then there really is no way to cope with a changing environment, apart from disrupting some of those perfect genes, and producing a less-than-perfect being.

It is distressing to admit, but the evidence is overwhelming that the bulk of the human genome (the soul of the cell) is about 80% junk. By junk, we mean to say that it serves no discernible function. So far as we can determine, if all that junk DNA were suddenly eliminated from our cells, we would not expect to suffer any detrimental consequences.

Purpose 2. Creating de novo genes

When we refer to evolution, we like to imagine that totally new genes are being created, producing wonderful new species. This is almost never the case. We shall see in Section 4.2, "The Biological Process of Speciation," that new species start out very much like the old species from which they branched. We won't explain this next assertion until we get to Chapter 4, but it turns out that new species emerge without any new species-specific genes, and without any variant genes other than those that are found in the gene pool of the parent species.

New genes do appear in the gene pools of species, but infrequently. As we've already pointed out, most of the genes in the human genome are homologous to ancient genes, and perform functions that are similar to the functions performed in ancestral organisms. This doesn't leave a whole lot of room for new genes. However, every old gene was a new gene at one time and there must be mechanisms for creating something new. Let's disregard those genes that are acquired by horizontal gene transfer or from retroviruses, or from some other form of biological thievery. Such genes seem new to the acquisitive organism, but they are really just old genes shoved into a different cell.

For many decades, it was assumed that all new genes develop from duplications of existing genes. The idea here is that a duplication of a functional gene permits the natural selection of gene variants that serve to expand the functionality of the original gene, or that serve to select small changes to the gene that provide some similar or related function in the cell. The evidence supporting this theory is that when we look at all of the genes in our genome, we find that more than 97% of our genes have a close sequential match to some other gene that is already in our genome [68]. When we look at the ancestry of these near-duplicates, they seem to match genes that are present through multiple levels of our phylogenetic lineage. The simplest explanation for these findings is that a new gene appearing in an ancestral species was replicated, and that the replicates, or parts thereof, evolved through descendant classes of organisms, yielding the genes that currently reside in our genome.

A second set of observations reinforces the notion that new genes come from duplicated ancient genes; that being the evolution of gene families [57, 69]. As discussed in Section 1.3, "Our Genes, for the Most Part, Come from Ancestral Species," gene families consist of paralogous genes with similar or complementary activities and traceable through a long phylogenetic lineage.

Before we leave the point that new genes come from duplicated genes, let's consider just one more set of observations; this being the oft-repeated association between polyploidization and the radiation of new species. Occasionally, species experience a polyploidization event, wherein the stable number of chromosomes in the species doubles its normal chromosome count. For example, a genus wherein all the organisms have 10 chromosomes in its haploid set, now finds organisms with 20 chromosomes. Following polyploidization in a stem lineage, we may sometimes observe a radiation of new species, all having the polyploid chromosome number. Polyploidization has been proposed as the major genetic force speeding the evolution of flowering plants, for which there are about 300,000 known species, and in ray-finned fish, for which there are over 20,000 species [70–72]. Insofar as polyploidization is the ultimate type of gene duplication, these observations strengthen our hypothesis that new genes arise from duplicated genes (Fig. 2.6).

The assumption that new genes derive from old genes, exclusively following gene duplication events, has been discredited by the observation of recently evolved genes having no homologs to ancient genes [73]. Whole genome sequencing has uncovered genes that are species-specific; hence, not inherited from ancestral organisms and not belonging to gene families. These species-specific genes were, at first, called "orphan genes" (i.e., having no parental origin), and are now called "de novo genes" (i.e., new to a species).

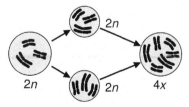

FIG. 2.6 Polyploidization produces copies of each diploid pair of chromosomes. The schematic indicates a doubling of ploidy, producing a 4-*n*-ploid cell. If the polyploidization event is stable and persistent in progeny, then the species has in effect acquired an extra set of chromosomes that can mutate and participate in the evolutionary process without necessarily harming the fitness of the organism. Ancient polyploidization events have occurred in fish and in angiosperms (flowering plants). *Source: Wikipedia, and entered into the public domain.*

It is hypothesized that de novo genes arise from noncoding gene sequences that mutate to include a start codon sequence that can initiate RNA transcription and protein synthesis. We assume that most such events are either deleterious or of no advantage to the cell. However, if the new protein fragment serves some useful purpose for the cell, then the new coding sequence may be conserved [74]. The emergence of de novo genes is not particularly rare [75]. Scientists have found 142 genes that are present in some populations of flies, but not in others, a sign that these genes arose de novo [73]. The emergence of de novo genes is yet another example of the value of junk DNA. [Glossary Open reading frame]

Purpose 3. Genomic regulation

We have inferred that at least some portion of the noncoding regions of the genome must be functional. In point of fact, some noncoding sequences are highly conserved. According to the Law of Sequence Conservation (Section 2.2 of this chapter), highly conserved sequences always have a function. As noted previously, some of these conserved noncoding regions are transcription enhancers [34]. The epigenome regulates the genome, but the epigenome does not monopolize the genome regulation business. We will see that noncoding regions of DNA have a variety of different regulatory functions that will be discussed in Section 3.5, "Pathologic Conditions of the Genomic Regulatory Systems." [Glossary Enhancer]

– How much of the genome is junk?

When we speak of junk DNA, we are referring to DNA that serves no immediate function. If we confine our attention to DNA that has a definable function during the life of an organism, we cannot dismiss the conclusion that the vast majority of DNA is noncoding, has no function, and is certifiably junk [52].

Humans and marbled lungfish are relatively close cousins, both being members of Class Dipnotetrapodomorpha (see Section 6.2, "Vertebrates to Synapsids"). When we compare marbled lungfish, having a 133 billion base pairs, to humans, having 3 billion base pairs, we cannot help but conclude that nearly all of the genome of the marbled lungfish is junk. When we compare the human genome with that of another animal, Kudoa iwatai, a myxozoan with a genome size of 22.5 megabases, we get the idea that nearly all of the human genome is junk. The humiliation never seems to end. When we search for the minimal number of genes necessary to sustain the life of a terrestrial organism, a mere 182 genes seems to suffice [76]. This tells us that the excess 25,000 or so genes in humans are devoted to nonessential activities (e.g., watching football on TV, purchasing lottery tickets, and shopping for stiletto-heeled shoes).

Let's be generous and assume that human genome sequences that are known to be conserved must be functional; otherwise they would not have been conserved. Conserved sequences represent 5% of the human genome. In addition, about 4% of the human genome is under some degree of selective pressure. Generously totaling the two, an estimated 9% of the human genome is not junk [77].

Glossary

Aging Alternate spelling, "Ageing". A chronic degenerative process that occurs in cells that have lost the ability to divide, while retaining their functional obligations. Such cells include neurons, chondrocytes (i.e., cartilage cells), muscle cells, and cells of the eye lenses. Cells that maintain the ability to divide, indefinitely, such as epithelial lining cells of ducts, mucosal surfaces, glands, and *epidermis* do not suffer from the degenerative changes associated with aging cells. Empirically, the colon of a centenarian may look much like the colon of a 20-year-old, and nobody dies from an old colon or an old liver.

For the purposes of this book, aging is considered a disease, differing from other diseases only in its inevitability.

Allele One of a pair of matched genes on paired chromosomes, wherein each of the matched genes may vary in sequence from the other. In most cases one allele comes from the father, the other from the mother. One biological exception to biparental alleles may arise with uniparental disomy. Uniparental disomy occurs when the offspring's germline has a chromosome pair derived from one parent, with no contribution from the other parent. Uniparental disomy of a chromosome or of a part of a chromosome can also occur as an acquired feature in somatic (i.e., nongermline) cells and is a frequent alteration found in cancers. There are several mechanisms whereby somatic cell mitosis can lead to an acquired uniparental disomy; these involve faulty chromosome migration with or without accompanying translocations when the chromosomes migrate to daughter cells [78]. The acquired, or somatic form of uniparental disomy does not actually involve the participation of two parents (as seen in uniparental disomy of germline cells), and is often referred to by the less mechanistic, but more precise term, "copy neutral loss of heterozygosity." Because only one allele is represented in the uniparental disomic gene, there is loss of heterozygosity. Because cells have the normal number of copies of expressed genes producing a normal gene expression level, the result is "copy neutral."

Association vs cause An association between events A and B is the finding that A and B often occur together. A causal relationship between A and B is the finding that A initiates an event that results in the occurrence of B (or vice versa).

It is easy for scientists to confuse associations with causes. For example, the renowned statistician, R.A. Fisher (1890–1962), who should have known better, concluded that the causal link between smoking and lung cancer was mere speculation. He suggested that a precancerous condition (not caused by smoking) could irritate the bronchial tree to the extent that affected individuals might seek the cooling and soothing comfort of cigarettes to self-medicate, thus prompting the statistical illusion that smoking causes cancer when the opposite is true (the cancerous process causes smoking). Fisher's unhelpful analysis demonstrates the importance of sequential timing data in medical research. If A precedes B, then we can assume that B is not the cause of A. We now know, through experimental research, that smoking precedes the development of precancerous lesions and that precancerous lesions do not form unless there is a preceding cause, such as smoking. [Glossary Bronchogenic carcinoma, Lung cancer]

It can be difficult or impossible to assign causality to an association. In general we look to see if the associated factor always segregates with the disease (i.e., is found in every instance of disease and is found in no instances of the control population), and if the association always precedes the development of disease and does not occur as a late finding in chronic or fully developed instances of the disease. Such findings do not establish proof of causality, but they can be considered supportive evidence.

For example, scientists have long wondered whether tau proteins, a hallmark of several neurodegenerative diseases, are causal factors, or just associated bystanders. Frontotemporal dementia and parkinsonism linked to chromosome 17 (FTDP-17) is an inherited disease characterized by mutations in the gene coding for Tau. In this disease, a specific Tau inclusion is always found in the pathologic lesions and never in control subjects. Most importantly, the appearance of the Tau inclusion always precedes the development of neural degeneration; hence, the inclusion does not occur as a consequence of neural degeneration. These observations have led researchers to believe that altered Tau protein plays a causal role in the pathogenesis of FTDP-17 [79]. Nonetheless, associations never prove hypotheses. It is best to think of associations as clues; not answers.

Bioinformatics The science of the curation and analysis of biological data. The field of bioinformatics focused on genomic data for several decades. Presently, the field has expanded its purview into epigenomics, proteomics, metabolomics, and so on. Not to be confused with biomedical informatics.

Copy number It is possible to produce a rare genetic disease without actually producing a mutation in a gene; simply changing the number of genes may be sufficient [80]. Charcot-Marie-Tooth disease is an inherited neuropathy caused by duplication of a 1.5 megabase segment on chromosome 17. No altered protein is produced. The root cause of the clinical phenotype is the increased gene dosage.

Cytopenia A reduction in the normal number of cells of a particular type. The term is usually applied to hematopoietic cells (i.e., marrow-derived blood cells). Thrombocytopenia is a cytopenia of platelets (i.e., thrombocytes). Granulocytopenia is cytopenia of granulated white cells. Pancytopenia refers to a reduction of all the different types of cells of hematopoietic lineage.

De novo mutation De novo is a term derived from Latin, meaning "new." De novo mutation diseases arise from disease-causing mutations found in the germline of a child that were not present in the germline of the parents (i.e., mutations that were not inherited from a parent's germline).

A careful reading of the above definition tells us that a new mutation that is present in a parental gamete fits the definition of "de novo" insofar as the new mutation in the gamete was not present in the germline of the parent (i.e., not present in all the somatic cells of the parent). The implication here is that a de novo mutation can be passed from a father or a mother to the offspring, without being inherited from the parents. The notion that an individual can receive a noninherited disease from a mutation present in a parental cell is always confusing, unless we fully understand the definitions of "de novo" and "germline."

In addition, a new mutation arising in the fertilized zygote of the offspring, or a new mutation in an embryonic cell of the offspring will also fit the definition of a de novo mutation. When the mutation occurs in an embryonic cell (e.g., morula, blastula), then the resulting mutation will be expressed as a mosaic in the developed organisms (i.e., present in some somatic cells and not in others). The cells of the offspring that contain the mutation will be the clonal descendants of the original newly mutated embryonic cell.

Differentiation The human body contains several hundred types of cells endowed with a set of functions and morphologic features that distinguish each cell type from all of the other cell types. The process by which each type of cell becomes different from the other (several hundred) cell types is called differentiation. Because every cell type, regardless of its particular properties, has the same genome as every other cell type, we can infer that some factors other than genome sequence controls the process of differentiation. In mammals, differentiation is determined by the epigenome, which controls the levels of expression of genes in a cell, which in turn determines how a gene looks and behaves. The epigenomic modifications are heritable from a cell to its progeny (e.g., when a hepatocyte replicates, it produces hepatocytes and not any other cell types), and this would lead us to expect that when a cell of a given cell type replicates, it will give rise to another cell of the same cell type. This, in fact is the case. The process by which undifferentiated cells give rise to a lineage of cells with progressively greater levels of differentiation, until a fully differentiated cell type is achieved, is referred to as stem cell maturation, or lineage maturation.

Enhancer A site on DNA that binds to trans-acting protein factors, to enhance transcription. Unlike promoters, enhancers do not need to be close to their target genes. Enhancers are one of many regulators of gene expression [81].

Etiology The cause of a disease. One of the themes of this book is that diseases do not have an etiology; diseases have a pathogenesis. Still, it is convenient to assign one cause to a disease, if for no reason other than abbreviating explanations. If the word "etiology" is to be used at all, it should probably refer to whatever is generally believed to be the root cause of the disease.

Exome sequencing Also known as targeted exome capture, exome sequencing is a laboratory technique wherein only the exons (the sections of DNA that code for proteins) are sequenced, sparing analysts from dealing with the noncoding regions of DNA [82]. In the human genome, there are only about 180,000 exons, accounting for about 1% of the genome, and about 85% of sites where disease-causing mutations occur [82].

Facultative intracellular organism An organism that is capable of living, and reproducing inside cells. Most of the facultative intracellular organisms are bacteria. There are a number of single-celled eukaryotic organisms that have states of their life cycles in which they alternate from intracellular to extracellular existence (e.g., the trypanosomes and the plasmodia), but such organisms follow a strict life cycle dictating their biological identities in and out of cells. The term "facultative," indicating a circumstantial or optional adaptation, doesn't seem to apply to the eukaryotes.

Genetic instability The process whereby the genome accumulates genetic alterations (e.g., SNPs, GSVs) over time. Low levels of unrepaired DNA damage are an inescapable feature of living cells. The older the cell, the more mutations might be found [83]. Many cancers have a high rate of genetic instability. Only those mutations that arise in germ cells can be passed onto progeny and contribute to the germ pool of the species [2].

As a general rule, cancer cells have high rates of mutation, with rates that may exceed a 100-fold higher than normal cells, and this is thought to contribute to the high degree of genetic instability that is almost always found in cancer cells [6].

Genetic surplus disorder Mutations that increase the size of an organism's genome or that produce an increased dosage of one or more genes, or of portions of a chromosome, can produce rare diseases [84]. Examples are: Charcot-Marie-Tooth disease, an inherited neuropathy caused by a duplication of a segment of chromosome 17 [80]; and Down syndrome, caused by an extra chromosome 21.

Genomic disorder Although all genetic diseases are technically genomic diseases, the term "genomic disease" is usually reserved for disorders arising from the loss or gain of portions of the DNA in chromosomes (i.e., usually involving several megabases, and never single nucleotide mutations). Disorders in which there is an increase or decrease of portions of DNA that normally occur as multiple copies (i.e., copy number losses and copy number gains) are included in the genomic diseases. Most of the genetic disorders arise as sporadic, de novo conditions caused by chromosomal rearrangements that may result in interstitial or terminal deletions or duplications. Meiotic nondisjunction events in phenotypically normal carriers of balanced translocations may be passed on as a disorder of gene dosage in their offspring [85].

Genomic structural variation Abbreviation: GSV. A variation in the structure of chromosomes, usually involving more than a single nucleotide (i.e., not a point mutation and not a SNP). GSVs include alterations in karyotype or cytogenetic alterations observable with special techniques, as well as changes too small to see with a microscope, such as small deletions, insertions, SNPs, and larger insertions, inversions, and translocations. GSVs would also include duplications and other copy-number alterations as well as gene conversions [57]. GSVs among different individuals in the human population occur frequently, and may account for more phenotypic variations in the human population than do SNPs [59].

Germline The germline consists of the cells that derive from the fertilized egg of an organism. All of the somatic cells (i.e., the cells composing the body), as well as the germ cells of the body (oocyte and spermatozoa) arise from the same germline. The extraembryonic cells (e.g., placental cells) have the same germline as the somatic cells. An inherited condition can be described as being in the germline; in every cell that derives from the fertilized egg.

The word "germline" has confused many students, who use the term "germline cell" interchangeably with "germ cell" or with "germ cell line." In fact, if you search for a definition of germline on the web, you're apt to find the somewhat recursive definition that the germline constitutes all of the cells in the offspring that are produced by the germ cell lines of the parents. True, but apt to mislead.

It is best to think of a germline mutation by its functional definition, a mutation found in every cell in an organism; and not by the mechanism whereby the mutation came to occur (i.e., passed from a parental germ cell).

Germline mutation A germline mutation is a mutation found in every cell of the body. The word "germline" was perhaps a poor choice of words. "Germline" is used to indicate that the mutation must have been present in the earliest conceptus (i.e., the zygote). It should not be confused with germ cells or germinal cells or the germ cell lineage, all of which pertain to the cell line that gives rise to mature germ cells (i.e., sperm cells in males and oocytes in females).

Haplotype A set of DNA polymorphisms that tend to be inherited together, often as a result of their close proximity on a chromosome. The term is often used to describe a set of SNPs that are statistically associated with one another, on a chromosome. Another term, "haplogroup" refers to a subpopulation of individuals that share a common ancestor and a haplotype.

Homozygosity Condition that occurs when only one allele of a gene is expressed in cells. This may occur when both of the inherited alleles of a gene (the maternally-derived allele and the paternally derived allele) are identical to each other, or when both alleles came from one parent (e.g., uniparental disomy). It may also result when the expression of one of the inherited alleles is unattained or lost in which case homozygosity is said to result from the loss of heterozygosisty.

Horizontal gene transfer The direct transfer of genetic material between organisms, by mechanisms other than by reproduction (i.e., other than the transfer of DNA from parents to offspring). The very first eukaryotic ancestors derived their genetic material by horizontal gene transfer from prokaryotes (bacteria and archaean organisms), viruses, and possibly from other now-extinct organisms that might have preceded the eukaryotes. The early eukaryotes almost certainly exchanged DNA between one another, and we see evidence of such exchanges in modern single-celled eukaryotes and fungi [86].

To an unknown extent horizontal gene transfer occurs throughout the animal kingdom. For example, tardigrades, a microscopic animal, have a genome one-sixth of which was derived from bacteria, archaeans, plants, and fungi [46].

It should be noted that many of the most significant evolutionary advances came from interspecies gene acquisitions. The primordial mitochondrion that helped to create the first eukaryotic cell was an acquisition from a bacteria. The very first chloroplast in the most primitive precursor of the plant kingdom was an acquisition from a cyanobacteria. The big jump in adaptive immunology came with acquisition of the RAG1 gene. This gene enabled the DNA that encodes a segment of the immunoglobulin molecule to rearrange, thus producing a vast array of protein variants [87]. The RAG1 gene, which kicked off adaptive immunity in animals, was derived from a transposon, an ancient DNA element that was acquired through horizontal gene transfer or through infection from another living organism or from a virus; not by random base mutations.

Inheritance Inheritance, to the layman, indicates that a trait or disease occurring in an offspring can be attributed to some genetic condition or predisposition that is present in one or both of the parents. To the geneticist, the term "inheritance" has a much more narrow meaning, and the difference in these two interpretations can lead to a great deal of confusion. It's best to take the time to explain the medical meaning of inheritance and how the precise meaning of the term influences our understanding of the pathogenesis of genetic diseases.

For the geneticist, inheritance always indicates the passage of affected genes from the germline of a parent to the germline of the offspring. This means that the inherited genes are present in every cell of the parent. When a genetic abnormality is present only in the fertilized oocyte or the fertilizing sperm, then we do not speak of inheritance as the cause of the abnormality in the offspring; the mutation that

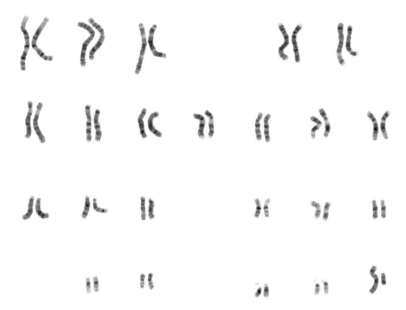

FIG. 2.7 Diploid karyotype prepared from a chromosome spread of a cell from a human male. Each chromosome is paired with the Y chromosome paired to the X chromosome. *Source: Wikipedia, from the US government public domain image.*

is new to a parent's gamete is said to be de novo (i.e., new, and not inherited). Hence, Down syndrome (trisomy 21) is a condition that often comes from the mother (i.e., when there is a genetic defect in the oocyte) but which is seldom inherited from the mother.

The medical definition of inheritance is an anachronism that has persisted from the early days of genetics when there was no way to examine the cellular pathogenesis of disease. In the case of Down syndrome, if both the mother and the father did not have Down syndrome, and if there was no history of Down syndrome in the family, then the infant with Down syndrome must not have inherited the condition. It is worth pointing out that this last statement is not always true. Cases of so-called translocation Down syndrome are inherited from a parent who does not have Down syndrome. In these cases, one parent carries a balanced translocation between chromosome 21 and another chromosome, producing no gain or loss of chromosomal material in the affected parent. If this translocation is passed to the next generation, it can become unbalanced, producing extra genetic material from chromosome 21 in the offspring, with the consequent development of Down syndrome.

Karyotype From the Greek root, karyon, meaning nucleus, the karyotype is a standard shorthand that describes the chromosomal complement of a cell (Fig. 2.7).

The normal karyotype of a human male diploid somatic cell is 46 XY, a somewhat confusing way to express that there are two sets of 23 chromosomes, which includes one haploid X chromosome (the female sex chromosome) and one haploid Y chromosome (the male sex chromosome). The normal female karyotype is 46 XX. Abnormalities in karyotype are described using The International System for Human Cytogenetic Nomenclature (ISCN).

Mendelian inheritance A pattern of inheritance observed for traits inherited from the mother or the father. The modes of Mendelian inheritance are autosomal dominant, autosomal recessive, sex-linked dominant, and sex-linked recessive. The most comprehensive listing and discussion of the Mendelian diseases has been collected, for many decades, in Mendelian Inheritance in Man, currently available online [88]. The number of Mendelian diseases varies depending on how they are counted (e.g., a smaller number if counted by disease phenotype; a larger number if counted by genotypic subtypes), but it is generally accepted that there are at least 7000 documented Mendelian diseases that occur in humans.

Microdeletion Microdeletions are cytogenetic abnormalities that typically span several megabases of DNA. Microdeletions are too small to be visible with standard cytogenetics, but they can often be detected with FISH (fluorescent in situ hybridization). All of the microdeletion syndromes are rare diseases, and they typically arise as de novo germline aberrations (i.e., not inherited from mother or father, in most instances). Conditions that occur rarely and sporadically to produce a uniform set of phenotypic features in unrelated subjects, may be new cases of microdeletion syndromes [89]. DiGeorge syndrome is a typical microdeletion disease, with a germline 22q11.2 deletion encompassing about 3 million base pairs on one copy of chromosome 22, containing about 45 genes. Neurofibromatosis I sometimes occurs as a microdeletion syndrome involving a region of chromosome 17q11.2 that includes the NF1 gene. Microdeletion disorders are a subtype of contig disorders (i.e., contiguous gene disorder).

Microsatellite Also known as simple sequence repeats (SSRs), micorsatellites are DNA sequences consisting of repeating units of 1–4 base pairs. Microsatellites are inherited and polymorphic. This means that within a population there will be wide variation in the number of repeats at any microsatellite locus.

Microsatellite instability When there is a deficiency of proper mismatch repair (a type of DNA repair), DNA replication is faulty, and novel microsatellites (repeating short DNA sequences) may appear in chromosomes. This phenomenon is called microsatellite instability, and it is observed to occur at high frequency in cells obtained from various types of cancer. Microsatellite instability is a characteristic feature of colon cancers occurring in Hereditary Non-Polyposis Colorectal Cancer syndrome.

Monogenic disease Same as single-gene disease. The term is used in cases wherein a mutation in a single gene is the root cause of a disease, and in which no mutations in any other genes are involved.

Knowing what we now know about the pathogenesis of disease, the term "monogenic disease" is somewhat misleading. All diseases develop in steps and many different genes, pathways, and biological events may contribute to the pathogenetic process. Because we continue to discover "carriers" of monogenic disease genes who never develop the disease, we can infer that such diseases are not truly monogenic. It is easy to imagine that specific polymorphisms of multiple genes, in addition to the "monogenic disease gene" may be necessary to produce the disease phenotype.

It is worth remembering that any disease attributed to the loss of function of a single gene can be mimicked by selective epigenetic silencing. We don't call such conditions "zero-genic diseases." It is safest to refrain from assigning cardinality genetic diseases until we fully understand their pathogeneses.

Mutagen A chemical that produces alterations in the genetic sequence of DNA molecules. Most mutagens are carcinogens, and most carcinogens are mutagens. Contrariwise, there are examples of carcinogens that are not mutagens and examples of mutagens that are not carcinogens. These somewhat inconsistent findings reflect the fact, repeated throughout this book, that carcinogenesis is a multistep process occurring over time. This being the case, we would expect carcinogenic agents with different properties (i.e., mutagenic, nonmutagenic) to have effects on different steps of carcinogenesis.

Myeloproliferative disorder A blood disorder characterized by a clonal expansion of a hematopoietic cell population. When cells of lymphoid lineage proliferate in the blood, the analogous term, lymphoproliferative disorder, is applied. Many of these disorders have characteristic single-gene mutations [90].

Neurofibromatosis Any of several inborn conditions associated with neurofibromas (a benign neoplasm of Schwann cells). Common usage assigns "neurofibromatosis" to type 1 neurofibromatosis, or Von Recklinghausen Disease, which is associated with multiple cafe au lait spots (large epidermal nevi) plus neurofibromas. The so-called central form of neurofibromatosis, now called neurofibromatosis type 2, is characterized by bilateral acoustic schwannomas, and meningiomas. Skin neurofibromas, as seen in Neurofibromatosis type 1, are not a prominent feature of neurofibromatosis type 2. Juvenile subcapsular cataracts are often found in neurofibromatosis type 2.

Next-generation sequencing Refers to a variety of new technologies that sequence genomes quickly and cheaply [91].

Nomenclature A nomenclatures is an authoritative listing of terms intended to cover all of the concepts in a knowledge domain. A nomenclature is different from a dictionary for three reasons: [1] the nomenclature terms are not annotated with definitions, [2] nomenclature terms may be multiword, and [3] the terms in the nomenclature are limited to the scope of the selected knowledge domain. Many nomenclatures group synonyms under a group code. For example, a food nomenclature might collect submarine, hoagie, po' boy, grinder, hero, and torpedo under an alphanumeric code such as "F63958." Optimally, the canonical concepts listed in the nomenclature are organized into a hierarchical classification [92, 93].

Non-Mendelian inherited genetic disease Refers to genetic diseases that do not exhibit the patterns of inheritance described by Mendel. Specifically, diseases with an autosomal dominant inheritance pattern would be expected in half the offspring of a couple in which one parent was affected. A recessive inheritance pattern would be expected in one-quarter of offspring in which each parent carried one recessive gene.

Polygenic diseases (i.e., in which altered variants must occur in multiple genes for the expression of disease) are never Mendelian. In addition, monogenic diseases that might be expected to display a Mendelian inheritance pattern may become non-Mendelian due to a variety of confounding factors (e.g., allelic heterogeneity, epistasis, expressivity, genetic heterogeneity, incomplete dominance, lethal genotypes, penetrance, phenocopies, and pleiotropy). As a general rule, Mendelian inheritance is monogenic; non-Mendelian inheritance is polygenic; de novo diseases are not inherited (i.e., not Mendelian and not non-Mendelian). Diseases that have Mendelian patterns of inheritance tend to be less common than diseases with polygenic inheritance.

Noncoding mutational diseases We are all familiar with the concept of a genetic disease, wherein a mutation in a gene serves as the root cause of a disease. We are much less familiar with the idea that a mutation in a noncoding region of the genome (i.e., a nongene) can also produce a disease.

Most single nucleotide mutations in noncoding regions of the genome cannot cause human disease, because noncoding mutations are, at worst, regulatory. We know this because noncoding nucleotides, by definition, do not code for proteins. If they aren't coding for proteins, but they exert some biological effect on the cell, then that effect must be regulatory in nature. Regulatory effects caused by mutations in noncoding regions generally produce mild changes in the level of expression of coding genes (i.e., they do not turn off the expression of genes). Hence, a mutation in a single noncoding site is seldom sufficient to produce a disease.

Nonetheless a few examples of conditions caused by mutations in noncoding regions can be found, as shown:

- A form of frontotemporal dementia and/or amyotrophic lateral sclerosis (FTDALS1) is caused by a heterozygous hexanucleotide repeat expansion (GGGGCC) in a noncoding region of the C9ORF72 gene on chromosome 9p21.
- Lactase persistence is associated with noncoding variation in the MCM6 gene (601806) upstream of the lactase gene, in a region that appears to act as a cis element capable of enhancing differential.
- Variants of the OCA2 gene play a role in determining blue versus nonblue eye color and blond versus brown hair. Noncoding variants in the HERC2 gene 200 kb downstream of OCA2 are thought to alter the expression of the OCA2 gene, and thus indirectly influence eye and hair color.
- Chronic tubulointerstitial nephropathy can be caused by a 5656A-G transition in mitochondrial DNA. This adenine is the single noncoding nucleotide separating the structural genes of 2 tRNAs.
- Hyperferritinemia-cataract syndrome is caused by heterozygous mutation in the iron-responsive element in the 5-prime noncoding region of the ferritin light chain gene) on chromosome 19q13.

Noninherited genetic disease A significant but unquantified portion of genetic diseases is non-inherited; occurring as a new (de novo) mutation in the zygote or in the early germ cells of the developing embryo.

Obligate intracellular organism An obligate intracellular organism cannot survive for very long outside its host. Obligate intracellular organisms live off cellular products produced by the host cell. This being the case, it would be redundant for such organisms to maintain all of the complex cellular machinery that a free-living organism must synthesize and maintain. Consequently, obligate intracellular organisms adapt simplified cellular anatomy, often dispensing with much of the genome, much of the cytoplasm, and most of the organelles that were present in their ancestral classes, prior to their transition to intracellular (i.e., parasitic) life.

Open reading frame Abbreviation: ORF. A sequence within the genome that can be translated (via an mRNA intermediate) into protein. As such, an ORF will have a start codon (usually AUG) and a stop codon (usually UAA, UAG, UGA).

Parasite Any organism that lives and feeds in or on its host. In common usage, the term "parasite" was often reserved for animals that are parasitic in humans and in other animals. Usage has since been extended to include the once-called one-celled animals (i.e., members of the class formerly known as Protozoans). We now know that members of Class Protozoa belong to any of several distinct major classes of organisms, and are not one-celled animals. As we learn more and more about classes of organisms, and their lifestyles, the term "parasite" seems to have diminishing biologic specificity and utility.

Parthenogenesis Reproduction from an ovum without fertilization by a sperm. Occurs in plants, invertebrates, and some vertebrates but does not occur naturally in mammals.

Polycythemia Polycythemia is an increase in the number of circulating red blood cells. An increase in red blood cells can occur as a response to a physiologic stimulus (e.g., chronic anoxia, high-altitude living). The term polycythemia vera, or "true" polycythemia, is reserved for an increase in the number of red blood cells due to a clonal expansion of hematopoietic stem cells. The underlying mutation occurs in the JAK2 gene [94]. Mutations in JAK2 are associated with a variety of myeloproliferative conditions, including myelofibrosis, and at least one form of hereditary thrombocythemia [17, 94].

Polymorphism The term "polymorphism" can have several somewhat different meanings in various fields of biology. Herein, polymorphism refers to genetic polymorphism, indicating that variants of a gene occur in the general population. A polymorphism is usually restricted to variants that occur with an occurrence frequency of 1% or higher. If a variant occurs at a frequency of less than 1%, it is considered to be sufficiently uncommon that it is probably not steadily maintained within the general population. All commonly occurring polymorphisms are assumed to be benign or, at worst, of low pathogenicity; the reasoning being that natural selection would eliminate frequently occurring polymorphisms that reduced the fitness of individuals within the population. Nonetheless, different polymorphisms may code for proteins with some deficits in functionality.

Polyploidy A condition in which a cell or organism contains more than the usual complement of chromosomes, by some multiple of two. Polyploidization often results from the so-called endoduplication in which the customary doubling of chromosome number during mitosis is retained in a single cell, rather than divided among two cells.

On occasion, such duplication events stabilize to produce an organism with double the typical complement of chromosomes for the species. These events can occur more than one time, producing species having very large numbers of chromosomes. Polyploidization is particularly common in plants.

For example, the black mulberry (*Monus nigra*) has 308 chromosomes.

Polyploidization confers an evolutionary advantage by providing "spare" chromosomes suitable for rapid evolution, and some of the most impressive evolutionary radiations (i.e., appearances of many species arising in a relatively short period of time) have occurred after polyploidization events [70–72].

Proto-oncogene An oncogene in an inactive form.

Reassortment Often confused or used interchangeably with "recombination," but reassortment is generally reserved for a viral event wherein two similar segmented viruses exchange part of their genomes during the coinfection of a host cell. Reassortment seems to be the major mechanism accounting for new influenza virus strains.

Single-gene disease Synonymous with monogenic disease/disorder. A disease for which an aberration of one gene leads to the disease. Such disorders exhibit Mendelian inheritance patters (recessive or dominant). In the case of recessive single-gene disorders, the two alleles of the single gene are affected. A single-gene disorder may have genetic heterogeneity. For example, there may be many different genes that are capable of causing the disorder (as, e.g., Familial forms of dilated cardiomyopathy can result from mutations in any one of the following genes: LDB3 gene, TNNT2 gene, SCN5A gene, TTN gene, DES gene, EYA4 gene, SGCD gene, CSRP3 gene, ABCC9 gene, PLN gene, ACTC gene, MYH7 gene, PSEN1 gene, PSEN2 gene, gene encoding metavinculin, gene encoding fukutin, TPM1 gene, TNNC1 gene, ACTN2 gene, DSG2 gene NEXN gene, MYH6 gene, TNNI3 gene, SDHA gene, BAG3 gene, CRYAB gene, LAMA4 gene, MYPN gene, PRDM16 gene, MYBPC3 gene, TNNI3 gene, and GATAD1 gene. Nonetheless, each case is caused by an aberration of only one of those listed genes; hence familial dilated cardiomyopathy is considered a single-gene disorder. In addition, different mutations for a single gene may lead to the disease. If a single gene causes a disease, regardless of the number of disease-causing variants in the gene, it would be considered a single-gene disorder. Also, in single-gene disorders, there may be multiple genes that modify the clinical phenotype. Regardless, if a disease is characterized by a known mutation in a single gene that is sufficient for the expression of the disease, then the disease is considered a single-gene disorder (even when other gene variants may contribute to the clinical phenotype).

Single nucleotide polymorphism Abbreviation: SNP, single nucleotide polymorphism. Locations in the genome wherein different individuals are known to have single base differences in DNA sequence. It is currently estimated that there are more than 35 million SNPs in the human population, and a SNP occurs about once in every 300 nucleotides [95]. When SNPs are sought specifically in genes that are potential drug targets, their frequency appears even higher; 1 of every 17 nucleotides. By general acquiescence, the term SNP is reserved for variations that are found in at least 1% of the general population. The term single nucleotide variant (SNV) is applied to any observed variations among individuals, regardless of its frequency of occurrence in the general population. As DNA samples are sequenced for more and more individuals, the number of SNVs increases. Presumably, SNVs may occur anywhere along the 3 billion nucleotides that sequentially line the genome. Assuming that there is, at each nucleotide location, one particular nucleotide that dominates in the general population, then there would be 3 possible nucleotide variants at each location, producing a theoretical limit of 9 billion SNV location/nucleotide pairs. SNPs are just one form of genetic polymorphism among individuals. Other polymorphisms would include alterations in chromosome number, small alterations within a chromosome, deletions of stretches of DNA, insertions of DNA, and a host of subtle complex variations wherein parts of chromosomes are translocated elsewhere within the same chromosomes or to other chromosomes.

Skin cancer Any cancer arising from the epidermis (the multilayered blanket of squamous cells that covers the body), adnexa (e.g., hair, sweat glands, sebaceous glands), or integument (dermis and subcutis). Basal cell carcinoma and squamous carcinoma account for the greatest number of skin tumors. Each year in the US, squamous carcinoma of skin plus basal cell carcinoma of skin account for more than a million new cancers. This number exceeds the yearly occurrence of all other types of cancer, from all other anotomic sites in the body, combined. Because these tumors are seldom fatal and are so frequent, they are neither recorded in cancer registries, nor included in statistical compilations of cancers affecting the US population. Melanoma is another type of skin cancer and is potentially lethal, if not treated in an early stage of growth. Though melanoma occurs with a much lower frequency than either squamous cell carcinoma or basal cell carcinoma, it accounts for the greatest number of deaths due to skin tumors.

Somatic mutation A mutation occurring in the soma (i.e., the body, specifically, the nongerm cells of the body) and not present in the germline cells that give rise to the cells of the soma. Some diseases associated with acquired somatic mutations include aplastic anemia [96], myeloproliferative disorders [97], paroxysmal nocturnal hemoglobinemia [11], and virtually all cancers occurring in adults.

Stem cell A cell that divides to produce two different types of cells: another stem cell plus a cell that is more differentiated than the original stem cell. According to the stem cell theory of development, all of the differentiated cells of the body derive from an ancestry of stem cells, leading back to the zygote. In most cases, fully differentiated cells are incapable of further division and are sometimes referred to as postmitotic cells.

Translocation Exchanges of sections of a chromosome between two unpaired chromosomes.

Trinucleotide repeat disorder Rare diseases characterized by trinucleotide repeats in DNA. About half of the studied trinucleotide repeat disorders have repeated CAG sequences. CAG codes for glutamine, so the repeat CAG sequences code for a polyglutamine amino acid protein fragment. Examples of polynucleotide repeat disorders are: dentatorubropallidoluysian atrophy, fragile x syndrome, Friedreich ataxia, Huntington disease, myotonic dystrophy, several forms of spinocerebellar ataxia, and spinobulbar muscular atrophy.

Von Recklinghausen disease Eponym for neurofibromatosis type 1. Should not be confused with Von Recklinghausen disease of bone, also known as osteitis fibrosa cystica, which has no pathological relationship to neurofibromatosis.

X-chromosome The female sex chromosome. Genetically normal human males have one X chromosome and one Y chromosome. Genetically normal human females have paired X chromosomes, one inherited from the mother, and one from the father. In each somatic cell, one X chromosome is active, and the other X-chromosome is inactive. The inactive X-chromosome can often be visualized in cytologic preparations as a small dense clump of heterochromatin hugging the nuclear membrane. Hence, human females are X-chromosome mosaics; composed of cells expressing one of two different X-chromosomes.

How one X-chromosome is selected for inactivation, while the other is spared, is something of a mystery. There is evidence to suggest that the selection process is nonrandom [98, 99].

X-linked inheritance Refers to genetic conditions resulting from mutations on the X-chromosome. Most X-linked conditions are expressed preferentially in males and are inherited (in males) from the mother. The reason for this common pattern is that the X-chromosome in males always comes from the mother, and because males having only one X-chromosome, cannot compensate for an X-chromosomal mutation with a second X-chromosome.

There are exceptions to the rule. Because females have two X-chromosomes, they are twice as likely to harbor an X-chromosome mutation than are males. If the mutation on either X-chromosome is dominant, then the X-linked condition will express itself more often in females than in males. Examples of dominant X-linked diseases are:

```
Aicardi syndrome
Alport syndrome
Giuffre-Tsukahara syndrome
Goltz syndrome
Incontinentia pigmenti
Joao-Goncalves syndrome
Rett syndrome
X-linked dominant protoporphyria
X-linked hypophosphatemia
```

In addition, there are also examples of X-linked diseases that are typically expressed with greater clinical severity in females than in males. One such example is craniofrontonasal dysplasia (CFND) syndrome.

Y-chromosome The male sex chromosome. Genetically normal human males have one X chromosome and one Y chromosome. The Y chromosome is always inherited from the father. Hence, Y chromosome variants serve as clues to paternal lineage.

References

[1] Nagel ZD, Chaim IA, Samson LD. Inter-individual variation in DNA repair capacity: a need for multi-pathway functional assays to promote translational DNA repair research. DNA Repair (Amst) 2014;19:199–213.

[2] Crow JF. The high spontaneous mutation rate: is it a healthrisk? Proc Natl Acad Sci U S A 1997;94:8380–6.

[3] Nachman MW, Crowell SL. Estimate of the mutation rate per nucleotide in humans. Genetics 2000;156:297–304.

[4] Roach JC, Glusman G, Smit AF, Huff CD, Hubley R, Shannon PT, et al. Analysis of genetic inheritance in a family quartet by whole-genome sequencing. Science 2010;328:636–9.

[5] Oller AR, Rastogi P, Morgenthaler S, Thilly WG. A statistical model to estimate variance in long term low dose mutation assays: testing of the model in a human lymphoblastoid mutation assay. Mutat Res 1989;216:149–61.

[6] Jackson AL, Loeb LA. The mutation rate and cancer. Genetics 1998;148:1483–90.

[7] Whittaker JC, Harbord RM, Boxall N, Mackay I, Dawson G, Sibly RM. Likelihood-based estimation of microsatellite mutation rates. Genetics 2003;164:781–7.

[8] Martincorena I, Roshan A, Gerstung M, et al. High burden and pervasive positive selection of somatic mutations in normal human skin. Science 2015;348:880–6.

[9] Lynch M. Rate molecular spectrum, and consequences of human mutation. Proc Natl Acad Sci U S A 2010;107:961–8.

[10] Bianconi E, Piovesan A, Facchin F, Beraudi A, Casadei R, Frabetti F, et al. An estimation of the number of cells in the human body. Ann Hum Biol 2013;40:463–71.

[11] Johnston JJ, Gropman AL, Sapp JC, Teer JK, Martin JM, Liu CF, et al. The phenotype of a germline mutation in PIGA: the gene somatically mutated in paroxysmal nocturnal hemoglobinuria. Am J Hum Genet 2012;90:295–300.

[12] Migliaccio AR, Migliaccio G, Dale DC, Hammond WP. Hematopoietic progenitors in cyclic neutropenia: effect of granulocyte colony-stimulating factor in vivo. Blood 1990;75:1951–9.

[13] Gibbons RJ, Pellagatti A, Garrick D, Wood WG, Malik N, Ayyub H, et al. Identification of acquired somatic mutations in the gene encoding chromatin-remodeling factor ATRX in the alpha-thalassemia myelodysplasia syndrome (ATMDS). Nat Genet 2003;34:446–9.

[14] Yoshizato T, Dumitriu B, Hosokawa K, Makishima H, Yoshida K, Townsley D, et al. Somatic mutations and clonal hematopoiesis in aplastic anemia. N Engl J Med 2015;373:35–47.

[15] Zhao R, Xing S, Li Z, Fu X, Li Q, Krantz SB, et al. Identification of an acquired JAK2 mutation in polycythemia vera. J Biol Chem 2005;280:22788–92.

[16] Mead AJ, Rugless MJ, Jacobsen SEW, Schuh A. Germline JAK2 mutation in a family with hereditary thrombocytosis. New Engl J Med 2012;366:967–9.

[17] Barosi G, Bergamaschi G, Marchetti M, Vannucchi AM, Guglielmelli P, Antonioli E, et al. JAK2 V617F mutational status predicts progression to large splenomegaly and leukemic transformation in primary myelofibrosis. Blood 2007;110:4030–6.

[18] Berman JJ. Neoplasms: principles of development and diversity. Sudbury: Jones & Bartlett; 2009.

[19] Berman J. Precision medicine, and the reinvention of human disease. Cambridge, MA: Academic Press; 2018.

[20] Benvenuti S, Arena S, Bardelli A. Identification of cancer genes by mutational profiling of tumor genomes. FEBS Lett 2005;579:1884–90.

[21] Cahill DP, Lengauer C, Yu J, Riggins GJ, Willson JK, Markowitz SD, et al. Mutations of mitotic checkpoint genes in human cancers. Nature 1998;392:300–3.

[22] Forbes SA, Bindal N, Bamford S, Cole C, Kok CY, Beare D, et al. COSMIC: mining complete cancer genomes in the catalogue of somatic mutations in cancer. Nucleic Acids Res 2011;39(Database issue):D945–50.

[23] Greenman C, Stephens P, Smith R, Dalgliesh GL, Hunter C, Bignell G, et al. Patterns of somatic mutation in human cancer genomes. Nature 2007;446:153–8.

[24] Sonnenschein C, Soto AM. Somatic mutation theory of carcinogenesis: why it should be dropped and replaced. Mol Carcinog 2000;29:205–11.

[25] Eyre-Walker A, Keightley PD. The distribution of fitness effects of new mutations. Nat Rev Genet 2007;8:610–8.

[26] Francioli LC, Polak PP, Koren A, Menelaou A, Chun S, Renkens I, et al. Genome-wide patterns and properties of de novo mutations in humans. Nat Genet 2015;47:822–6.

[27] Acuna-Hidalgo R, Veltman JA, Hoischen A. New insights into the generation and role of de novo mutations in health and disease. Genome Biol 2016;17:241.

[28] Kong A, Frigge ML, Masson G, Besenbacher S, Sulem P, Magnusson G, et al. Rate of de novo mutations and the importance of father's age to disease risk. Nature 2012;488:471–5.

[29] Green PM, Saad S, Lewis CM, Giannelli F. Mutation rates in humans. I. Overall and sex-specific rates obtained from a population study of hemophilia B. Am J Hum Genet 1999;65:1572–9.

[30] Veltman JA, Brunner HG. De novo mutations in human genetic disease. Nat Rev Genet 2012;13:565–75.

[31] Woese CR. Bacterial evolution. Microbiol Rev 1987;51:221–71.

[32] Forterre P. The two ages of the RNA world, and the transition to the DNA world: a story of viruses and cells. Biochimie 2005;87:793–803.

[33] Alfoldi J, Lindblad-Toh K. Comparative genomics as a tool to understand evolution and disease. Genome Res 2013;23:1063–8.

[34] Lindblad-Toh K, Wade CM, Mikkelsen TS, Karlsson EK, Jaffe DB, Kamal M, et al. Genome sequence, comparative analysis and haplotype structure of the domestic dog. Nature 2005;438:803–19.

[35] Engle SJ, Womer DE, Davies PM, Boivin G, Sahota A, Simmonds HA, et al. HPRT-APRT-deficient mice are not a model for lesch-nyhan syndrome. Hum Mol Genet 1996;5:1607–10.

[36] Novarino G, Akizu N, Gleeson JG. Modeling human disease in humans: the ciliopathies. Cell 2011;147:70–9.

[37] Burt A. Perspective: sex, recombination, and the efficacy of selection—was Weismann right? Evolution 2000;54(2):337–51.

[38] Trombetta B, Sellitto D, Scozzari R, Cruciani F. Inter- and intraspecies phylogenetic analyses reveal extensive x-y gene conversion in the evolution of gametologous sequences of human sex chromosomes. Mol Biol Evol 2014;31:2108–23.

[39] Veerappa AM, Padakannaya P, Ramachandra NB. Copy number variation-based polymorphism in a new pseudoautosomal region 3 (PAR3) of a human X-chromosome-transposed region (XTR) in the Y chromosome. Funct Integr Genomics 2013;13:285–93.

[40] Faustino NA, Cooper TA. Pre-mRNA splicing and human disease. Genes Dev 2003;17:419–37.

[41] Sorek R, Dror G, Shamir R. Assessing the number of ancestral alternatively spliced exons in the human genome. BMC Genomics 2006;7:273.

[42] Koonin EV. Evolution of genome architecture. Int J Biochem Cell Biol 2009;41:298–306.

[43] Baldauf SL. An overview of the phylogeny and diversity of eukaryotes. J Syst Evol 2008;46:263–73.

[44] Collins L, Penny D. Complex spliceosomal organization ancestral to extant eukaryotes. Mol Biol Evol 2005;22:1053–66.

[45] Nowell RW, Almeida P, Wilson CG, et al. Comparative genomics of bdelloid rotifers: insights from desiccating and nondesiccating species. Tyler-Smith C, ed. PLoS Biol 2018;16:e2004830

[46] Boothby TC, Tenlen JR, Smith FW, Wang JR, Patanella KA, Nishimura EO, et al. Evidence for extensive horizontal gene transfer from the draft genome of a tardigrade. Proc Natl Acad Sci U S A 2015;112:15976–81.

[47] Pereira SL, Reeve JN. Histones and nucleosomes in Archaeae and Eukarya: a comparative analysis. Extremophiles 1998;2:141–8.

[48] Armstrong L, Lako M, Dean W, Stojkovic M. Epigenetic modification is central to genome reprogramming in somatic cell nuclear transfer. Stem Cells 2006;24:805–14.

[49] Balhorn R. The protamine family of sperm nuclear proteins. Genome Biol 2007;8:227.

[50] Regev A, Lamb MJ, Jablonka E. The role of DNA methylation in invertebrates: developmental regulation or genome defense? Mol Biol Evol 1998;15:880–91.

[51] Feltman R. The mysterious genes of carnivorous bladderwort reveal themselves. The Washington Post; 2015. February 23.

[52] Koonin EV. How many genes can make a cell: the minimal-gene-set concept. Annu Rev Genomics Hum Genet 2000;1:99–116.

[53] Chandley AC, Short RV, Allen WR. Cytogenetic studies of three equine hybrids. J Reprod Fertil Suppl 1975;23:356–70.

[54] Leitch IJ. Genome sizes through the ages. Heredity 2007;99:121–2.

[55] Yu J, Hu S, Wang J, Wong GK, Li S, Liu B, et al. A draft sequence of the rice genome (*Oryza sativa* L. ssp. *indica*). Science 2002;296:79–92.

[56] Zybina EV, Zybina TG. Polytene chromosomes in mammalian cells. Int Rev Cytol 1996;165:53–119.

[57] Kim HL, Iwase M, Igawa T, Nishioka T, Kaneko S, Katsura Y, et al. Genomic structure and evolution of multigene families: "flowers" on the human genome. Int J Evol Biol 2012;2012:917678.

[58] Collinson MN, Fisher AM, Walker J, Currie J, Williams L, Roberts P. Inv(10)(p11.2q21.2), a variant chromosome. Hum Genet 1997;101:175–80.

[59] Korbel JO, Urban AE, Affourtit JP, Godwin B, Grubert F, Simons JF, et al. Paired-end mapping reveals extensive structural variation in the human genome. Science 2007;318:420–6.

[60] Nobrega MA, Zhu Y, Plajzer-Frick I, Afzal V, Rubin EM. Megabase deletions of gene deserts result in viable mice. Nature 2004;431:988–93.

[61] Liu Y, Koyuturk M, Maxwell S, Xiang M, Veigl M, Cooper RS, et al. Discovery of common sequences absent in the human reference genome using pooled samples from next generation sequencing. BMC Genomics 2014;15:685.

[62] Siegel JJ, Amon A. New insights into the troubles of aneuploidy. Annu Rev Cell Dev Biol 2012;28:189–214.

[63] Fryns J, Lukusa TP. Monosomies. In: Encyclopedia of life sciences. Hoboken, NJ: John Wiley & Sons; 2005.

[64] Snape K, Hanks S, Ruark E, Barros-Nunez P, Elliott A, Murray A, et al. Mutations in CEP57 cause mosaic variegated aneuploidy syndrome. Nat Genet 2011;43:527–9.

[65] Jacquemont S, Boceno M, Rival JM, Mechinaud F, David A. High risk of malignancy in mosaic variegated aneuploidy syndrome. Am J Med Genet 2002;109:17–21.

[66] Graur D, Zheng Y, Price N, Azevedo RB, Zufall RA, Elhaik E. On the immortality of television sets: "function" in the human genome according to the evolution-free gospel of ENCODE. Genome Biol Evol 2013;5:578–90.

[67] Graur D, Zheng Y, Azevedo RBR. An evolutionary classification of genomic function. Genome Biol Evol 2015;7:642–5.

[68] Britten RJ. Almost all human genes resulted from ancient duplication. PNAS 2006;103:19027–32.

[69] Levitt M. Nature of the protein universe. Proc Natl Acad Sci U S A 2009;106:11079–84.

[70] Soltis DE, Albert VA, Leebens-Mack J, Bell CD, Paterson AH, Zheng C, et al. Polyploidy and angiosperm diversification. Am J Bot 2009;96:336–48.

[71] Otto SP, Whitton J. Polyploid incidence and evolution. Annu Rev Genet 2000;34:401–37.

[72] Thompson JN, Nuismer SL, Merg K. Plant polyploidy and the evolutionary ecology of plant/animal interactions. Biol J Linn Soc 2004;82:511–9.

[73] Zimmer C. The continuing evolution of genes. The New York Times; 2014. April 28.

[74] Schmitz JF, Bornberg-Bauer E. Fact or fiction: updates on how protein-coding genes might emerge de novo from previously non-coding DNA. F1000Res 2017;6:57.

[75] Neme R, Tautz D. Phylogenetic patterns of emergence of new genes support a model of frequent de novo evolution. BMC Genomics 2013;14:117.

[76] Ball P. Smallest genome clocks in at 182 genes. Nature October 12, 2006.

[77] Palazzo A, Gregory TR. The case for junk DNA. PLoS Genet 2014;10:e1004351.

[78] Tuna M, Knuutila S, Mills GB. Uniparental disomy in cancer. Trends Mol Med 2009;15:120–8.

[79] Delacourte A. Tauopathies: recent insights into old diseases. Folia Neuropathol 2005;43:244–57.

[80] Lupski JR, de Oca-Luna RM, Slaugenhaupt S, Pentao L, Guzzetta V, Trask BJ, et al. DNA duplication associated with Charcot-Marie-Tooth disease type 1A. Cell 1991;66:219–32.

[81] Heintzman ND, Hon GC, Hawkins RD, Kheradpour P, Stark A, Harp LF, et al. Histone modifications at human enhancers reflect global cell-type-specific gene expression. Nature 2009;459:108–12.

[82] Choi M, Scholl UI, Ji W, Liu T, Tikhonova IR, Zumbo P, et al. Genetic diagnosis by whole exome capture and massively parallel DNA sequencing. Proc Natl Acad Sci U S A 2009;106:19096–101.

[83] Dolle MET, Snyder WK, Gossen JA, Lohman PHM, Vijg J. Distinct spectra of somatic mutations accumulated with age in mouse heart and small intestine. PNAS 2000;97:8403–8.

[84] Roberts J. Looking at variation in numbers. The Scientist; 2005. March 14.

[85] McDermid HI, Morrow BE. Genomic disorders on 22q11. Am J Hum Genet 2002;70:1077–88.

[86] Fitzpatrick DA. Horizontal gene transfer in fungi. FEMS Microbiol Lett 2012;329:1–8.

[87] Kapitonov VV, Jurka J. RAG1 core and V(D)J recombination signal sequences were derived from Transib transposons. PLoS Biol 2005;3:e181.

[88] Omim. Online Mendelian inheritance in man. Available from: http://omim.org/downloads [viewed June 20], 2013.

[89] Harmon A. The DNA age: searching for similar diagnosis through DNA. The New York Times; 2007. December 28.

[90] Tefferi A, Gilliland DG. Oncogenes in myeloproliferative disorders. Cell Cycle 2007;6:550–66.

[91] Behjati S, Tarpey PS. What is next generation sequencing? Arch Dis Child Educ Pract Ed 2013;98:236–8 2013.

[92] Berman JJ. Tumor classification: molecular analysis meets Aristotle. BMC Cancer 2004;4:10.

[93] Berman JJ. Tumor taxonomy for the developmental lineage classification of neoplasms. BMC Cancer 2004;4:88.

[94] Zhang L, Lin X. Some considerations of classification for high dimension low-sample size data, Stat Methods Med Res 2011; Available from:http://smmsagepubcom/content/early/2011/11/22/0962280211428387long [viewed 26.01.13].

[95] Genetics Home Reference. National Library of Medicine, Available from:http://ghr.nlm.nih.gov/handbook/genomicresearch/snp; 2013 [viewed 06.07.13].

[96] Solomou EE, Gibellini F, Stewart B, Malide D, Berg M, Visconte V, et al. Perforin gene mutations in patients with acquired aplastic anemia. Blood 2007;109(12):5234–7.

[97] Naramura M, Nadeau S, Mohapatra B, Ahmad G, Mukhopadhyay C, Sattler M, et al. Mutant Cbl proteins as oncogenic drivers in myeloproliferative disorders. Oncotarget 2011;2:245–50.

[98] Fey MF, Liechti-Gallati S, von Rohr A, Borisch B, Theilkas L, Schneider V, et al. Clonality and X-inactivation patterns in hematopoietic cell populations detected by the highly informative M27 beta DNA probe. Blood 1994;83:931–8.

[99] Busque L, Mio R, Mattioli J, Brais E, Blais N, Lalonde Y, et al. Nonrandom X-inactivation patterns in normal females: lyonization ratios vary with age. Blood 1996;88:59–65.

3

Evolution and Embryonic Development

OUTLINE

Section 3.1 The Tight Relationship Between Evolution and Embryology

History doesn't repeat itself, but it rhymes.

Attributed variously to Mark Twain and to Joseph Anthony Wittreich

If we want to understand how modern medicine is influenced by evolution, then we must take some time to study human embryology. The reason for this is that the steps of embryologic development in the human are tightly coupled to the milestones of evolutionary development. In this section, we will learn why this assertion is true, and we will review some examples that demonstrate how this assertion applies to the study of human disease.

First, let's put to rest the outdated concept that the genome is the blueprint for the construction of the human body. A blueprint is a schematic, containing detailed descriptions of the pieces of a construction, with a diagram showing how the pieces are fit together to form the final product. When we look at human development, from the earliest stages of the embryo to the eventual appearance of a mature adult, we see nothing remotely resembling what we would expect to see if the genome were a blueprint. There are no identifiable "pieces" and no indication that anything is being fastened together to form a whole organism. Instead, we see clearly that human development follows successive transformations of living stages. This process in no way resembles the construction from a blueprint. A much better analogy is that the genome operates like a recipe book in which each successive recipe depends on the successful completion of the preceding recipe. In this case, there is a recipe for making a morula from a zygote, a recipe for making a blastocyst

Evolution's Clinical Guidebook. https://doi.org/10.1016/B978-0-12-817126-4.00003-5

from a morula, a recipe for making an inner cell mass from a blastocyst, a recipe for making a gastrula from an inner cell mass, and so on.

When we look at various phylogenetic classes of eukaryotic organisms, we see an historical retelling of the sequential recipes for embryonic development.

```
Class Eukaryota: Marked by the preembryonic zygote,
Class Metazoa: Marked by the formation of the blastulated embryo,
Class Eumetazoa: Marked by the formation of a two-layered embryo,
Class Bilateria: Marked by the formation of the three-layered embryo,
Class Deuterostomia: Gastrulation and formation of an enterocoelom,
Class Chordata: Marked by the formation of an embryonic backbone,
Class Amniota: Marked by the evolution of egg-laying, on land,
Class Eutheria: Marked by the evolution of the placenta.
```

In Chapter 5, "Phylogeny: Eukaryotes to Chordates," as we review the lineage of eukaryotic classes, we will see that each new class-defining embryologic innovation is preserved in the embryologic development of its descendant classes. [Glossary Evo-devo]

How do we know that phylogenetic innovations (i.e., evolutionary advancements) are always incorporated as embryological steps? It is really quite simple. Adult organisms are the product of developmental processes that occurred in utero. It is exclusively through embryonic development that adult organisms acquire their distinguishing anatomic and physiologic features. Hence, every evolutionary feature expressed in adult organisms (e.g., feathers on birds, placentas in mammals, fins on fish) must have arisen in the embryo.

As an example, let's consider Class Bilateria, animals with axial symmetry. Humans are bilaterians, with two legs, two arms, two eyes, and so on. Likewise, starfish are members of Class Bilateria, with axial symmetry and five arms. But wait. How is it possible for a five-armed organism to have bilateral symmetry. Shouldn't a symmetrical starfish have four arms, or maybe six arms? (Fig. 3.1).

We must remember that phylogeny has everything to do with sequential embryologic processes, and very little to do with what we observe in adult organisms. Adult organisms often display secondary changes due to developmental events that occur after a fundamental embryologic process has occurred. Yes, the adult starfish violates axial symmetry, but the embryonic starfish has lovely symmetry, and it is the embryo that expresses the phylogenetic innovations of its ancestors (Fig. 3.2).

Likewise, humans are descendants of Class Bilateria. When we look at adult humans, we see some indication that we arose from a bilateral embryo (e.g., two kidneys, two lungs, two lobes to our brains), but we also see asymmetrical organs (one spleen, one liver, and a wandering alimentary tract that criss-crosses the body's central axis). If we confine our observations to the early human embryo, we see bilateral symmetry of every anlage (i.e., every predecessor to an adult organ). We can infer that all the twists, turns, and organ fusions came later (Fig. 3.3).

FIG. 3.1 A red-knobbed starfish with five arms, openly violating axial symmetry. *Source: Wikipedia, released into the public domain by its author, Arpingstone.*

Rule: Every sequential step in human development represents an evolutionary advancement achieved by our ancestral species.

Some readers must be thinking that this "Rule" is just a rephrasing of the discredited and often vilified maxim that "Ontogeny recapitulates phylogeny." We must digress a moment to explain the meaning and history of this catchy but largely discredited phrase, and why it does not apply herein. [Glossary Phylogeny]

Near the end of the nineteenth century, one of the greatest evolutionary biologists to have ever graced this world made a small blunder that ravaged his scientific reputation and provided creationists with enough ammunition to attack the theory of evolution for the next century. The biologist was Ernst Haeckel (1834–1919), and his blunder can be summarized in three words: "Ontogeny recapitulates Phylogeny." As you might imagine, Haeckel's original assertions came under fire by his colleagues, leading to reinterpretations and revisions of the original flawed work.

What is the meaning of "Ontogeny recapitulates phylogeny?" There are many interpretations of this three-word bombshell, but let's try to keep it simple. Ontogeny is embryonic development. During embryonic development, the anlagen of organs appear, and their anatomic arrangement is determined. Phylogeny is the evolutionary descent and diversification of species from common ancestors. The slogan "ontogeny recapitulates phylogeny" asserts that during the embryologic development of an organism, it will reproduce all of the embryologic organisms in its ancestral lineage. For example, as the human embryo

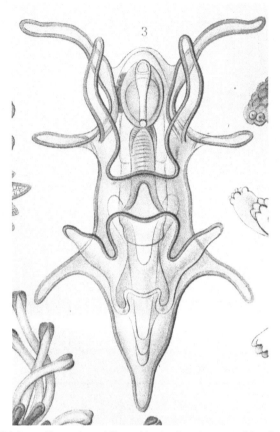

FIG. 3.2 Diagram of a starfish larva, illustrating a bilateral body plan. *Source: Wikipedia, from a drawing by Ernst Haeckel published in "Kunstformen der Natur," 1904e.*

develops, it will begin as a simple eukaryote (the zygote) and progress to a ball of cells (recapitulating the proposed ancient gallertoids that preceded animals) and then a blastula (recapitulating an embryonic metazoan) and then a two-layered organism (recapitulating an embryonic eumatazoan), followed by a tripoblast (recapitulating an embryonic bilaterian), and so forth until a human fetus emerges. [Glossary Anlagen, Gallertoid]

If ontogeny truly recapitulated phylogeny, we could reconstruct the entire phylogenetic ancestry of any living species, simply by capturing every phase of its embryologic development. In particular, somewhere in human development, we could expect to find an embryo that had the same form as the embryo of the ancestral organism for all living apes.

Haeckel's theory holds up moderately well, at least for the early embryo. In some species, observations seem to provide strong confirmation. For example, amphibians evolved from sarcopterygian fish. Every species of amphibia passes through a "fishy" stage of

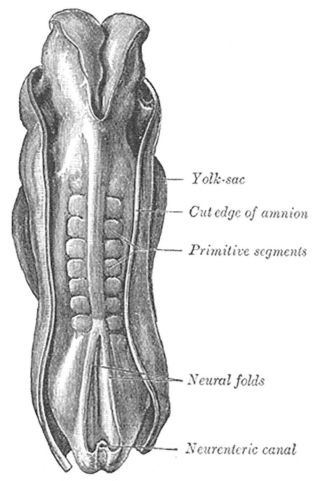

Yolk-sac

Cut edge of amnion

Primitive segments

Neural folds

Neurenteric canal

FIG. 3.3 Drawing of dorsum of human embryo, 2.11 mm in length. In early embryonic stages, human anatomy has bilateral symmetry, as we would expect from any member of Class Bilateria. *Source: 20th US edition of Gray's Anatomy of the Human Body, originally published in 1918.*

development before emerging as fully developed frogs, toads, salamanders, or caecilians. This being the case, why has Haeckel's theory been discredited? There are fundamentally two very large errors in the proposition that "ontogeny recapitulates phylogeny."

−1.   Evolutionary innovations are ontogenetically nonsequential.

There is no law that limits evolutionary innovations to a strictly temporal order. For example, Class Eutheria (also known as Class Placentalia) is characterized by the evolution of the placenta. The placenta is not an add-on at the end of ontogeny. The placenta begins to form at the earliest stage of embryonic development. Hence, one of the defining embryologic features of eutherians, a class of organism positioned way down the ancestral lineage of animals, is found by examining the very earliest embryo. If ontogeny recapitulated phylogeny, we would expect the first appearance of the placenta to occur late in embryonic development, at the stage that recapitulates the first eutherian.

Today, it's easy for us to imagine new mutations arising in any gene, potentially affecting any stage of embryonic development. In Haeckel's hay-day, DNA and the genetic code were unknown, and genes were a vague abstraction.

–2. Embryologic processes occur concurrently, and the pace of those processes will vary among species.

We think of embryologic development as a stepwise affair: first the zygote, then the blastula, then the gastrula, and so on. We lose sight of the fact that sequential processes can branch, with each branch of the process developing simultaneously with other branches. Moreover, the pace at which different branches of embryonic development progress may vary among different organisms of the same ancestral lineage. This tells us that even if the order of events is the same in every species, variations in speed will cause some branches of development to fall "out of step" with other branches, when we compare different species. Hence, there can be no expectation that any temporal stage of embryonic development will have a corresponding temporal stage of development in an ancestral embryo. This poses as a violation of "ontogeny recapitulates phylogeny."

Putting aside his professional difficulties arising from the assertion that ontogeny recapitulates phylogeny, Haeckel was one of the greatest naturalists in the history of science. Today, Haeckel is best remembered for his remarkable legacy of beautiful and detailed drawings of classes of organisms (Fig. 3.4).

A quotation variously attributed to both Mark Twain and to Joseph Anthony Wittreich, reminds us that "History doesn't repeat itself, but it rhymes." So true. Every species is unique, and embryologic events occurring within a species cannot faithfully recapitulate the events occurring in ancestral species. Nonetheless, we can recognize that embryologic development is not freshly invented by each species. In an age of bioinformatics, we continue to find that the traits that are most useful in classifying organisms are ancestral features that develop in the embryo and are observed in the adult. Let's revise Haeckel's law as the following:

Rule: Ontogeny does not repeat phylogeny, but it rhymes.

The observation that embryologic advances serve as the fundamental basis for phylogenetic classification reminds us that evolution is a process that operates on the embryo, not the adult organism. As humans, we spend our lives as adult organisms, and we tend to overlook the fact that the vast majority of genomic function is devoted to the task of producing the embryo. The cellular activities related to adult behavior (e.g., walking, gathering food, and eating) are impressive, but not nearly so much as the activities of embryonic development. We see evidence of the overwhelming dominance of the embryonic life form in those organisms that spend the majority of their lives passing through stages of development, with the adult organism appearing for a brief time, primarily for the purpose of procreating (e.g., butterflies and cicadas). It is in the embryo and the fetus that germ layers are created, cell types are programmed, metabolic pathways are established,

FIG. 3.4 An Ernst Haeckel diagram of sea anemones, one of many paintings and drawings by the 19th-century naturalist. *Source: Haeckel's "Kunstformen der Natur," 1904.*

and organs are formed. The embryo deals the cards, and the adult organism plays the hand that it is dealt. [Glossary Germ layers]

Rule: Natural selection favors the embryo, not the adult.

Having referred to the embryo repeatedly in this chapter, we should take a moment to reflect about what it means to be an embryo. Everyone who devotes any attention to the debate around abortion must be aware that it has become commonplace, among anti-abortion advocates to refer to embryos as babies. Regardless of your position on the abortion issue, it can be stated with certainty that an embryo is a fundamentally different biological entity from a newborn baby, although it is obvious that the latter develops from the former. For medical scientists, it is important to understand the special place of the embryo in development, insofar as the embryo ties together our phylogeny (the history of our evolutionary ancestry) with our pathology (the history of the events and pathways that lead to the development of human diseases). We will see in later chapters that our understanding of the relationship of phylogeny and pathology is key to our understanding of modern medicine. Hence, we need to fully understand the process of embryonic development, and there is no better place to start than explaining why an embryo is not a baby. Let's look at a few specific points. [Glossary Embryo vs fetus]

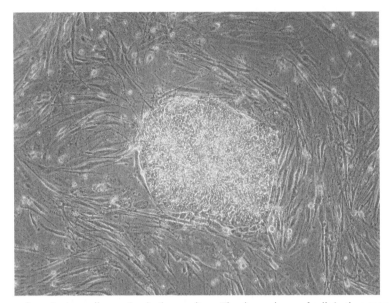

FIG. 3.5 Human embryonic stem cells growing in tissue culture. The dense clump of cells in the center are stem cells. The flat cells in the periphery of the image are a layer of mouse fibroblast cells that are intended to help the embryonic stem cell colony thrive. *Source: Wikipedia, from an image entered into the public domain by its author, Ryddragyn.*

−1.   Any cell type can be converted to an embryo-forming cell.

Historically, embryonic stem cells have been treated with great awe by scientists; and rightly so. The embryonic stem cell is capable of developing into any and all human cell types, and, under proper conditions, can be implanted into the uterus of an animal to develop into a fully formed organism. [Glossary Embryonic stem cell, Multipotent stem cell, Pluripotent stem cell, Totipotent stem cell]

We need to remember that the embryonic stem cell has the same genome as any of the differentiated cells of the body. In the past few decades, we have come to learn that the primary difference between an embryonic stem cell and a fully differentiated cell of the body (i.e., a somatic cell) is the erasure of the epigenome. Genomic erasure is a process that we will be discussing in further detail in the next section of this chapter [1]. For now, suffice it to say that differentiation (i.e., the process by which the morphologic and physiologic properties of a mature cell type are acquired) is controlled by the epigenome. A hepatocyte is a hepatocyte because its modified epigenome controls the expressions of the cell's genes in such a manner as to produce the cellular features of a hepatocyte. We might guess that if all those hepatocyte-specific modifications to the epigenome were somehow erased, then the cell would revert to an uncommitted state, much like a totipotent stem cell. Such is the case [2–5] (Fig. 3.5).

−1.   The embryo has anlagen; the developed organism has organs.

Anatomists typically divide gestation into three periods: preembryonic, embryonic, and fetal. The preembryonic period, which corresponds to the first 2 weeks following fertilization or the first month following the mother's last menstrual period, is the time in which the zygote (fertilized ovum) travels to the uterus, implants, begins to build the extraembryonic tissues, and forms a blastocyst. The cells of the preembryo are uncommitted and totipotent. In the third week following fertilization (fifth week since the last menstrual period), the three layers of the embryo form, and the embryonic cells differentiate to the extent that they are not totipotent. During the embryonic phase, which corresponds to the 4th to 9th weeks after fertilization (or the 6th to the 11th week since the last menstrual period), all the germ layers are completed, and all the anlagen of organs have begun to develop.

The fetal period, which begins at week 10 after fertilization (and week 12 since the last menstrual period), marks the time when the organs and cell types that characterize the fully developed human are recognizable. From this point onwards, it is not unreasonable to consider the fetus as a preformed human or even as a tiny baby, although it is customary to wait until the time when the fetus is viable outside the womb (about 26 weeks after the last menstrual period) before applying such politically charged terms.

Human development is not fundamentally different from the development of any other animal. All animals (i.e., all members of Class Metazoa) develop from an embryo, and all embryos develop in a similar fashion, even the embryos of insects. Insects develop through a series of distinctive morphologic stages, called instars. Growing instars molt to become the next instar. The development of the fly *Drosophila melanogaster* begins very soon after fertilization, with the first instar at 24 h. The second instar is seen at 48 h, and the 5th at 118 h. The pupa appears on the 6th day. When the pupa molts on the 9th day, the adult insect emerges. During each stage, structures from the preceding stage may be reorganized or otherwise eliminated. The adult is the only stage that exhibits the familiar body structures that we commonly associate with insects: antenna, legs, wings, and so on. These adult structures arise from undifferentiated organizing cells that form imaginal discs. Each of the imaginal discs is committed to form a specific structure (i.e., the wing disc produces a wing, and the antenna disc produces an antenna). Because insects develop through a succession of morphologically distinctive living forms, all of which lack the external structural features characteristic of the adult form, it is easy to see that the instar is not a small version of the adult insect.

Similarly, in the human embryo, we do not see tiny organs. We see committed tissues that undergo a series of developmental steps that eventually lead to the creation of a small organ, in the fetus. For example, the human kidney develops from a focus of mesoderm that eventually forms a primitive mesonephros, which is the anlage tissue for the subsequent development of the metanephros, which is the anlage for the development of the kidney. The embryo does not contain a tiny little kidney. Recognizable organs do not emerge until the fetal period. [Glossary Mesoderm].

–2.   Male and female organs both arise from the same embryonic anlagen.

Adult males are phenotypically different from adult females in many ways, but the most significant differences pertain to organs. Women have a uterus and a vagina and a clitoris. Males do not. Males have a prostate and a penis. Women do not. What accounts for these differences? Are there separate anlagen for female organs and for male organs? Actually, no. Female organs develop from the exact same anlagen as do the male organ. The gender-neutral anlagen respond to hormones and other gender-specified regulators to produce female organs in women and male organs in men.

Just to demonstrate, and without going into the anatomic details, here is a short and incomplete list of embryologic anlagen, alongside the adult tissue into which it develops, in men and in women.

– Embryonic gonad develops into the testis in men, the ovary in women.
– Paramesonephric duct develops into the appendix testis and prostatic utricle in men, and the Fallopian tubes, uterus, cervix, and vagina in women.
– Labioscrotal folds develop into the scrotum in men and the labia majora in women.
– The genital tubercle develops into the penis in men and the clitoris in women.

The point here is that every embryonic anlage develops into some adult structure, even if it is a nonfunctional tissue remnant, in both men and women. This accounts for superfluous nipples in men. The embryonic structure is not a tiny version of the adult male or female organ. It is a collection of precursor cells, equivalent in males and in females, just waiting for instructions.

Because men and women share the same embryonic anlagen, they are both at-risk for developing the same types of cancers, and these cancers can resemble the anlagen or the tissues derived from the anlagen. Thus men may develop breast cancer. Women may develop Leydig cell tumors of the ovary (exhibiting cytologic features of the Leydig cells in testes). Men and women may develop germinomas (known as seminomas in men and dysgerminomas in women). There are a host of rare and somewhat arcane tumors of males and females that are presumed to derive from the same embryologic anlagen [6]. [Glossary Neoplasm, Seminoma]

–3.   The types of cells in embryos are different from the types of cells in adult organisms.

We should not succumb to the belief that embryonic cells are just somewhat less differentiated versions of the cells that populate the fully developed human. Abundant evidence indicates that embryonic cells are fundamentally different from adult cells, responding very differently to toxins, and other injuries than cells in adults, and exhibiting patterns of cellular behavior that are sharply different from the behavior of cells in adult humans.

For example, a mutation of the gene Sas-4, coding for a centriolar protein in flies, inhibits the synthesis of centrioles in the fly embryo. Remarkably, these embryos mature normally, and mutant flies develop into morphologically unremarkable adults. Soon after birth, these flies die, because adult flies need their cilia for a wide variety of vital physiologic functions [7]. This experiment provides us with some notion of the fundamental differences between an embryo and a fully developed organism.

As another example, mutated oncogenes tend to produce developmental disorders (not cancers) when they occur in embryonic cells. Oncogenes produce cancers (not developmental disorders) when they occur in adult cells. In particular, germline BRAF mutations are associated with developmental syndromes such as Noonan, LEOPARD, and cardiofaciocutaneous syndromes [8]. BRAF mutations in somatic cells are associated with a variety of cancers, notably melanoma. In addition, neonates and children develop totally different types of tumors than those observed in adult humans, indicating that the first cells involved in prenatal carcinogenesis are fundamentally different from the cells involved in adult carcinogenesis. [Glossary Developmental disorder]

Let's summarize the points emphasized in this section.

- The genome is a recipe book, with a rich historical background, written in roughly chronological order (mode and tempo notwithstanding). By studying embryologic development, we can read the genome and understand its historical evolution.
- Phylogeny (the ancestral lineages of living organisms) does not recapitulate ontogeny (the development of the embryo and the fetus), but all of the major evolutionary advancements that characterize phylogenetic classes are found in embryonic development, and subsequently expressed in the adult organisms.
- Natural selection is mostly about the embryo; the adult organism is a secondary factor in evolution.
- Embryos are not tiny little babies. There are no tiny organs in the embryo, and the cells composing the embryo are different from the cells composing the tissues and organs of the baby. Embryos have the same relationship to babies as larval stages of animals have to the adult forms of the same species.

Section 3.2 The Epigenome and the Evolution of Cell Types

In mammals the genome is shaped by epigenetic regulation to manifest numerous cellular identities.

David A. Khavari, George L. Sen, and John L. Rinn [9]

The epigenome, at its simplest, consists of cell-type-specific chemical modifications to DNA that do not affect the sequence of nucleotides that comprise the genome. There are many kinds of nonsequence modifications that are included in the "epigenome." One of the best studied epigenomic modifications is DNA methylation. The most common form of methylation in DNA occurs on Cytosine nucleotides, most often at locations wherein Cytosine is followed by Guanine. These methylations are called CpG sites. CpG islands are concentrations of CpG sites. There are about 29,000–50,000 CpG islands in the human genome [10]. We infer that the epigenome must have arisen early in the evolution of eukaryotes, because plants (the nonmetazoan branch of multicellular organisms that develop from embryos) seem to have all the epigenetic control systems observed in metazoans [11]. [Glossary CpG island]

A simple way to think about the respective roles of genome and epigenome is as follows: **The genome establishes the identity of an organism; the epigenome establishes the identity of the individual cell types within the organism.**

At a minimum, the epigenome consists of the nonsequence modifications to DNA that control the expression of genes. These modifications include DNA methylations, histones, and nonhistone nuclear proteins. Beyond this minimalist definition, there are expanded versions of the definition that would include any conformational changes in DNA that influence gene expression, as well as protein interaction that influence gene expression. As used in this book, the terms "epigenome" and "epigenetics" apply exclusively to nonsequence alterations in chromosomes that are heritable among somatic cell lineages.

Rule: The epigenome produces the different cell types of the body.

There are at least 200 different cell types in the adult body. We infer that all the different cell types of the body are determined by the epigenome, for the following reasons [12]:

- –1. Every cell type in the body has the same genome as every other cell type. Hence, the differences between one cell type and another are nongenetic.
- –2. Cells of a fully differentiated cell type maintain that cell type throughout their lifespan. Hence, cell type is maintained through a regulatory function of the cell.
- –3. Cell types of a given lineage produce other cells of the same lineage (e.g., a dividing hepatocyte produces two hepatocytes; never one hepatocyte and one keratinocyte). Hence, the cell type of a cell is heritable.
- –4. Hence, the epigenome, which happens to be the nongenetic and somatically heritable cellular regulatory system, maintains cell types.

As we would expect, consistent differences in epigenomic patterns of methylation are found in each type of cell. Nonetheless, we have not yet reached the point where we can accurately identify a cell type by its epigenome [13].

Rule: The epigenome is heritable.

The epigenome can be easily modified with methylating agents or with hypomethylating agents. In an often-cited study, the diet of mice during gestation was supplemented with genistein resulting in an inherited shift in coat color of the offspring (from agouti to pseudoagouti) and an associated increased methylation in a regulatory site upstream of the Agouti gene [14, 15]. Hence, methylation, via an environmental additive, can produce a heritable epimutation.

The patterns of methylation are replicated among cells of the same type (e.g., neuron to neuron, liver cell to liver cell). Hence, the epigenome is somatically inherited. Alterations in methylation patterns are referred to as epimutations. Epimutations may persist in those specialized cells that are descended from an epimutated cell [16].

Rule: Thanks to the epigenome, many genetic diseases are organ-specific.

If a genetic defect is present in the germ line of an organism, and thusly present in every cell of the organism, then why are genetic diseases organ-specific, affecting some organs and not others?

The superficial answer to this question is simple: genetic defects need to be expressed in cells before they cause disease. If a cell of a particular type does not express the defect (e.g., does not produce proteins altered by the genetic defect or does not use metabolic pathways affected by the genetic mutation), then the disease phenotype will not be expressed. On a deeper level, we see that the epigenome is ultimately responsible for determining which genes are expressed and which genes are not expressed, in any give cell type. Hence, the epigenome determines the organ specificity of genetic diseases carried in the germline.

In addition, we can imagine that any disease resulting from a loss-of-function mutation in a gene could be mimicked by a germline epimutation (a heritable change in the epigenome) that suppresses the gene. For example, an epimutation that silences the MLH1 gene, without the involvement of a gene mutation, may account for some cases of hereditary nonpolyposis colorectal cancer (HNPCC) [17]. [Glossary Epimutation, Hereditary nonpolyposis colorectal cancer syndrome, Lynch syndrome]

Rule: The epigenome accounts for phenotypic differences among genetically identical organisms.

It is a common observation that monozygotic twins look alike at birth, often growing into early adulthood as a pair of strikingly similar individuals. As the decades go by, identical twins begin to diverge in their appearances and in the diseases they develop [18, 19]. Moreover, phenotypic differences among genetically identical animals occur in the absence of any observable environmental differences [19].

Monozygotic twins are born with nearly identical epigenetic patterns of DNA methylation and histone acetylation. This near-identity of epigenomes persists in the early years, but in later years, monozygotic twins have widely divergent patterns of DNA methylation and histone acetylation. These divergent patterns are accompanied by discordances in gene expression [20].

It is tempting to conclude that changes in the epigenome, arising during life, account for the disparities among the diseases affecting identical twins. Hypothetically, it is reasonable to infer that if two individuals with the same genome were to have very different epigenomes, then we should not expect their phenotypes to be identical (i.e., they would not look like twins).

Erasure

There are moments in the life of every metazoan organism when it needs to be free of all epigenomic constraints. For example, the zygote (the cell resulting from a spermatocyte fertilizing an oocyte) must serve as a totipotent cell, producing all of the embryonic and extraembryonic cell types, through a process of successive epigenomic modifications, eventually yielding the complete set of differentiated cell types that constitute the fully developed organism. To do so, the zygote must be fully undifferentiated, with no epigenomic modifications that would restrain its role as the root cell of the developing organism. To attain an undifferentiated state, the zygote initiates an active demethylation process, stripping itself of epigenomic markers. This process, which begins almost immediately following fertilization is referred to as erasure [1].

In early stage embryos (about day 7 postfertilization in the mouse embryo), a new cell type appears, in the endoderm, known as the primordial germ cell. These cells migrate as a group to the genital ridges where they will dwell in the developing gonads to become the oocyte-generating cells in females and the spermatocyte-generating cells in males. Because the primordial germ cells are a differentiated type of cell, committed to a particular function, they attain specific epigenomic markings typical of their cell type. These germ cells will be producing a distinctive form of cell (i.e., the haploid gametocyte), through a type of division that is only employed when haploid cells are created (i.e., meiosis). In Section 1.2, "Bootstrapping Paradoxes," we discussed the alternating roles played by diploid and haploid structures, and explained that every metazoan organism is somewhat of a hybrid, containing both haploid and diploid life forms. In a sense, the primordial germ cells are the progenitor cells of the haploid organisms, while the zygote is the progenitor of the diploid organism. In both cases, erasure precedes the development of the organism, so that epigenomic controls can be inscribed on a "blank slate" genome. Hence, the primordial germ cell, like the zygote, undergoes its own erasure process [21, 22]. After erasure, epigenomic modifications appear in the early oocytes and spermatocytes (e.g., oogonia and spermatogonia) in a pattern that is appropriate for each of these two committed cell types [22]. The erasure process has one more cycle. When the mature spermatocyte is formed, an additional round of erasure, this time affecting the paternal epigenetic program of the spermatogonia, is executed; thus preparing the spermatocyte for fertilization [23]. A similar round of erasure occurs in late oogenesis, in women [24].

We have omitted, for the moment, an evolved exception to the typical process of erasure that is present in mammals and their descendant classes, and in no other animals. This exception is called imprinting, and it has uniquely influenced the evolution, embryologic development, and modes of disease inheritance of mammals. We will be discussing imprinting, and its evolutionary consequences, in Section 6.3, "Mammals to Therians." [Glossary Imprinting]

The purpose of cell types

There are hundreds of different types of cells in the human body, but how did they evolve? For that matter, why is it important for organisms to have specialized cells? We can guess that the very first service provided by one type of cell for another was compartmentalization of pathways, allowing the metabolic products of one cell to diffuse to another cells, where the first cell's product became the second cell's substrate. Under primitive living conditions, reciprocating metabolic relationships may have existed in a prebiotic world wherein cells were mere bubbles in rocks conducting chemical reactions with whatever material washed inside. As replicating, free-floating cells emerged, we can imagine cells taking on the complementary roles of effector and responder. Such cellular arrangements, working in close proximity with one another, may have formed the first examples of different cell types.

A very good example of the need for cell types is found in the intimate and enduring marriage between photoreceptor cells (e.g., rods and cones) and retinal pigment epithelial cells [25]. It seems that every animal with eyes, regardless of the optical mechanism whereby light is focused, is equipped with photoreceptor cells in close proximity to retinal pigment epithelial cells. Eyes with photoreceptor cells in proximity to pigmented cells were present in the very first metazoans, as an aid to predation. The chemistry of vision employs photoexcited receptors that produce retinal, which is transported, as a sort of waste product, out of the photoreceptor cells, where it is absorbed by the retinal pigment epithelial cells and reisomerized to 11-cis retinal, and transported back to the photoreceptor cells to participate in another round of visual excitation. Aside from that, the retinal pigment epithelial cells phagocytize photoreceptor outer segments and transport a host of nutrients back to the photoreceptor cells so that they rejuvenate their outer segments. [Glossary Epithelial cell]

The illuminating story of the photoreceptor and the retinal pigment epithelial cells reminds us that cells require some form of specialization if they are to survive in a competitive and dangerous world. Every organism needs some regulatory method by which groups of cells perform a special function, in a particular manner, whenever needed. Functional specialization results in altered phenotypes (i.e., the specialized cell looks different from the unspecialized cell), and we see phenotypically distinguished cell types in simple bacteria. For example, many bacteria have distinctive vegetative and spore cell types, as exemplified by Clostridia (Fig. 3.6).

Likewise, unicellular eukaryotic organisms have life cycles in which functionally nonequivalent (i.e., specialized) cell types are sequentially produced. For example, the malarian life cycle involves an infective sporozoite that matures in liver cells as schizonts, which rupture and release merozoites, that invade red cells, and produce gametocytes that are sucked by the female Anopheles mosquito wherein they develop as sporozoites (Fig. 3.7).

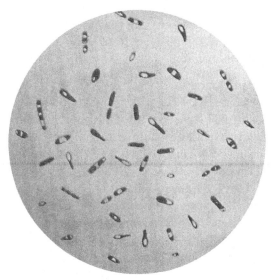

FIG. 3.6 Photomicrograph of a Gram-stained culture of *Clostridium feseri*. The poorly stained round structures are its endospores. *Source: a public domain image from the US Center for Disease Control and Prevention.*

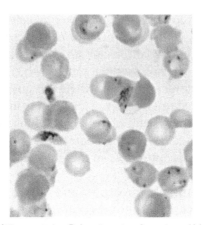

FIG. 3.7 Photomicrograph of blood (Giemsa-stained) showing *ring-forms* in red blood cells, and macrogametocytes of *Plasmodium falciparum. Source: US Center for Disease Control and Prevention public domain word.*

The epigenome, along with a retinue of processes that regulate the genome, was probably present in some of the very earliest single cell organisms and is certainly present in present-day bacteria [26, 27], and in all eukaryotic organisms, including single-celled eukaryotes, animals, plants, and fungi. In any human, there are a little more than 200 well-recognized cell types (with an indeterminate number of currently uncharacterized cell types), and all these different cell types have the same genome, with the same DNA sequence. This should be a clue that new cell types are not generally created by new mutations. New cell types are created by subtle changes in the transcription of the genome, perhaps in differences of just a few transcription factors or epigenomic methylation sites. There are exceptions. In Section 4.5, "Viruses and the Meaning of Life," we will see that retroviruses contributed the genes that played the key role in the development of several evolved cell types, including the syncytiotrophoblast, and the primary immunocytes of the adaptive immune system.

Cell type as the basis for phylogeny

Major evolutionary advances, the kind of advances that establish new classes of organisms, are confined to specific cell types. For example, the evolution of metazoans awaited the acquisition of epithelial cells that could tightly bind to one another, forming cystic structures delimited by a water-tight epithelial layer. Without epithelial cells held together by desmosomes and tight junctions, there could be no blastocyst, and without the blastocyst, there could be no Class Metazoa. Without the acquisition of the ectodermal cell, there could be no embryonic bilayer and no Class Eumetazoa. Without the acquisition of the mesodermal cell, there were be no triple-layered embryo and hence no Class Bilateria. Skipping forward, without the acquisition of a notochordal cell, there can be no notochord, and consequently no Class Chordata. Without the acquisition of the syncytiotrophoblast, there can be no Class Theria. We can play this game all day. [Glossary Ectoderm]

Our ability to associate the phylogenetic lineage of organisms with the acquisition of specific cell types indicates that evolution is just as much the story of epigenomic evolution as it is the story of genetic evolution. How so? New genes arising in a species are present in the germline (i.e., in the genome of every cell of the organism). It is the epigenome that coordinates genetic expression, and it is the epigenome that creates a new cell type from evolved genes.

Diseases arise within specific cell types, and cell types are carried through an ancestral lineage. Therefore, we can examine the evolution of disease within the ancestral classes of organisms to learn the identity of the disease-specific cell type. The subject of orthodiseases (i.e., genetic diseases in classes of organisms that share orthologous genes with humans) captures this idea, and we shall discuss the orthodiseases in detail in Section 8.3, "New Animal Options." [Glossary Clade, Orthodisease]

We should stop to note that linking phylogeny to cell type is a radical departure from the time-honored tradition of organizing classes of organisms by bones. Traditionally, we associate successive classes of organisms by the skeletal changes that accompanied their earliest appearances as fossilized remains, in chronologically stratified rocks. Due credit must go to the paleontologists who, knowing that the truth lies buried in rocks, painstakingly explored the relationships among classes of organisms, based on their examinations of fossilized bones. Hence, our current classification of animals is full of osteologic landmarks (e.g., Class Gnathostomata, from the Greek for jawed mouth; Class Synapsida from the Greek for fused arch). When new cell types appear, we can expect to find a set of anatomic accommodations that change the skeletal morphology, resulting in the kinds of class-specific fossil findings recorded by paleontologists. For example, Class Craniata named for the cranium present in all species of the class, and their descendants, is a product of the acquisition of neural crest cells, which produce the flat bones of the head, composing the cranial vault, from which Class Craniata takes its name. [Glossary Craniata]

Nonetheless, diseases have specific cell types; not specific skeletal anatomies. Hence, medical researchers might find a cell type taxonomy more helpful than our current osteologic taxonomy. At the moment, the paleontologists hold sway over taxonomy. **In the future, as we synthesize an approach to phylogeny that incorporates our growing understanding of evolutionary genetics, and disease biology, we may be able to produce a new taxonomy of organisms that is closely tied to the phylogenetic acquisition of cell types.**

Cell types evolve from other cell types

Because all cells come from cells, we can infer that every cell type of the body evolved from modifications of preexisting cell types. In some cases, we can successfully reconstruct the evolution of a new cell type. For example, venom producing cells of snakes evolved from salivary gland cells. This evolutionary feat was achieved by both anatomic and chemical modifications. Digestive enzymes, normally synthesized by salivary glands, were modified to serve as toxins. The anatomic route of salivary gland ducts was subtly redirected to the fangs [28]. Most highly venomous snakes have triangular heads, while nonvenomous snakes have tubular heads. The difference is accounted for by the hypertrophied modified salivary glands, producing bilateral neck bulges in some of the deadliest snakes (Fig. 3.8).

As another example, electrocytes, the cells that compose electrocytic organs in electric eels, mantas, and several species of fish, are modified muscle cells. In gymnotiform and mormyrid fish, genes for the NaV1.4a sodium channels accumulated mutations that modulate channel kinetics to produce high voltages [29]. Stacks of electrocytes act like a voltaic pile to produce a high voltage pulse of electricity. It has been suggested that animals with electric organs may serve as useful animal models for the study of human channelopathies. [Glossary Channelopathy]

In general, once a cell has differentiated to a particular cell type, it does not change to become another cell type. For example, a squamous cell does not become a glandular cell, and a hepatocyte never becomes a pancreatic cell. Exceptions to this general rule are recognized by the term "transdifferentiation," indicating a biological process whereby one cell, of a particular cell type, converts into another cell type.

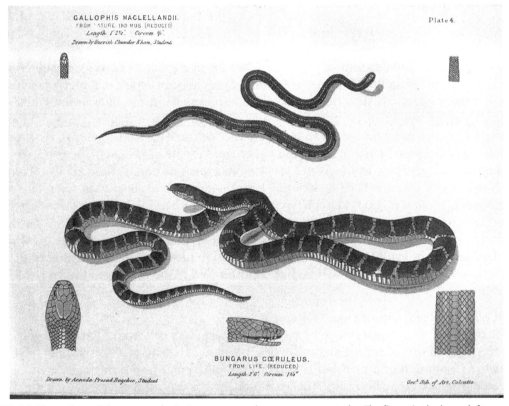

FIG. 3.8 Illustrations of the common krait, *Bungarus caeruleus*, a venomous snake. The figure in the lower left corner demonstrates the triangular head of the snake, produced by bilateral bulging venom glands, an evolutionary modification of salivary glands. *From: Joseph Ewart's "The poisonous snakes of India," published in 1878 and now held in the public domain.*

In mammals, transdifferentiation may be a phenomenon that does not occur naturally in living organisms. What may appear to be naturally occurring transdifferentiation in tissues is more easily explained as the replacement of one existing cell type by another existing cell type that happens to be occupying the same tissue. For example, in the esophagus, the squamous epithelium is sometimes interrupted by zones of mucus-secreting cells. In this instance, it would be wrong to say that the squamous cells transdifferentiated into mucus-secreting cells. What actually happens is that mucus stem cells generate mucus-secreting cells that gradually replaced the squamous cells.

Over the past decade, evidence has mounted indicating that differentiated cells can, under special circumstances, dedifferentiate to stem cells, and the stem cells can redifferentiate as some alternate differentiated cell type. Hence, one cell type eventually becomes another cell type, but not through a direct biological process of transdifferentiation. Under naturally occurring situations, there seems to be the requirement of an intermediate cell type, like so [30, 31]:

```
differentiated cell ->
undifferentiated stem cell ->
alternate differentiated cell
```

Current evidence indicates that direct transdifferentiation can be induced in cultured cells, with the use of transcription factors. For example, hepatocytes can be induced to become neurons, without dividing, and without reverting to a pluripotent intermediate cell [32]. Similarly, mouse fibroblasts can be converted to hepatocytes [33]. It is now well established that cells of various cell types can be transdifferentiated as cells of other cell types with the judicial use of transcription factors [34]. These facile manipulations of cell type, using transcription factors, prompt us to wonder where these lines of investigation are leading. It has been suggested that variations in transcription factor activity account for most of the phenotypic variations among members of a species, and may even account for most of the differences between one species and another closely related species [35, 36]. In vivo cellular reprogramming methods are currently under study [37]. This being the case, might there come a day when phenotypically new breeds and species of animals are created through the application of transdifferentiating transcription factors? Food for thought.

The cell theory of disease implies diseases affect particular cell types.

Back in 1858, Rudolph Virchow formalized the cellular basis of human disease, in his influential work, "*Die Cellularpathologie in ihrer Begrundung auf physiologische und pathologische Gewebelehre.*" [38]. He proposed that every disease begins with a population of affected cells, and that we can understand the pathogenesis of diseases by focusing our attention on those specific cells. Simple enough, but Virchow's assertion runs into some difficulty when it encounters inherited disorders. The genetic diseases are characterized by germline gene alterations that are present in every nucleated cell of the body. Nonetheless, the clinical phenotypes that are produced by these omnipresent genes are typically restricted to one or a few organs or cell types. What is the mechanism whereby a defect that is present in every cell of the body is manifested in only a few organs?

It should come as no surprise that many different mechanisms account for this fascinating phenomenon. Here are a few:

—   Cell-type-specific gene expression

One obvious reason why certain tissues, and not others, are affected by genetic diseases relates to cell-type-specific gene expression. For example, congenital hypothyroidism is a primary disease of the thyroid. It cannot be a primary disease of any other organ because the thyroid gland is the only organ that produces thyroid hormones. The effects of hypothyroidism are found in those organs that respond to thyroid hormones. The primary site of disease and the secondarily affected organs are predetermined by normal endocrine physiology.

Sometimes cell-specific gene expression plays a transient role during development. In Blepharophimosis, Ptosis, and Epicanthus inversus Syndrome type I (BPES1) a FOXL2 mutation causes eyelid abnormalities and reduces the number of ovarian follicles in the fetal organism. Expression of the FOXL2 gene during development is found primarily in the ovaries and the eyelids, thus matching cell type specificity to the cells wherein the underlying gene mutation is expressed [39, 40].

The cell-type-specific control of gene expression can be highly complex. Cornelia de Lange syndrome is an inherited disease associated with many different developmental defects, including intellectual disability, skeletal abnormalities, small stature, and gastrointestinal dysfunction. The root genetic cause of Cornelia de Lange syndrome is a mutation in the Nipbl gene, which loads cohesin at promoter sites. Cohesin occupies different promoters on different types of cells, thus mediating cell-type-specific gene expression [41]. As you might expect, a defect in the process through which cell-type gene expression is controlled, will adversely influence the development of many different tissues. [Glossary *Trans*-acting, *Cis*-acting vs *trans*-acting, Promoter]

Of course, if we fully understood the mechanisms that control cell-type-specific gene expression, we might be able to recruit various types of cells to produce proteins that are normally restricted to another cell type. For example, if insulin secretion were deficient due to a developmental absence of pancreatic islets, then it might be advantageous to flip on insulin synthesis in hepatocytes or gut lining cells.

Cell type specificity of pathways

Processes that occur exclusively in one cell type will obviously produce a primary disease restricted to one tissue or organ. For example, every genetic disorder of hemoglobin synthesis produces a primary disease of red blood cells, the only cells in the body that synthesize hemoglobin. Because hemoglobin is an essential protein involved in oxygen exchange, deficiencies in hemoglobin will produce widespread secondary changes in many different organs. [Glossary Primary disease, Primary disease versus secondary disease, Secondary disease, Tertiary disease]

Rule: Mutations of pathways used by many different cells will produce disorders that are found in multiple organs.

For example, cystic fibrosis is caused by an inherited deficiency of the cystic fibrosis transmembrane conductance regulator (CFTR). CFTR regulates the movement of chloride and sodium ions across epithelial membranes. Ion exchange across epithelial tissues is deployed by many different kinds of cells, including virtually every type of duct lining cell, and every mucus-producing epithelial cell, such as the mucus producing cells of the lung, and gastrointestinal tract. In cases of cystic fibrosis, mucus-producing cells produce a thick, viscous product that cannot be easily cleared from ducts (e.g., pancreas) and organ conduits (e.g., bronchi, gastrointestinal tract, seminal vesicles). A defect in a widely necessary cellular function will produce primary disease in many different organs.

On occasion, mutations of pathways used by many different cell types will produce an isolated disease of a single cell type. For example a variant of nonsyndromic deafness is maternally inherited and is associated with a mutation in mitochondrial DNA. This mitochondrial DNA mutation is expressed in every cell of the body, with the exception of red blood cells (which lack mitochondria). One might assume that a mitochondrial defect would produce pathological changes in many different types of cells. Such is not the case. In the case of nonsyndromic hearing loss, mitochondrial mutations that result in deafness produce a highly specific loss of cells of the inner ear, without producing known deficits in other organs. As it happens, a stress pathway leading to cell death is preferentially expressed in inner ear neurons [42]. A mitochondrial mutation associated with nonsyndromic hearing loss, the A1555G mtDNA mutation, activates this pathway, causing cells to die exclusively in the inner ear, and producing progressive deafness [42].

–   Cell type disease specificity determined by the weakest link

There are many instances wherein a mutation affects pathways in many different cells, but produces a clinical phenotype in a small subset of affected tissues. In many of these instances, the clinical phenotype arises in cells that are least able to cope with the mutational effects.

Here are three examples of weakest link tissue specificity:

–1.   Vitamin B12 deficiency

Vitamin B12 is required for efficient DNA synthesis, and cell division. Cells that have the highest cell division rate are the same cells that are most affected by B12 deficiency. Bone marrow has a very high rate of cell division. As expected, B12 deficiency results in anemia, particularly in the red cell population, which develops a dual syndrome of anemia (i.e., reduced number of red cells in the peripheral blood) and megaloblastoid morphology (i.e., defect in maturation of red cells).

In addition to its weakest link toxicity for bone marrow cells, vitamin B12 exerts a specific toxic effect on the nervous system, via a cell-type-specific mechanism. Vitamin B12 is required for the last step of the pathway leading to succinyl CoA formation. A deficiency of vitamin B12 produces elevated levels of methylmalonic acidemia. Over time, methylmalonic acid incorporates into myelin and destabilizes the myelin sheath, required for the normal fast-conduction of impulses along neuronal axons. The resultant myelinopathy produces sensory and motor neuron deficiencies and subacute combined (i.e., posterior and lateral column) degeneration of the spinal cord. Alterations in mentation have also been observed.

–2.   Retinitis pigmentosa

Retinitis pigmentosa is one of the most genetically heterogeneous of the inherited diseases. It can occur as a solitary disease, or it can occur in syndromic conditions (e.g., Usher syndrome, which is characterized by losses of hearing and vision).

A nonsyndromic form of retinitis pigmentosa is caused by a mutation in any one of the four genes coding for splicing factors (i.e., PRPF31, PRPF3, PAP1, and PRPF8). These splicing factors are found in virtually every type of cell in the body. Why would deficits in any of these ubiquitous factors lead to one specific disease? It has been suggested that the protein processing demands in retinal cells are extremely high, due to the high turnover of rhodopsin molecule [43]. Hence, the same splicing defects that are tolerated in other types of cells are thought to be destructive to retinal photoreceptor cells. This particular form of retinitis pigmentosa is an example of weakest link cell specificity.

–3.   Leukoencephalopathy with vanishing white matter

Leukoencephalopathy with vanishing white matter is a leukodystrophy (i.e., a degenerative disease of the white matter of the brain), caused by loss of function mutation in any of five genes encoding subunits of the translation initiation factor EIF-2B. The gene product initiates protein synthesis in cells throughout the body, but its clinical expression is often isolated to two cell types in the brain: oligodendrocytes and astrocytes. These cells have a particularly high rate of protein synthesis, and it is this heightened requirement for the gene product that seems to render oligodendrocytes and astrocytes sensitive to the mutation. The disease is triggered or exacerbated by certain types of stress to the central nervous system. Susceptible individuals who carry the mutation may develop acute symptoms of leukodystrophy (cerebellar ataxia, spasticity, optic atrophy) after head trauma. [Glossary Translation factor]

It should be noted that a weakest link mechanism may apply in instances for which no differences in pathway requirements can be measured in the affected tissues. The reason that a weakest link cause for tissue specificity cannot always be determined by any measurable test is subtle, and best explained with an economic analogy. Suppose everyone in a population receives a weekly salary of $200. One person has expenses of $199 per week, and he manages to save $1 each week. Another person has expenses of $201 per week, and he is continually in debt. Over time, he loses his house and cannot provide for his own existence. The difference in requirements between the two individuals is just $2, an insignificant quantity that might evade an accountant's inspection. The lesson here is that a cell that is particularly susceptible to disease may have a pathway requirement not measurably different from the pathway requirement of cells that do not express disease susceptibility.

– Co-conditional factors

When we impose an increasing number of restrictions on a cellular process, we narrow the range of participating cell types. For example, in xeroderma pigmentosum, there is an inherited DNA repair deficiency in every cell of the body. Xeroderma pigmentosum cells are particularly sensitive to the toxic effects of DNA damage induced by ultraviolet light. Ultraviolet light cannot penetrate deeper than the skin. Hence, the clinical phenotype of xeroderma pigmentosum is restricted to the skin and the cornea. [Glossary Toxic response versus disease process, Mutator phenotype, Xeroderma pigmentosum]

Section 3.3 An Embryonic Detour for Human Diseases

Embryonic structure is no more free from the effects of new mutations, from adaptation, convergence, divergence, and the like than is structure in any other phase of the life cycle.

George Gaylord Simpson [44]

In the past few decades, we have identified an underlying or root mutation for several thousand rare genetic diseases. Unfortunately, knowing the root cause of an inherited or de novo genetic disease seldom allows us to predict the clinical outcome of the mutation [45]. It is not hard to understand why this is so. First off, diseases develop through a sequence of biological steps occurring over time. Knowing the first step in a disease process doesn't tell us all that much about the subsequent steps in pathogenesis.

Several additional biological realities muddle our ability to trace how a single mutation might cause a disease, one being that a mutation involving a single gene may result in many different diseases. In some cases, each of the diseases caused by the altered gene are fundamentally similar (e.g., spherocytosis and elliptocytosis, caused by mutations in the alpha-spectrin gene; Usher syndrome type IIIA and retinitis pigmentosa-61 caused by mutations in the CLRN1 gene). In other cases, diseases caused by the same gene may have no obvious relation to one another. For example, different mutations of the same gene, desmoplakin, cause the following diseases [46]:

– Arrhythmogenic right ventricular dysplasia 8
– Dilated cardiomyopathy with woolly hair and keratoderma
– Lethal acantholytic epidermolysis bullosa [47]
– Keratosis palmoplantaris striata II
– Skin fragility-woolly hair syndrome

We will see how a mutation in desmoplakin can cause disease in tissues as diverse as heart and epidermis when we explore the origins of metazoan (i.e., animal) evolution in Section 5.4, "Opisthokonts to Parahoxozoa."

Just as mutational changes in a single gene may be the root cause for any number of different diseases, we see that a single disease may be caused by mutations in any of several different genes; a condition known as genetic heterogeneity. [Glossary Genetic heterogeneity, Locus heterogeneity]

Let's look at an extreme example of genetic heterogeneity. Severe combined immunodeficiency disease (SCID) has different genetic causes, all producing immunodeficiencies of both arms of the adaptive immune system (i.e., B cells and T cells). The condition arises by any one of a number of different genetic mutations, involving totally different genetic pathways that all lead to a common phenotype, in infants. Here are the different mutations that are the root genetic causes of SCID [Glossary Combined deficiency, Immune system]:

- Mutations in the gene encoding the common gamma chain (of interleukin receptors)
- Defective adenosine deaminase, an enzyme involved in the breakdown of purines
- Mutations of the purine nucleoside phosphorylase gene, the protein product of which is a key enzyme in the purine salvage pathway
- Insufficiency of mitochondrial adenylate kinase 2
- Insufficiency of recombination activating gene products, necessary for the manufacture of immunoglobulins
- Loss of expression of major histocompatibility complex proteins
- Janus kinase-3 deficiency
- Mutation in Artemis gene, required for DNA repair and normal immunologic defenses

In SCID, the different pathogeneses are remarkably diverse. In cases such as this, the medical scientist must face a very difficult question. "How can mutations in any of these many different and biologically unrelated genes precipitate a series of events that eventually converge to the same disease?" As all of the mutations that cause SCID are present in the embryonic germline and produce disease in infants, we can infer that the various sequences of events leading to SCID must unfold in the embryo. Hence, the medical scientist is tasked with learning the embryonic pathways that lead to SCID.

It happens that over the past few decades, we have learned a great deal about the many different underlying causes of SCID. In the case of SCID associated with adenosine deaminase deficiency, it has been shown that adenosine deficiency interferes with the development of the fetal thymus gland. The thymus is a tissue that involutes throughout childhood, but plays a key role in the fetal development of the immune system. In this case, maldevelopment of the thymus seems to account for the combined (T cell and B cell) deficiencies associated with SCID.

As in so many biomedical investigations, we find that solving one mystery uncovers another, deeper mystery. It seems that all rapidly dividing cells require adenosine deaminase to eliminate deoxyadenosine, a cytotoxic molecule produced when DNA is catabolized. Immunocytes (T and B cells and their precursor cells) are rapidly dividing and require high levels of adenosine deaminase for their survival. These observations fit our simple explanation relating adenosine deaminase deficiency with SCID. Or do they?

Cell division occurs at a very high rate throughout the development of the embryo and the fetus. Why would a deficiency of adenosine deaminase specifically block thymic development, when every organ in the fetus is rapidly dividing? It doesn't make much sense, when you think about it.

The answer to this question might come from an understanding of the pathogenesis of some of the other forms of SCID. Individuals with deficiencies of several of the proteins that participate in a particular DNA repair mechanism (i.e., nonhomologous end joining DNA repair) account for several of the subtypes of SCID [48]. The process of lymphocyte differentiation involves the enzymatic breakage followed by the enzymatic repair of the so-called V(D)J recombination units that account for the remarkable diversity of recognition sites that characterize the adaptive immune system. Following recombination, these breaks must be repaired, or the T and B cells do not mature, and affected individual develops SCID [48]. When resting blood lymphocytes are treated with deoxycoformycin, an adenosine deaminase inhibitor, single-strand breaks accumulate [49]. Thus a relationship between deoxyadenosine deaminase deficiency and DNA repair deficiency is established, suggesting that this form of SCID develops much like the other forms of SCID; from a DNA repair deficiency.

In this case, our understanding of the pathogenesis of SCID, in several subtypes of the disease, has helped us to hypothesize a generalized mechanism for the pathogenesis of SCID, that involves an embryologic DNA repair pathway eventually leading to the arrested development of the fetal thymus.

One of the most perplexing aspects of genetic disease is observed when injuries to the embryo are expressed as diseases of the adult, leaving no observable trace of the underlying embryonic damage. Historically, one of the most dramatic examples of this phenomenon has been seen in the daughters of women exposed to diethystilbestrol (DES) during their pregnancies. Their female offspring suffered a high incidence of malformations and tumors of the genital tract [50, 51]. From a scientific standpoint, the three most perplexing features of transplacental tumors produced by in utero DES exposure were:

- –1. The tumors did not occur until the children were in their teens or were in early adulthood.
- –2. The tumors that arose were tumors of adult-type tissues, not primitive embryonic tissues.
- –3. The cell types of the tumors, being of differentiated cells, were not present in the embryos and fetuses at the time of DES exposure; hence the tumor arose from a type of cell that was not exposed to the carcinogen. [Glossary Carcinogen]

Similar observations have been reported in transplacental carcinogenesis studies in animals [52]. Taken together, these findings clearly indicate that tumors develop over years and decades, during which embryonic cell types directly exposed to a carcinogen may participate in the normal steps of embryonic and fetal development, to eventually differentiate as functional adult cell types. In the end, the type of cell observed in the developed tumors are different from the cell types in which the carcinogenic process originated. [Glossary Teratogenesis]

The phenomenon of transplacental carcinogenesis reminds us that pathogenesis is always a multistep process and that embryonic cellular development always proceeds in a forward direction, passing their collective mutations and cell injuries to their descendant cell types.

In summary, when we study genetic diseases that we diagnose in children or adults, we are always dealing with a disease process that began in the earliest embryo, and developed in conjunction with the development of the embryo and the fetus. Because embryonic development follows a phylogenetic recipe, we can expect that every genetic disease, and every metabolic pathway occurring in a genetic disease, can be modeled and studied in organisms that share the phylogenetic ancestry of humans.

Section 3.4 The Borderland of Embryology and Cancer

But - - once I bent to taste an upland spring *.
And, bending, heard it whisper of its Sea.

Ecclesiastes

In 1958, RA Willis published a landmark book entitled, "Borderland of Embryology and Pathology," in which he explored the theory that cancer was considered the forme fruste (Latin, wrong form) of embryologic development. For instance, a Wilms' tumor was considered a disorganized growth arising from a failed attempt at recapitulating the growth of the renal mesoderm (the metanephros). Hence, a synonym for Wilms tumor was "nephroblastoma." With the discovery of oncogenes, in the last few decades of the 20th century, the focus of cancer shifted to the task of elucidating the molecular steps leading from oncogene activation to the emergence of clinically developed cancers [53]. The concept of cancers as failed embryologic organs was abandoned. [Glossary Forme fruste]

Like many discredited biological theories, there is some merit to Willis' ideas, and the modern classification of cancers continues to pay its respects to the concept that cancers may embody the metabolic pathways and cellular phenotypes of embryologic cells [54–56].

Let's examine a few examples:

The hepatopancreas of arthropods, gastropods and fish

Mammals have a liver and a pancreas, and the two organs are anatomically and functionally distinct. In some animals, including arthropods, gastropods and fish, the pancreas and the liver are a single organ. Fish, like humans, belong to Class Craniata, indicating that the hepatopancreas is preserved in animals with a close lineage to mammals.

The observation that the hepatopancreas is a single organ in fish helps explain the close similarity between two tumors in humans: pancreatoblastoma and hepatoblastoma. Both tumors occur primarily in infants. Both tumors have a somewhat similar morphology, characterized by fetal-type acini and embryonic organ stem cells. Both tumors produce alpha-fetoprotein, a protein secreted in great quantities into the bloodstream by the normal fetal liver, and both tumors are associated with the inherited growth disorder,

Beckwith-Wiedemann Syndrome. Both tumors have loss of heterozygosity for markers on the 11p chromosome [57]. [Glossary Loss of heterozygosity]

The similarity between pancreas and liver extends to the adult. Hepatoid adenomas occasionally occur in the pancreas [58], and liver differentiation can be induced in hypoplastic mouse pancreas [59].

Infantile hemangioma

Well-differentiated neoplasms sometimes function much like normal cells. For example, in chronic myelogenous leukemia, the neoplastic neutrophils that circulate in the blood perform many of the physiologic activities that are observed in normal neutrophils.

In addition to behaving like normal, differentiated tissues, neoplasms sometimes behave like their developmental ancestors. For example, hepatocellular carcinomas typically secrete alpha-fetoprotein, an albumin-like protein normally secreted by fetal hepatocytes. Colon cancers will often secrete carcino-embryonic antigen, a protein that is synthesized normally during fetal life, but which is not synthesized to any degree by adult colonic epithelial cells. [Glossary Hepatoma]

Occasionally, we see neoplasms that yield atavistic cell types consistent with their particular developmental lineage. Congenital hemangiomas are a common lesion found at birth and occurring in about 10% of infants. These tumors can grow rapidly following birth, but they tend to regress over the next few months or years. Investigators have found that some congenital hemangiomas contained myeloid precursor cells (i.e., blood-forming stem cells) [60]. Furthermore, the endothelial cells that composed the hemangiomas co-expressed immunohistochemical markers for endothelium and for myeloid cells [60]. [Glossary Lesion, Regression]

Why would an infantile hemangioma have myeloid differentiation? Though adult myeloid cells (erythrocytes, neutrophils, monocytes, and lymphocytes) come from bone marrow elements, fetal myelopoiesis has a different anatomic source. The earliest myeloid cells in the embryo derive from primitive vessels called insulae sanguineae (i.e., blood islands) [61]. [Glossary Angiogenesis, Myelopoiesis]

It may seem out of the ordinary for a hemangioma to produce blood cells, but the behavior of a congenital hemangioma is consistent with its normal developmental lineage. A primitive vascular tumor behaves like its anlage, an embryonic blood island, producing capillary endothelium and hematopoietic cells.

Carcinosarcomas of uterus

An unusual type of tumor, known as a carcinosarcoma, occasionally arises in the uterus. This tumor is characterized by areas that have an epithelial morphology (resembling the polygonal cells lining the endometrium) and a sarcomatous lining (resembling the spindle-shaped muscle cells or fibrous tissue cells in the wall of the uterus). Most cancers occurring in humans are monomorphic (e.g., epithelial morphology or sarcomatous morphology, but not both). To say that a tumor can be both a carcinoma and a sarcoma is like saying that a eukaryote can be both a bikont and a unikont, or that a plant can be both a monocot and a dicot, or that a geometric shape can be both a circle and a square.

Hence, the occurrence of carcinosarcomas with a dual morphology has been something of a puzzle for generations of pathologists. Theories for the dual nature of uterine tumors have been proposed through the years, and include [Glossary Histopathology, Tissue block]:

- Collision tumor theory

Synchronous biclonal tumors produced a mixed carcinoma/sarcoma histology. This theory has been thoroughly discredited by the finding that carcinosarcomas are clonal neoplasms, each tumor population representing a colony of cells descending from a single, clonogenic cell [62]. The observation of clonality in carcinosarcomas is in line with the generalization, formed from observations on many different tumors, that cancers are clonal growths [63–69].

- Composition theory

The sarcomatous component is an exuberant, nonneoplastic overgrowth of connective tissue in reaction to the presence of the neoplastic carcinoma that invades the uterine musculature. This theory has been discredited by the same observations that discredited the collision tumor theory; that carcinosarcomas are clonal. Both the sarcomatous and the carcinomatous components of the tumor derive from the same clone. Hence, neither component can be a nonneoplastic reaction to the other component. [Glossary Connective tissue]

- Conversion theory

The tumor begins as an epithelial tumor (i.e., a carcinoma) but, for reasons unknown, converts to a sarcoma.

Actually, the dual nature of carcinosarcomas is easily solved, if we just take a moment to think about the embryologic development of the uterus. The uterus, like the kidney, is derived entirely from mesoderm. The uterus is formed from a duct that forms within the mesoderm adjacent to the coelomic cavity (the paramesonephric duct). This paramesonephric duct gives rise to all of the uterus, including the endometrial epithelium and the underlying specialized stroma, and all of the fibromuscular tissue composing the uterine wall.

Embryologists understand that the distinction between carcinoma and sarcoma is a biologically absurd dichotomy, because embryonic mesoderm can produce both types of cells. The uterus is derived entirely from an anlage from the mesoderm known as the paramesonephric duct. This paramesonephric duct gives rise to the endometrial epithelium and the fibromuscular tissue composing the uterine wall. Consequently, tumors of the endometrium can develop as epithelial tumors or spindled connective tissue tumors, or both. Incidentally, under experimental conditions, a single uterine carcinogen can produce tumors of epithelial type, or sarcomatous type, as we would expect [70]. [Glossary Adenocarcinoma, Mucosa, Mixed tumor]

Section 3.5 Pathologic Conditions of the Genomic Regulatory Systems

When debugging, novices insert corrective code; experts remove defective code.

Richard Pattis

Each of us employs a dizzying array of complex mechanisms to control our genes. Transcription (e.g., transcription factors, promoters, enhancers, silencers, pseudogenes, siRNA, miRNA, and competitive endogenous RNAs), posttranscription (splicing, RNA silencing, RNA polyadenylation, and mRNA stabilizers), translation (e.g., translation initiation factors and ribosomal processing), posttranslational protein modifications (e.g., chaperones in mammals, protein trafficking), and all of the chemical and structural modifiers of DNA that constitute the epigenome (e.g., chromatin packing, histone modification, base methylations, imprinting). Disruptions of any of these regulatory processes may produce disease in humans and other metazoans [17, 43, 71–79]. [Glossary Posttranslational protein modification, Vesicular trafficking disorders, MicroRNA, MiRNA, Pseudogene, Alternative RNA splicing, Regulatory DNA element, Regulatory RNA element, Spliceosome, and Ubiquitination]

Why do we have all these different regulatory mechanisms? Here are four plausible assertions. Take a moment and see if you can pick the correct answer, from the choices listed here.

- –1. Every genome regulatory system is essential in its own right
- –2. Each regulatory system has a different and complementary function
- –3. None of the regulatory systems are efficient or error-free, requiring backup regulators
- –4. It's easier and faster to regulate a gene than to create a new gene

Each of these choices has its merits, and it is impossible to say which is correct and which is incorrect, at least for now. However, it is hard to resist option 4, for its lazy and irresponsible approach to cellular physiology. Because it takes a very long time to evolve a new gene, we can appreciate that it is much easier to create multiple fixes, as needed, rather than creating brand new genes that are optimally suited for the job. If this were the case, and genomic regulators existed primarily for the purpose of "making do" with what we have, then we might expect that each of the many regulatory systems may not be absolutely essential, and that disruptions of any individual system may not have catastrophic consequences for the organism.

In point of fact, whenever we study the effects of disrupting any of the known genome regulatory systems, the results are underwhelming. Let's look at a few examples:

Epigenome disruptors

Earlier in this chapter, we discussed how the epigenome can be modified with methylating agents, such as genistein, to produce an inherited shift in coat color in offspring mice [14, 15]. Really, now, is that all we might expect from an agent that fundamentally alters the methylation patterns that control the genome? Wouldn't we expect genistein to be a potent poison, causing instantaneous death, as every cell in the body loses its phenotypic identity? Wouldn't we expect the genistein-fed mouse to transform into a formless clump of billions and billions of nameless, identical and undifferentiated cells?

Of course, this never happens. What we actually observe is that potent methylating agents and demethylating agents, both global disruptors of the epigenome, are famously well-tolerated agents [80]. Indeed, we have accumulated quite a bit of data on the human organism's response to epigenome disruptors, and the results are not what we might expect. For example, we have anticancer drugs such as azacytidine and decitabine that are potent DNA methyltransferase inhibitors, causing the epigenome to hypomethylate. The side effects of these two drugs are so mild that they are recommended for older individuals with acute myelocytic leukemia who have comorbidities that would render such patients particularly vulnerable to standard chemotherapeutic agents [81].

Although epigenetic defects have been associated with slowly developing diseases, including cancer, there seems to be very little acute toxicity associated with epigenomic disruptors [19, 79].

Mild effects of histone disruptors

Histones are the major proteins in chromatin (i.e., the material composing chromosomes). When histones are deacetylated, they tighten around DNA, reducing normal transcription by blocking transcription factors from attaching to their target sites.

Why do drugs like vorinostat, a histone deacetylase inhibitor that acts on the major protein that regulates gene transcription, have mild and nonspecific side effects (e.g., diarrhea, fatigue, nausea) [82]?

Mild effects of microRNA disruptors

MicroRNAs, also known as miRNA, are a short but abundant form of RNA that regulate gene expression by pairing with complementary sequences of mRNA. Such complementation usually reduces the rate of translation of mRNA into protein. It is estimated that humans have more than 1000 different molecular species of miRNA, targeting the majority of human mRNA species [83]. One miRNA may have hundreds of different mRNA targets, and one mRNA target may be regulated by any of several different miRNAs. The miRNAs seem to have a somewhat similar role to that of siRNAs (small interfering RNAs), insofar as they both reduce gene expression.

The miRNA regulatory system, though it has widespread effects on gene expression, seems to have very little effect on the differentiation or viability of cells. Several observations confirm this assertion.

- When dicer, a key enzyme in the synthesis of miRNA molecules, is deleted from mouse kidney cells, thereby drastically reducing levels of miRNA, there is no consequent reduction in the animal's viability. If anything, deleting dicer seems to improve the health of animals by increasing the kidney's resistance to ischemic damage [84].
- MiRNAs are found in animals, plants, fungi, and even viruses, but miRNAs seem to be absent in some eukaryotes, and present but nonfunctional in other eukaryotes, including yeasts [85].
- In a study of the effects of miRNA mutations in *Caenorhabditis elegans* germ cells, no abnormal phenotypes were observed following mutations induced in 95 different miRNAs [86].
- Although miRNAs can reduce the stability of many different mRNAs, the degree of suppression is not great, usually less than twofold [87].
- At present, there are very few diseases that have been associated with specific mutations of the genes coding for miRNAs. Chronic lymphocytic leukemia and B-cell lymphomas seem to be the exception [88]. Also, a form of autosomal dominant hearing loss is caused by mutations in the MIRN96 miRNA.

Mild clinical course of HOX gene diseases

We discussed the HOX genes back in Section 1.5, "Cambrian Explosion," speculating that the acquisition of these regulatory genes may have played a pivotal role in the rapid emergence of animal body plans.

A variety of extremely rare inherited syndromes may result from HOX mutations. Despite the key role of HOX in regulating embryologic development, the diseases resulting from HOX mutations are surprisingly mild in affected humans and other animals [89]. Some individuals with supposedly disease-causing HOX gene mutations have no discernible clinical symptoms.

Mild transcription factor diseases

Transcription factors are proteins that bind to specific DNA sequences to control the transcription of DNA to RNA. Because transcription factors alter the rate of synthesis of many different proteins, we expect to see a diversity in the ways that the disease can express itself in different organs. Furthermore, because each of the genes affected by an altered regulator gene is itself subject to variable expressivity, we can expect that genetic aberrations of transcription factors will produce complex phenotypes involving multiple organs.

The condition wherein one gene produces an array of apparently unrelated clinical effects is known as pleiotropism, and transcription factor mutations provide some of the classic examples of pleiotropic diseases. For example a mutation in the gene encoding transcription factor TBX5 causes Holt-Oram syndrome, characterized by the variable expression of hand malformations, heart defects, and other malformations.

As a general rule, transcription factor diseases, though they effect multiple systems, are relatively mild [90].

Nonsyndromic spliceosome diseases

Errors in normal splicing can produce inherited disease, and it is estimated that 15% of disease-causing mutations involve splicing [91, 92]. Examples of spliceosome diseases are spinal muscular atrophy and some forms of retinitis pigmentosa [71]. In both diseases, pathology is limited to a specific type of cell; retinal cells and their pigment layer in retinitis pigmentosa and motor neuron cells in the spinal muscular atrophy. [Glossary Nonsyndromic disease]

Because spliceosomes are conserved constituents of every eukaryotic cell, we might expect that mutations in spliceosomes would cause deficiencies in many different cell types, with multiorgan and multisystem failures. That this is not the case is somewhat of a mystery, and the catalyst for much speculation [71].

Genetic fine-tuning and common diseases

For a long time, from about 1955 to 1985, it was widely assumed that each of the common diseases (e.g., type 2 diabetes, atherosclerosis, stroke, hypertension, asthma, chronic obstructive pulmonary disease, and cancer) was caused by its own specific disease-causing mutation. It was the job of the research scientists to find the single mutation, for each of the common diseases, and to use that knowledge to discover new, miraculous cures. This fanciful idea was based, in no small part, on the early discoveries of single-gene causes for a large number of metabolic disorders. At the time, there was no compelling argument against finding a single-gene cause for each of the common diseases, including cancer.

In the last several decades of the 20th century, researchers have learned that the notion of a single gene, causing any common disease, was simply wrong. For every common disease, researchers were finding that there were many different associated gene variants that seemed to contribute to the disease phenotype [93–97]. In addition, a slew of additional factors (e.g., regulatory defects and environmental influences) were mixed into the brew. For all the common diseases, our hopes for a cure were looking grim. It was in the last two decades of the 20th century that new information arrived, changing the outlook from grim to probably hopeless. The news was that in common cancers, and in other common diseases, the key genomic mutations seemed to be occurring in noncoding regions of the genome; the very regions that should have had no role in the pathogenesis of disease [46, 98].

What can we learn from the observations that common diseases are polygenic, multifactorial, develop over a long period of time, and involve mutations of noncoding regions of the genome? The common diseases have something in common with the rare conditions resulting from defects in regulatory systems. In both cases, diseases are caused by factors that modify the activity of many different genes. We can easily imagine that noncoding regulatory sequences may modify gene expression levels without eliminating the activity of a gene. Variants in noncoding areas would be unlikely, in that case, to have much of an effect on the overall fitness of individuals [99]. We can imagine that multiple gene variants, along with multiple variants in noncoding regions, might, as a group, influence the development of common, chronic diseases. [Glossary Digenic disease, Heritability, Multifactorial, Oligogenic inheritance, Polygenic disease]

After reviewing the conditions associated with genome regulatory systems, and the diseases resulting from multiple mutations in coding and noncoding gene regions, we can infer the following:

Rule: Diseases resulting from errors in genome regulatory systems are generally mild. They may be isolated, involving a single organ or cell type, or syndromic, involving multiple tissues, and they may include cases in which genetically affected individuals will have no clinical manifestations from their regulatory deficits.

The reason why diseases of genomic regulation are mild is that genomic regulators are fine-tuning devices, and do not behave like on-off switches. As such, regulatory errors do not produce the kinds of catastrophic results observed in mutational diseases that eliminate the activity of important enzymes or fundamentally alter the function of key structural proteins.

Having made the point that disruptors of genomic regulation typically produce mild disease, in unpredictable organs, we should backstep just a bit to make a special exception for the brain. Despite our best research efforts, we know very little about neurologic development or cognition. We don't even understand the biology of consciousness, and how it differs, on a cellular level, from unconsciousness. The brain is pretty much a mystery. Compounding our confusion, there is growing evidence that the brain is particularly susceptible to conditions wherein genomic regulation is reduced [100–102]. In particular, several disorders of genomic regulation result in neurologic disorders [103]. These include hereditary sensory and autonomic neuropathy type 1 with dementia and hearing loss (HSAN1E), Rett syndrome [72], Prader-Willi syndrome and Angelman syndrome, and autosomal dominant cerebellar ataxia, deafness and narcolepsy (ADCA-DN) [103]. It has also been suggested that genomic regulators may lead to the development of common neurologic diseases including autism and psychosis [102].

Together, these finding suggest that the genomic regulatory processes play an important role in neurologic development and function. Although the relationship between epigenome and the brain is in its intellectual infancy, the topic is raised here simply to bring the reader's attention to a new and possibly important field of inquiry. Tread lightly, though. Study of the brain is always fraught with paradox. Ultimately, everything we think we know about the brain is told to us by our brains.

Glossary

Adenocarcinoma Malignant tumors of epithelial cells that are capable of forming glands. As a generalization that is almost always true, the most common tumors developing from glandular organs (e.g., pancreas, salivary glands, and breast glands) and from tissues lined by mucus-secreting cells (e.g., stomach, small intestines and large intestines) are adenocarcinomas (Fig. 3.9).

Adenocarcinoma is an example of a morphology-based diagnostic entity that has no pathogenetic uniformity or consistent embryologic origin. For example, an adenocarcinoma of lung derives from endoderm. An adenocarcinoma of uterus arises from mesoderm. An adenocarcinoma of skin (e.g., an adnexal carcinoma arising from sebaceous glands or hair root) is of ectodermal origin. Because the diagnosis "adenocarcinoma" includes tumors that are biologically unrelated to one another, it would be unlikely that a treatment designed for one variety of adenocarcinoma would

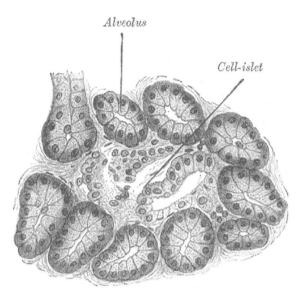

FIG. 3.9 Drawing of an idealized gland. Epithelial cells line a lumen *(empty space)* that connects to a duct. *Source: Wikipedia, from a figure in Henry Gray's "Anatomy of the Human Body," 1918.*

be effective against other varieties of adenocarcinoma, having a different pathogenesis. One of the goals of modern medicine is to develop a classification of diseases that is largely determined by metabolic pathways that have evolved for particular types of cells (Fig. 3.10).

Alternative RNA splicing A mechanism whereby one gene may code for many different proteins [104]. In humans, about 95% of genes that have multiple exons employ alternative splicing. It is estimated that 15% of disease-causing mutations involve splicing variants [91, 92]. Cancer cells are known to contain numerous splicing variants that are not found in normal cells [105, 106]. Normal cells eliminate most abnormal splicing variants through a posttranscriptional editing process. Alternative RNA splicing diseases may result from mutations in splice sites or from spliceosome disorders. An example of a splicing mutation disease is hereditary thrombocythemia, which is characterized by an overproduction of platelets. An activating splice donor mutation in the thrombopoietin gene leads to mRNAs that are more efficiently translated than transcripts that lack the mutation. This, in turn, causes the overproduction of the thrombopoietin, which induces an increase in platelet production [107].

Angiogenesis The formation of new vessels. Angiogenesis in the adult organism always refers to the growth of small vessels, not arteries and veins. The large vessels in the human body develop in utero. Tumor cells must receive oxygen from blood; hence, every invasive and growing solid tumor is capable of inducing angiogenesis. The so-called liquid tumors (i.e., leukemias and myeloproliferative syndromes) do not grow as solid masses, and these malignancies receive oxygen directly from the blood in which they circulate. Hence, angiogeneis would not be a constitutive property of leukemias. Because angiogenesis is a necessary biological step in the pathogenesis of all solid tumors, angiogenesis inhibitors such as bevacizumab, sorafenib, sunitinib, pazopanib, and everolimus, have been used in the treatment of various cancers. Angiogenesis inhibitors may also have value in the treatment of non-neoplastic disorders that are characterized by an overgrowth of new vessels [108–111]. The targeted treatment of many disorders having widely different clinical phenotypes, but with one pathogenetic step in common (angiogenesis in this case) exemplifies the importance of understanding disease pathogenesis. If we can find a pathogenetic step that is shared by many diverse diseases, then we might be able to develop a generalized treatment that provides generalized benefit to a related group of diseases.

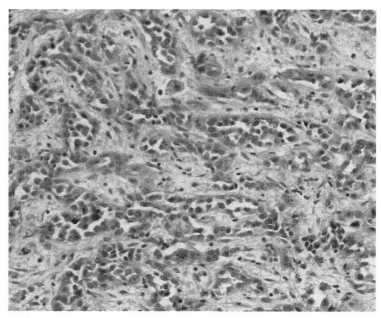

FIG. 3.10 Adenocarcinoma of stomach featuring insinuating glands of various sizes and shapes, each with a central lumen *(white areas),* lined by cancer cells, in a fibrous stroma. *Source: Wikipedia, from an image entered in to the public domain by its copyright holder.*

Anlagen Embryologic precursors for fetal organs and tissues. It is important to know that the anlagen of an organ is not a miniature version of the organ, and does not contain any of the differentiated cells that will eventually appear in the developed organ. The anlagen contains stem cells growing within an architectural framework that does not have the appearance of the organ that will eventually emerge. As an example, imaginal disc are the insect equivalent of analagen. Insect wings grow from imaginal discs that are biologically committed to develop into a wing. An insect wing does not grow from a tiny wing.

Carcinogen The term "carcinogen" refers to agents that cause cancer, but there is considerable controversy over how to apply the term. Some people use the term "carcinogen" to mean a chemical, biological, or physical agent that produces cancer in animals, without the addition of any other agents or processes. Sometimes, the term "complete carcinogen" is used to emphasize the self-sufficiency of the agent as the primary underlying cause of a cancer. Others in the field use the term "carcinogen" to include any agent that will increase the likelihood of tumor development. This definition would apply to agents that must be followed or preceded with other agents for tumors to occur, or agents that increase the number of cancers occurring in a population known to be at high risk of cancer due to an inherited condition.

Channelopathy Disorders of the electrical systems in humans, all of which depend on the depolarization and repolarization of electrical current (i.e., the flux of charged molecules), across ion channels (e.g., sodium channel, potassium channel, chloride channel, and calcium channel). Ion channels are found on the membranes of specialized cells. Disorders of these channels are termed channelopathies, and encompass a wide range of neural, cardiac, and muscular disorders and always play at least a contributing role in common seizures and arrhythmias.

***Cis*-acting versus *trans*-acting** *Cis*-acting regulatory elements are regions of noncoding DNA which regulate the transcription of nearby genes. The *Cis*-acting elements typically regulate gene transcription by functioning as binding sites for transcription factors (i.e., the proteins encoded by *trans*-acting regulatory elements). Polymorphisms (i.e., sequence variants) in these noncoding cis elements have strong effects on the phenotype of cells, by modifying the levels of gene expression. The best understood types of *cis*-acting elements are enhancers and promoters.

Trans-acting regulatory elements are genes that encode transcription factor proteins that modify the expression of distant genes. The transcription factors interact with a *cis*-acting regulatory elements adjacent to the regulated gene. Hence, *cis*- and *trans*-elements work in concert to regulate gene expression.

Clade   A class plus all of its descendant classes. A clade should be distinguished from a lineage, the latter being the list of a class's ascendant superclasses. Because a class can have more than one child class, a pictogram of a clade will often look like a branching tree. In a classification, where each class is restricted to one parent class, ascending lineages are represented as a nonbranching line of ancestors, leading to the root (i.e., top class) of the classification.

Combined deficiency   Occasionally a disease is encountered wherein defects in several proteins contribute to the clinical phenotype. Some of these diseases are due to deletions of segments of DNA, resulting in multiple genes being deleted, as a single chromosomal event.

Alternately, a combined deficiency may be caused by one gene that controls the synthesis of several different proteins. In combined factor V and factor VIII clotting factor deficiency, a defect in either the LMAN1 OR MCFD2 genes results in diminished transport of both factor V and factor VIII from the endoplasmic reticulum to the Golgi apparatus [112]. Hence, the posttranslational processing of both these factors is incomplete, and a combined deficiency results.

Connective tissue   Tissues that connect one tissue with another, keeping them from falling apart, and providing a firm, physical structure to the body. The connective tissues include bone, cartilage, and fibrous tissue.

CpG island   DNA methylation is a form of epigenetic modification that does not alter the sequence of nucleotides in DNA. The most common form of methylation in DNA occurs on Cytosine nucleotides, most often at locations wherein Cytosine is followed by Guanine. These methylations are called CpG sites. CpG islands are concentrations of CpG sites having a GC content over 50% and ranging from 200 base pairs to several thousand base pairs in length. There are about 29,000–50,000 CpG islands and most of these are associated with promoters [10]. Various proteins bind specifically to CpG sites. For example, MECP2 is a chromatin-associated protein that modulates transcription. MECP2 binds to CpGs; hence, alterations in CpG methylation patterns can alter the functionality of MECP2. Mutations in MECP2 cause RETT syndrome, a progressive neurologic developmental disorder and a common cause of mental retardation in females. It has been suggested that the MECP2 mutation disables normal protein-epigenome interactions [113].

Craniata   Craniata is the class of animals that have a cranium (skull) encasing a brain. Class Crianata is sometimes used in place of Class Vertebrata from which it is distinguished by the inclusion of several species that have skulls but lack vertebrae. Animals with a cranium always have a neural crest. Before the appearance of organisms of Class Craniata, there was no neural crest or the neural crest was primitive and incapable of producing all of the cell types and derivative tissues found in organisms of class Craniata. With the evolutionary arrival of the neural crest came all of the diseases that derive from neural crest (i.e., the neurocristopathies). Conversely, noncraniate organisms are not capable of developing any of the neurocristopathies.

Developmental disorder   Within the context of this book, a developmental disorder is a disease or malformation that results from some error in the development of the embryo, fetus, or young child. Outside of the context of this book, the term "developmental disorder" is used by psychiatrists, pediatricians, and related health care workers to refer to language disorders, learning disorders, motor disorders, and autism spectrum disorders. The term might encompass attention deficit disorder, a range of antisocial behaviors and schizophrenia that begins in childhood.

Digenic disease   Digenic diseases require mutations in each of two genes to produce the complete clinical phenotype. There are several rare diseases that are known or suspected to be digenic. Several different forms of Usher disease, combined retinitis pigmentosa and hearing loss, are digenic. A digenic cause of several forms of Long QT syndrome, a type of heart arrhythmia, has been reported to be digenic [114]. Kallman syndrome, a form of hypogonadotropic hypogonadism is suspected to be digenic [114]. Digenic diseases often have a variable clinical phenotype, even among family members with the disease. Mice with digenic diabetes have a non-Mendelian pattern of inheritance, as is typical of a polygenic familial disease [115]. As a group of disorders, the inherited digenic disease occupy an intermediate niche, between monogenic diseases and polygenic diseases.

Ectoderm There are three embryonic layers that eventually develop into the fully developed animal: endoderm, mesoderm, and ectoderm. The ectoderm gives rise to the skin epidermis and the skin appendages (e.g., hairs, sebaceous glands, and breast glandular tissue) and to the specialized neurectoderm from which the central nervous system arises.

Embryo versus fetus As pertains to humans, anatomists typically divide gestation into three periods: pre-embryonic, embryonic, and fetal. The preembryonic period, which corresponds to the first 2 weeks following fertilization or the first month following the mother's last menstrual period, is the time in which the zygote (fertilized ovum) travels to the uterus, implants, begins to build the extraembryonic tissues, and forms a blastocyst. The cells of the preembryo are totipotent. In the third week following fertilization (5th week since the last menstrual period), the three layers of the embryo form, and the embryonic cells differentiate to the extent that they are not multipotent (not totipotent). During the embryonic phase, which corresponds to the 4th to the 9th weeks after fertilization (or the 6th to the 11th week since the last menstrual period), all the germ layers are completed, and all the organs have begun to develop. By the time that the fetal period begins, at week 10 after fertilization (and week 12 since the last menstrual period), all the body and the organs and cell types of the fully developed human are recognizable.

When discussing the periods of embryonic development, it should be understood that there are no universally accepted definitions of the various phases of prenatal human development, or of the timing of such phases. You'll find references where a preembryonic stage is not recognized, with the embryonic organism beginning at the moment of fertilization. Other works will refer to the preimplantation organism as a conceptus; something separable from an embryo.

Most importantly, the timing of prenatal development is a source of great confusion. Obstetricians and pregnant women almost universally mark the beginning of pregnancy with the date of the first day of the last menstrual period. Doing so yields an expected date of delivery 40 weeks later. If you mark the beginning of prenatal development with the moment of conception (typically 2 weeks following the date of the last menstrual period), then the length of gestation is 2 weeks shorter (i.e., 38 weeks). In real life, only about 4% of women actually deliver on the date predicted. Whenever we discuss the timing of embryologic events, it is important to clarify whether we are measuring from the date of the last menstrual period, or whether we are measuring from the presumptive date of fertilization, as a 2 week discrepancy in the first 2 or 3 months will account for a large difference in stage of embryonic development.

Embryonic stem cell A line of tissue-cultured cells that was cloned from a cell taken from the inner cell mass of a blastocyst (i.e., not from the trophectoderm and not from the thin-walled cystic area of the blastocyst). The inner cell mass consists of pluripotent cells that soon differentiate into an embryo composed of three germ layers. Embryonic stem cells in tissue culture can differentiate into any of the cell types that occur in the human body, under proper laboratory conditions.

One of the most surprising advances in cell biology has been the discovery of a methodology by which differentiated cells obtained from a fully developed human, can be induced to become embryonic stem cells; thereby bypassing the traditional dependence on blastocysts as a source of embryonic stem cells. It seems that four factors (OCT4, SOX2, NANOG, and LIN28) can reprogram human somatic cells to become embryonic stem cells, capable of differentiating into all three germ layers [116].

Epimutation The condition in which the normal epigenetic modifications, that characterize differentiated cell types, have changed in individual cells, and their clonal progeny. Epigenomic instability, like genomic instability, is a near-constant feature of tumor progression. Because cellular differentiation is under epigenetic control, the loss of tumor cell differentiation observed with tumor progression is presumably due to epigenetic instability. Likewise, cancer cells that have an unstable epigenome may inactivate or activate a variety of disease genes in surprising ways. For example, epigenetic instability may produce individual cells within a tumor that have inactivation of the Werner syndrome gene, the same gene that causes a premature aging syndrome when it occurs in the germline cells of an organism [117]. In similar fashion, individual cells in a tumor may have epigenetic inactivation of the lamin A/C gene, the same gene that when inactivated in germline cells causes a form of cardiomyopathy [75, 118].

Epithelial cell Epithelial cells are polyhedral units that are held tightly together, by specialized junctions. Epithelial cells form the mucosal lining of ducts, glands, and most pavemented surfaces. Tumors of epithelial lining cells account for well over 90% of the cancers that occur in humans, including the most commonly occurring forms of skin, lung, colon cancers, prostate, breast, and pancreatic cancers.

Evo-devo The study of the evolution of developmental processes.

Forme fruste From the French, crude, or unfinished form; plural formes frustes. A term used by diagnosticians and applied to difficult cases wherein a patient presents with some of the features of a recognized disease or syndrome, but who does not quite fit the accepted diagnostic criteria. The clinical presentation is said to be the forme fruste (i.e., wrong, incomplete, or unfinished form) of the disease. In some cases, studying the forme fruste of a disease may help us understand the classic form of a disease. For example, geneticists reported a child who presented with renal angiomyolipoma, a rare tumor sometimes found in patients with tuberous sclerosis. Several years later, the same patient developed cystic disease in the contralateral kidney, often associated with another genetic condition known as polycystic kidney disease. Genetic analysis demonstrated a contiguous gene deletion involving both the TSC2 gene for tuberous sclerosis and the PKD1 gene for polycystic kidney disease. The patient's phenotype was the forme fruste of two rare diseases, but genetic analysis proved that the presentation fit a contiguous gene syndrome [119].

Gallertoid Animals are thought to have evolved from simple, spherical organisms floating in the sea, called gallertoids. The living sphere was lined by a single layer of cells enclosing a soft center in which fibrous cells floated in extracellular matrix. As the gallertoids evolved to extract food from the seabed floor, they flattened out. The modern animals most like the gallertoids are the placozoans, discovered in 1833, plastered against the wall of a seawater aquarium. These organisms are just under a millimeter in length and are composed of about 1000 epithelial cells. With the exception of being flat, rather than round, they resemble the gallertoids, with an outer lining of cuboidal cells, and an inner gelatinous matrix holding a suspension of fibrous cells.

Genetic heterogeneity In the context of genetic diseases, the term refers to diseases that can be expressed by any one of multiple allelic variants in a gene (allelic heterogeneity) or by any one of multiple different genes that carry disease-producing alleles (locus heterogeneity). When a rare diseases demonstrates genetic heterogeneity, we are provided with an opportunity to learn the common pathogenesis that stems from errors in different genes.

In general, the more common the genetic disease, the more heterogeneous it is, for example, retinitis pigmentosa. This genetic disease is relatively common, with a worldwide prevalence of 1 in 4000 individuals. Retinitis pigmentosa is remarkably heterogeneous and can be caused by at least 4000 different mutations involving any of at least 100 different genes. Retinitis pigmentosa may have autosomal dominant, autosomal recessive, or x-linked inheritance. It can occur in the absence of a family history (i.e., as a de novo mutation). Its clinical phenotype is highly variable, and it can occur by itself, or as part of a syndrome. For example, Usher Syndrome combines retinitis pigmentosa and deafness. Usher syndrome, the most common form of syndromic retinitis pigmentosa, is itself clinically and genetically heterogeneous. Basically, the more genetic heterogeneity in a disease, the more genetic opportunities are there for the disease to occur, increasing the incidence of the disease.

Genetic heterogeneity should be contrasted with the concept of genetic pleiotropism, in which one gene may be responsible for several different functions or disorders.

Germ layers Bilaterians (i.e., a subclass of Class Animalia) are tripblastic, meaning that the early embryo contains three layers from which all of the tissues of the developed organism will arise. These three layers are ectoderm, endoderm, and mesoderm.

During embryogenesis, several additional germ layers arise (e.g., neuroectoderm and neural crest), but these layers are derivative of the three primary germ layers and not traditionally included when we count the germ layers.

The three primary germ layers traditionally exclude trophectoderm, the layer of cells that gives rise to the extraembryonic tissues that arise early in embryogenesis but which are not incorporated into the developed organism (e.g., placenta and amnion).

Also, the term "germ layer" must never be confused with "a layer of germ cells" (i.e., a layer of gametes and their precursors).

Hepatoma Cancer of liver cells (i.e., hepatocytes) and synonymous with hepatocellular carcinoma. The unqualified term, "liver cancer," generally refers to hepatocellular carcinoma, although there are dozens of types of liver cancer other than hepatoma.

Hereditary nonpolyposis colorectal cancer syndrome Abbreviation: HNPCC. A hereditary cancer syndrome characterized by an increased risk of colorectal cancer, endometrial cancer, and several other types of cancers of the ovary, stomach, small intestine, hepatobiliary tract, upper urinary tract, brain, and skin. The syndrome is associated with mutations that impair DNA mismatch repair.

Heritability A statistical inference of the degree of population variation in a phenotypic trait that can be assigned to inheritance. A crude way of rephrasing this is that heritability is the fraction of a trait that is due to inheritance, and not to environment, as determined for a population.

Heritability has been studied in cancers wherein researchers compared the occurrence of individual types of tumors in individuals with some predisposing inherited conditions compared with individuals with no known hereditary predisposition [120]. The highest hereditary fractions at individual tumor sites were seen for retinoblastoma (37.2%). For most tumors, the heritability fraction was much lower. The overall estimate of heritability for all types of cancers was estimated as 4.2% [120].

Histopathology Many diseases can be diagnosed by examining biopsied specimens, using a microscope, and this process is referred to as histopatholgy ("histo" referring to tissue, and "pathology" referring to the study of disease). Sampled tissues are fixed in formalin and embedded in paraffin (a waxy substance). Thin slices of the paraffin-embedded tissues are mounted on glass slides and stained so that the cellular detail can be visualized under a microscope. A histopathologic diagnosis is based on finding the specific cellular alterations that characterize diseases.

Immune system In humans, there are three known host defense systems that recognize and destroy foreign organisms: intrinsic, innate, and adaptive.

Imprinting Early in mammalian embryogenesis, the pattern of epigenetic modifications (e.g., methylations) inherited from the paternal and maternal gametes is erased, forcing the embryo to develop its own unique pattern of methylations. This process of epigenome erasure is necessary; otherwise, the embryonic germline would have a differentiated epigenome, and the normal process of gradual epigenetic modifications, applied throughout embryogenesis, could not occur. Erasure is not a totally thorough process. There are about 100 known genes that retain their parental epigenetic patterns. Retention of parental epigenetic patterns is known as imprinting.

Lesion Any tissue that is visibly affected by a disease process.

Locus heterogeneity Also known as nonallelic heterogeneity, occurs when mutations in different genes can produce the same disease. For example, mutations in c-KIT or PDGFR-alpha can lead to GIST tumors. Mutations in the gene encoding the protein hamartin or the gene encoding the protein tuberin can produce the disease tuberous sclerosis. Carney complex can be caused by mutations in the PRKAR1A gene on chromosome 17q23-q24, or it may be caused by a mutations in chromosome 2p16. Locus heterogeneity is a special case of the broader concept of Genetic heterogeneity.

Loss of heterozygosity Most genes come in two copies, the copy produced from the maternally derived chromosome and the copy produced by the paternally derived chromosome. These copies are called alleles, and in many cases, the two alleles are subtly different from one another. In this case, the gene is heterozygous. If one of the two alleles is inactivated or lost within a cell, only one of the alleles will be expressed, and this is referred to as loss of heterozygosity. An apparent loss of heterozygosity may also occur with uniparental disomy (i.e., when a zygote receives two copies of a chromosome, or of part of a chromosome, from one parent and no copy from the other parent), producing two identical alleles, or when a repair error replaces one allele with its alternate, producing two identical alleles.

Lynch syndrome Eponymous equivalent of hereditary nonpolyposis colorectal cancer syndrome.

Mesoderm The embryonic germ layer that lies between Ectoderm and Endoderm and which gives rise to the mesenchyme, which consists of most of the connective tissue, muscles, and bones of the body. The kidneys and the uterus are organs that are derived entirely from mesoderm.

MiRNA Same as microRNA.

MicroRNA Also known as miRNA. Small but abundant species of RNA that regulate gene expression by pairing with complementary sequences of mRNA. Such complementation usually reduces the rate of translation of mRNA into protein. It is estimated that humans have more than 1000 different molecular species of microRNA, targeting the majority of human genes [83]. One miRNA may have hundreds of different mRNA targets, and one mRNA target may be regulated by any of several different miRNAs. MicroRNAs seem to play a regulatory role somewhat similar to that of siRNAs (small interfering RNAs).

Mixed tumor There is a wealth of accumulated evidence supporting the hypothesis that cancers are clonal growths (i.e., growths that arise as a clone or colony from a single founding cell). If so, we can infer that whenever we encounter cancers that are comprised of more than one morphologic cell type, then these differing cell types must have arisen from the same founder cell; and presumably have the same histologic lineage. If we encounter a mixed tumor of salivary gland (also known as pleomorphic adenoma of salivary gland) that exhibits epithelial cells and stromal cells in a myxoid stroma, then we customarily presume that the epithelial cells and the spindle have the same endodermal origin. When we encounter mixed adenocarcinoma-carcinoid tumors, then we infer that the adenocarcinoma cells and the carcinoid cells must have the same endodermal lineage. When we encounter a mixed medullary-papillary carcinoma thyroid, we infer that the cells composing the medullary morphology (i.e., thyroid c-cells or calcitonin producing cells) and the cells composing the papillary morphology (i.e., thyroid follicle cells) have the same endodermal origin, supporting molecular studies that would suggest an endodermal origin for the C-cells of the thyroid [121, 122]. Thus, we can use our knowledge of human tumor biology to tell us something about the embryologic origin of cell types [53].

As a caveat, we should mention that the clonal nature of cancer should not be construed as proof that a cancer develops as a 1-step process wherein a single cell (the founder) mutates into a cancer cell. There is much evidence supporting the assertion that cancer, like all diseases, develops through a series of biological steps, over time, usually involving many different cells. It is best to think of the founding cell of a cancer as the product of a process occurring over time, that may have produced many different potential founding cells, but which finally yielded a single cell whose progeny composed the emergent cancer.

Mucosa Refers to the surface layer of epithelial cells, the basement membrane on which the epithelial cells sit, and the thin layer of connective tissue that sits between the basement membrane and an underlying thin muscle layer (the muscularis mucosa).

Multifactorial The term "multifactorial" has several vague and unhelpful meanings, in the context of pathogenesis. Does it mean that multiple factors must act in concert to produce disease, or does it mean that a disease can be produced by any one of many possible factors, or does it mean that the development of the disease is complex and occurs over time, as many different factors eventually contribute to the disease phenotype, or does it mean that no single factor can rationally account for the occurrence of the disease? More often than not, the term "multifactorial" serves only to obfuscate reality.

Multipotent stem cell Synonymous with pluripotent stem cell. A cell that can divide to produce differentiated cells of more than one cell type. Multipotent stem cells must be distinguished from totipotent stem cells, which can produce differentiated cells of any embryonic germ cell layer and of extraembryonic origin (e.g., trophoblasts).

Mutator phenotype One of the hallmarks of cancer is genetic instability. There are examples of cancer cells that have thousands of genetic mutations. It is hypothesized that during carcinogenesis, cells acquire a mutator phenotype that increases the rate at which genetic aberrations occur in cells, thus raising the likelihood that a clone of cells will emerge with mutations that confer a malignant phenotype [123, 124]. An alternate hypothesis suggests that the normal rate of random mutations in cells is sufficient to provide cancer-associated mutations, and a mutator phenotype is not a necessary condition for carcinogenesis [125].

Myelopoiesis The process by which myeloid stem cells eventually produce the normal contingent of fully differentiated nucleated cells that circulate in our blood (e.g., neutrophils, eosinophils, basophils, lymphocytes, and monocytes). Erythropoiesis is the corresponding process, by which mature red blood cells are produced by erythroid stem cells.

Neoplasm Neoplasm means "new growth," and is a near-synonym for "tumor." Neoplasms can be benign or malignant. Leukemias, which grow as a population of circulating blood cells and which do not generally produce a visible mass (i.e., do not produce a tumor), are included under the general term "neoplasm." Hamartomas, benign overgrowths of tissue, are typically included among the neoplasms, as are the precancers (precursor lesions cancers), which are often small and scarcely visible.

Nonsyndromic disease A disease that effects a single organ or function, unaccompanied by abnormalities of other organs or physiologic systems. Congenital deafness and deafness in early childhood usually appears as part of a syndrome, possibly involving facial structures or nerves. A pediatrician may use the term nonsyndromic deafness for emphasis, when deafness occurs without other accompanying pathologies.

Oligogenic inheritance In the context of genetic diseases, occurs when the expression of several genes (i.e., not one gene, and not many genes) produces a disease phenotype. If two genes are required, the term "digenic disease" applies. Macular degeneration may qualify as a common disease with oligogenic inheritance. The combination of a few gene variants may account for 70% of the risk of developing age-related macular degeneration [126, 127], the third leading cause of blindness worldwide [128]. Other oligogenic rare diseases are Bardet-Biedl syndrome [129], and Williams-Beuren syndrome [130].

Orthodisease Orthodiseases are conditions observed in nonhuman species that result from alterations in genes that are homologous to the genes known to cause diseases in humans.

Phylogeny A method of classifications based on ancestral lineage. The classification of terrestrial organisms is a phylogenetic classification.

Pluripotent stem cell Cells that can divide and produce differentiated cell types from any of the three embryonic layers (endoderm, ectoderm, and mesoderm.) Pluripotent stem cells differ from totipotent stem cells as they do not yield cells of extraembryonic type (e.g., trophoblasts). Pluripotent stem cells can be induced from cultured fibroblasts treated using a cocktail of transcription factors [2, 3].

Polygenic disease A disease whose underlying cause involves alterations in multiple genes. In general, the development of polygenic diseases is highly dependent on environmental modifiers that trigger bouts of disease, that enhance or reduce susceptibility to disease, or that sometimes serve as the apparent root cause of the disease.

As an example, consider a patient with no known underlying medical condition who is stung by a bee and immediately succumbs to anaphylactic shock. It is tempting to say that the root cause is the bee sting, but we know that most individuals who are stung by a bee do not develop an anaphylactic response. Clearly, some underlying condition must have predisposed the patient to develop shock. You want to blame the patient's genes, but if there was no parental history or familial history of anaphylaxis, then it would be hard to put the blame on an inherited gene. In such instances, we look toward a polygenic explanation, wherein multiple variants of gene expression together produce a physiological condition that predisposes the individual to anaphylactic shock. Of course, we cannot be certain that we are correct until we identify all of the modified genes and demonstrate the biological mechanism by which they exert their effect. That's a very tall order. In the meantime, we work under the tentative assumption that we are correct. That's science.

Posttranslational protein modification Proteins freshly translated from a mRNA template, typically undergo a variety of modifications before they are fully functional. Errors in the posttranslational process, including timing errors (i.e., the proper sequence of events that lead to the finished product) can have negative consequences. An example of a rare disease caused by a defect in a posttranslational process is congenital disorder of glycosylation type IIe, caused by homozygous mutation in a gene that encodes a component of a Golgi body protein that is involved in posttranslational protein glycosylation; the COG7 gene [131]. This rare disease produces a complex disease phenotype in infants, with multiple disturbances in organs and systems plus various anatomic abnormalities.

Primary disease The term "primary disease," usually refers to an inherited form of a disease. The acquired, noninherited form of the disease is usually referred to as the "secondary disease." For example, primary ciliary dyskinesia is an inherited disease of cilia in which the normal movements of ciliated epithelia are impeded, leading to the accumulation of cellular debris and mucous, and thus producing chronic otitis, chronic sinusitis, chronic bronchitis, and pneumonias. Secondary ciliary dyskinesia is caused by a toxin (e.g., cigarette smoke) or respiratory infection that impairs ciliary activity, producing the same clinical sequelae as are observed for primary ciliary dyskinesia. It happens that cilia come in two categories: primary and motile. Primary ciliary dyskinesia affects the motile, nonprimary cilia. Hence, the term "primary ciliary dyskinesia" excludes all diseases of primary cilia, of which there are many. All of the recognized diseases of primary cilia are inherited conditions, and thus are "primary diseases of primary cilia," and the class of these conditions excludes "primary ciliary dyskinesia," which is a condition of nonprimary cilia. It's all very confusing, but it gets much worse. The term "primary" may refer to a disease that is limited to one organ or that arises from a disease process that is not secondary to any other disease process. Hence, we must endure two terms that seem to contradict one another: "primary cardiac amyloidosis" and "systemic primary amyloidosis." The first term refers to amyloidosis that occurs exclusively in the heart. The second term, "systemic primary amyloidosis" refers to amyloidosis that does not occur secondary to any other condition, but which will be found everywhere in the body [132]. We also see "primary" occurring as the adverb, "primarily" in which case it may have the layman's meaning of "mostly," or it may have the medical meaning, "as a primary condition." The sentence "Diseases occurring in adults are primarily secondary diseases," happens to be true, but it would be best if it were never uttered.

Primary disease versus secondary disease A primary disease is a condition caused by a sequence of conditions or events, none of which, being themselves a disease.

A disease that occurs as a result of having a primary disease is a secondary disease. For example, when finger clubbing (hypertrophic osteoarthropathy) occurs in the setting of underlying diseases of the cardiovascular and pulmonary systems, we call this "secondary hypertrophic osteoarthropathy, indicating that the finger clubbing arises secondary to some other disease.

When hypertrophic osteoarthropathy occurs as an inherited, familial syndrome, we refer to it as primary hypertrophic osteoarthropathy, indicating that it does not occur secondary to some predisposing concurrent disease.

In many other cases, the term "secondary" is used synonymously with "acquired," to indicate that the disease is not inherited and was caused by some acquired event. You'll notice that there are two forms of "secondary" in common usage. One form of secondary refers to a disease caused by some other condition. Another form of "secondary" refers to noninherited form of a disease [133]. Having two definitions for "secondary" is confusing, but troubles deepen when the inherited and secondary forms of a disease are biologically unrelated. We see this in some forms of primary and secondary erythromelalgia. Primary erythromelalgia is a rare genetic channelopathy whose root cause is a mutation in a sodium channel gene (SCN9A), producing neuropathic pain and redness of the extremities. When burning and redness of the extremities occurs due to microthrombi occurring in small arterioles, we call this secondary erythromelalgia. Secondary erythromelalgia has a totally different pathogenesis than primary erythromelalgia [134], and this will pose a problem when searching for a cure for erythromelalgia. There is no reason to expect a secondary erythromelalgia of the microthrombotic type to respond to a treatment developed for primary erythromelalgia due to an inherited mutation in a SCN9A.

As a final comment, the terms "primary" and "secondary" mean something quite different to oncologists than they do to other healthcare professionals. Within the cancer field, the term "primary" refers to the first site of growth of a cancer. For example, an adenocarcinoma primary in lung refers to a cancer that arose from cells present in the lung. For oncologists, a "secondary" tumor is a second tumor that arises after a primary tumor, and sometimes as the result of treatment of the primary tumor. For example, a sarcoma arising inside the field of radiation of a retinoblastoma, would be considered a secondary cancer. You can see that an oncologist might speak of the primary site of a secondary cancer.

The term "Secondary disease" and "Primary disease" are used in so many different ways that they have both lost much of their meaning and should probably be abandoned.

Promoter The DNA site that binds RNA polymerase plus transcription factors, to initiate RNA transcription. Promoter mutations cause disease in subsets of patients affected by beta-thalassemia, Bernard-Soulier syndrome, pyruvate kinase deficiency, familial hypercholesterolemia, and hemophilia, among others [135].

As a general rule, promoter mutations, as well as all mutations that effect genomic regulatory systems, cause disease by reducing the quantity of a normal protein; not by producing altered protein and not by eliminating gene products. Because the drop in protein production may be small, promoter diseases may produce less extreme clinical syndromes than gene defects that eliminate a protein entirely.

Pseudogene A gene that does not code for protein. Theories explaining the origin of pseudogenes are many. Some pseudogenes presumably devolved from genes that acquired mutations that rendered the genes nonfunctional. Other pseudogenes may have been reverse-transcribed into DNA via RNA retrotransposons. Pseudogenes are identified from genomic sequence data using computational algorithms that search for stretches of DNA that have some sequence similarities to functional genes, along with sequences that might render the gene nonfunctional (e.g., premature stop codons, frameshift mutations, and a Poly-A tail).

It is currently believed that transcribed pseudogene sequences (i.e., pseudogene RNA) is one of a class of competitive endogenous RNA species that compete for microRNA binding sites and consequently diminish the repressive actions of microRNA on target expression. Hence, pseudogenes moderate microRNA activity and provide some level of gene expression enhancement. There are, at a minimum, several thousand pseudogenes in the genome, and some genes, such as actin, may have numerous pseudogenes [136, 137]. At present, pseudogenes are thought to play a role in the dysregulation of cancer cells, and in cell defects found in neurodegenerative disorders [100, 138].

Regression The act of going back to a previous state. "Regression" is an example of a Janus term; a term that can have opposite meanings, depending on its context. When "regression" is applied to tumor growth, it means that the tumor is shrinking (i.e., returning the patient to an earlier state, when the tumor was smaller). For oncologists and their patients, regression is a very good thing. Outside of the field of oncology, the term "regression" usually applies to patients who are losing ground to their disease. For example, for a child who suffers from a neurologic disease, regression may take the form of losing the ability to understand spoken language or losing the ability to walk. Thus, the regressing child is returning to a more dependent and more infantile state. Thus, outside the field of cancer treatment, regression is a very bad thing.

Likewise, disease progression indicates that the disease is worsening. If we say that we are seeing progress in the treatment of a disease, we indicate that the patient is improving.

Regression and progression are examples of Janus terms; words that mean their opposite, depending on the context in which they appear. Another medical Janus word is "divide." When you divide doses, you reduce the size of each dose, while maintaining the equivalent total dosage. When a cell divides, it doubles itself to become two cells. Saying that a cell has divided is the same as saying that the cell has multiplied (the opposite of division).

As an aside, Janus words are not restricted to medical lexicography. If you say, "The stars are out tonight," you probably mean that the stars are visible. If you say, "Put the light out," you are indicating that the light should be made invisible.

Regulatory DNA element Sites in DNA that bind to other molecules (e.g., transcription factors and RNA polymerase) to regulate transcription. Promoters and enhancers are types of regulatory DNA elements.

Regulatory RNA element Transcribed RNA can influence the subsequent transcription and/or translation of other RNA species. The various RNA regulatory elements include: antisense RNA (including *cis*-natural antisense transcript and *trans*-acting siRNA), long noncoding RNA, microRNA, piwi-interacting RNA, repeat-associated siRNA, RNAi, small interfering RNA, and small temporal RNA.

Mutations of regulatory RNA elements may cause disease. For example, miR-96 is expressed exclusively in the inner ear and the eye. Mutations in the miR-96 precursor molecule may cause a rare form of autosomal dominant hearing loss [139].

Secondary disease A disease that results from having a primary disease. For example, an individual with diabetes is prone to developing conditions related to vascular insufficiency, such as ischemic necrosis of toes or feet. Diabetes is the primary disease and vascular insufficiency condition is the secondary disease. It seems like a simple concept, but type two diabetes may arise secondary to obesity. This would put obesity as the primary disease, diabetes as the secondary disease, and vascular insufficiency as the tertiary disease. Individuals with vascular insufficiency have reduced ambulation, and may lead sedentary lives, leading to obesity. In such case, the causal rankings produce circularity, with obesity -> diabetes -> vascular insufficiency -> obesity. When the circle is complete, who can say which disease is primary or secondary or tertiary or quaternary? You may as well just spin the wheel.

Seminoma A tumor of male germ cells, almost always arising in the testis. The female equivalent of seminoma (i.e., the tumor arising from female germ cells) is the ovarian dysgerminoma.

Spliceosome In eukaryotes, DNA sequences are not transcribed directly into full-length RNA molecules, ready for translation into a final protein. There is a pretranslational process wherein transcribed sections of DNA, the so-called introns, are spliced together, and a single gene can be assembled into alternative spliced products. Alternative splicing is one method whereby more than one protein form can be produced by a single gene [71]. Cellular proteins that coordinate the splicing process are referred to, in aggregate, as the spliceosome. Errors in normal splicing can produce inherited disease, and it estimated that 15% of disease-causing mutations involve splicing [91, 92]. Even the unicellular eukaryotes have spliceosomes [140]. Examples of spliceosome diseases are spinal muscular atrophy and some forms of retinitis pigmentosa [71].

Teratogenesis The biological process that leads to a developmental malformation. Current thinking would suggest that teratogens (the agents of teratogenesis) act by killing specific types of embryonic or fetal cells at vulnerable moments in development. Theory suggests that normal development requires specific cells fulfilling specific functions at specific times. Interruption of this orchestrated process may result in developmental abnormalities.

Tertiary disease A term that is hardly ever used, though you might expect it to be, if the terms "primary disease" and "secondary disease" made any logical sense. When the term "secondary disease" is used to describe conditions arising from some other existing disease (i.e., the primary disease), we would expect to see tertiary diseases that arise from secondary diseases. Employing proof by induction, there must be a quaternary disease and a quinary disease and so on. The lack of any such terms in medical terminology indicates our inability to think much beyond the simplest sequences of disease development.

One of the important conceptual advances of modern medicine in its implicit dependence on multistep pathogenesis and these steps would include genetic "root causes" of disease as well as successive stages of disease, some of which may have their own clinical phenotypes. This being so, it is probably wise to abandon terms such as primary, secondary, and tertiary, and replacing them with statements that describe how one step in the development of a particular disease may relate to some other step in the development of some other disease or condition.

Tissue block All tissues removed from patients (e.g., by surgeons during operations, by dermatologists who sample small skin lesions) are brought to the pathology department where they are examined grossly (i.e., by the unaided eye) and histologically (i.e., by microscopic examination). Samples of the received tissues are fixed in formalin and then processed to produce a paraffin-infiltrated tissue encased in a block of paraffin. These blocks, sometimes referred to as cassettes (because plastic cassettes hold the paraffin block) are used as the source of thin tissue sections that can be mounted and stained on glass slides. Pathologists reach a diagnosis based on a correlation of clinical history with gross and microscopic examinations of tissue.

Totipotent stem cell A stem cell that can produce, after cell divisions, differentiated cells of any type. This would include cells of any of the three embryonic layers (ectoderm, endoderm, and mesoderm), germ cells, and cells of the extraembryonic tissue (e.g., trophoblasts). A totipotent stem cell is different from a pluripotent stem cell, the latter of which cannot produce extraembryonic cells and cannot produce germ cells.

Toxic response versus disease process A toxic response is a pathological condition that is produced and sustained by the intrusion of a chemical or physical agent. When the offending agent is removed, the toxic response typically diminishes or disappears. A disease is a pathologic process that, whichever its cause, becomes self-sustaining; and cannot be immediately stopped by the withdrawal of a particular chemical or physical agent. A disease runs its own course, which may be acute or chronic, self-limited or permanent.

***Trans*-acting** In molecular biology, a *trans*-acting agent is usually a regulatory sequence of DNA, that acts through an intermediary molecule (i.e., protein or RNA), on some other location of the chromosome or on some other chromosome. A *cis*-acting agent element does not operate through an intermediary molecule. Typically, *cis*-acting elements are DNA sequences that serve as regulatory binding sites.

Translation factor Also called initiation factor or, more precisely, translation initiation factor. These factors initiate protein synthesis by forming a complex with mRNA and ribosomal RNA. Mutations in the initiation factor EIF2B gene result in leukoencephalopathy with vanishing white matter. Pathogenic mutations may arise in any one of five genes encoding subunits of EIF2B [141]. Surprisingly, each of these five genes, that contribute to the eventual synthesis of one aggregate protein, are located on five different chromosomes.

Ubiquitination Ubiquitin is a protein found "ubiquitously" in eukaryotic cells. Ubiquitination is a process involving ubiquitin in which proteins that need to be broken down are tagged for removal.

Vesicular trafficking disorders Alternately known as protein trafficking disorder, cargo disorder, and vesicular transport disorder. After a protein molecule is translated from mRNA, a complex set of posttranslational events must occur for the protein to serve its intended purpose. The protein must be modified (e.g., glycosylated), shaped (e.g., folded), transported from the endoplasmic reticulum into a series of subcellular locations (e.g., Golgi apparatus and cargo vesicle), and delivered to its ultimate location. Such posttranslational steps are often divided into disorders of posttranslational modification (e.g., congenital disorders of glycosylation) and protein transport disorders [142]. Much of what we know of these disorders were based on examining transport gene mutants in yeast cells [143].

Xeroderma pigmentosum An inherited disorder in which affected persons cannot efficiently repair DNA damage produced by UV light. Persons with xeroderma pigmentosum are extremely sensitive to the acute (causing sunburn) and chronic (causing skin cancer) effects of sunlight. As such, xeroderma pigmentosum is an example of the sequential participation of genes and environment in carcinogenesis. First comes the genetic mutation producing a specific type of DNA repair deficiency. Later, exposure to ultraviolet light produces DNA mutations that are not effectively repaired. Some of these mutations, occurring in cells directly exposed to UV light, set into motion another set of biological events, which may, or may not, result in the development of a skin cancer.

References

[1] Armstrong L, Lako M, Dean W, Stojkovic M. Epigenetic modification is central to genome reprogramming in somatic cell nuclear transfer. Stem Cells 2006;24:805–14.

[2] Okita K, Ichisaka T, Yamanaka S. Generation of germline-competent induced pluripotent stem cells. Nature 2007;448:313–7.

[3] Takahashi K, Tanabe K, Ohnuki M, Narita M, Ichisaka T, Tomoda K, et al. Induction of pluripotent stem cells from adult human fibroblasts by defined factors. Cell 2007;131:861–72.

[4] Tanimoto Y, Iijima S, Hasegawa Y, Suzuki Y, Daitoku Y, Mizuno S, et al. Embryonic stem cells derived from C57BL/6J and C57BL/6N mice. Comp Med 2008;58:347–52.

[5] Takahashi K, Murakami M, Yamanaka S. Role of the phosphoinositide 3-kinase pathway in mouse embryonic stem (ES) cells. Biochem Soc Trans 2005;33:1522–5.

[6] Willis RA. Borderland of embryology and pathology. London: Butterworth; 1958. In mid-twentieth century, many pathologists believed that neoplasms were a pathologic form of tissue embryogenesis. For instance, a Wilms' tumor was a disorganized growth arising from tissue recapitulating growth of the renal mesoderm (hence the descriptive name, "nephroblastoma"). This excellent and scholarly book discusses the similarities between embryologic growth and neoplastic growth.

[7] Basto R, Lau J, Vinogradova T, Gardiol A, Woods CG, Khodjakov A, et al. Flies without centrioles. Cell 2006;125:1375–86.

[8] Sarkozy A, Carta C, Moretti S, Zampino G, Digilio MC, Pantaleoni F, et al. Germline BRAF mutations in Noonan, LEOPARD, and cardiofaciocutaneous syndromes: molecular diversity and associated phenotypic spectrum. Hum Mutat 2009;30:695–702.

[9] Khavari DA, Sen GL, Rinn JL. DNA methylation and epigenetic control of cellular differentiation. Cell Cycle 2010;9:3880–3.

[10] Bogler O, Cavenee WK, Zhang W, Fuller GN. Methylation and genomic damage in gliomas. In: Genomic and molecular neuro-oncology. Sudbury, MA: Jones and Bartlett; 2004. p. 3–16.

[11] Pikaard CS, Mittelsten Scheid O. Epigenetic regulation in plants. Cold Spring Harb Perspect Biol 2014;6:a019315.

[12] Holliday R. Paradoxes between genetics and development. J Cell Sci 1990;97:395–8.

[13] Liu H, Liu X, Zhang S, Lv J, Li S, Shang S. Systematic identification and annotation of human methylation marks based on bisulfite sequencing methylomes reveals distinct roles of cell type-specific hypomethylation in the regulation of cell identity genes. Nucleic Acids Res 2016;44:75–94.

[14] Dolinoy DC, Weidman JR, Waterland RA, Jirtle RL. Maternal genistein alters coat color and protects Avy mouse offspring from obesity by modifying the fetal epigenome. Environ Health Perspect 2006;114:567–72.

[15] Wolff GL, Kodell RL, Moore SR, Cooney CA. Maternal epigenetics and methyl supplements affect agouti gene expression in Avy/a mice. FASEB J 1998;12:949–57.

[16] Lancaster AK, Masel J. The evolution of reversible switches in the presence of irreversible mimics. Evolution 2009;63:2350–62.

[17] Martin DIK, Cropley JE, Suter CM. Epigenetics in disease: leader or follower? Epigenetics 2011;6:843–8.

[18] Chatterjee A, Morison IM. Monozygotic twins: genes are not the destiny? Bioinformation 2011;7:369–70.

[19] Wong AHC, Gottesman II, Petronis A. Phenotypic differences in genetically identical organisms: the epigenetic perspective. Hum Mol Genet 2005;14:R11–8.

[20] Fraga MF, Ballestar E, Paz MF, Ropero S, Setien F, Ballestar ML, et al. Epigenetic differences arise during the lifetime of monozygotic twins. Proc Natl Acad Sci U S A 2005;102:10604–9.

[21] Miyoshi N, Stel JM, Shioda K, et al. Erasure of DNA methylation, genomic imprints, and epimutations in a primordial germ-cell model derived from mouse pluripotent stem cells. Proc Natl Acad Sci U S A 2016;113:9545–50.

[22] Allegrucci C, Thurston A, Lucas E, Young L. Epigenetics and the germline. Reproduction 2005;129:137–49.

[23] Zheng J, Xia X, Ding H, Yan A, Hu S, Gong X, et al. Erasure of the paternal transcription program during spermiogenesis: the first step in the reprogramming of sperm chromatin for zygotic development. Dev Dyn 2008;237:1463–76.

[24] Sun F, Fang H, Li R, Gao T, Zheng J, Chen X, et al. Nuclear reprogramming: the zygotic transcription program is established through an erase-and-rebuild strategy. Cell Res 2007;17:117–34.

[25] Simo R, Villarroel M, Corraliza L, Hernandez C, Garcia-Ramirez M. The retinal pigment epithelium: something more than a constituent of the blood-retinal barrier: implications for the pathogenesis of diabetic retinopathy. J Biomed Biotechnol 2010;2010:190724.

[26] Sanchez-Romero MA, Cota I, Casadesus J. DNA methylation in bacteria: from the methyl group to the methylome. Curr Opin Microbiol 2015;25:9–16.

[27] Pereira SL, Reeve JN. Histones and nucleosomes in Archaeae and Eukarya: a comparative analysis. Extremophiles 1998;2:141–8.

[28] Kardong KV. The evolution of the venom apparatus in snakes from colubrids to viperids and elapids. Mem Inst Butantan 1982;46:106–18.

[29] Markham MR. Electrocyte physiology: 50 years later. J Exp Biol 2013;216:2451–8.

[30] Jopling C, Boue S, Belmonte JCI. Dedifferentiation, transdifferentiation and reprogramming: three routes to regeneration. Nat Rev Mol Cell Biol 2011;12:79–89.

[31] Rishniw M, Xin HB, Deng KY, Kotlikoff MI. Skeletal myogenesis in the mouse esophagus does not occur through transdifferentiation. Genesis 2003;36:81–2.

[32] Marro S, Pang ZP, Yang N, Tsai MC, Qu K, Chang HY, et al. Direct lineage conversion of terminally differentiated hepatocytes to functional neurons. Cell Stem Cell 2011;9:374–82.

[33] Huang P, He Z, Ji S, Sun H, Xiang D, Liu C, et al. Induction of functional hepatocyte-like cells from mouse fibroblasts by defined factors. Nature 2011;475:386–9.

[34] Lee TI, Young RA. Transcriptional regulation and its misregulation in disease. Cell 2013;152:1237–51.

[35] Zheng W, Zhao H, Mancera E, Steinmetz LM, Snyder M. Genetic analysis of variation in transcription factor binding in yeast. Nature 2010;464:1187–91.

[36] Kasowski M, Grubert F, Heffelfinger C, Hariharan M, Asabere A, Waszak SM, et al. Variation in transcription factor binding among humans. Science 2010;328(5975):232–5.

[37] Abad M, Mosteiro L, Pantoja C, Canamero M, Rayon T, Ors I, et al. Reprogramming in vivo produces teratomas and iPS cells with totipotency features. Nature 2013;502:340–5.

[38] Virchow R. Die Cellularpathologie in ihrer Begrundung auf physiologische und pathologische Gewebelehre. Berlin: Hirschwald; 1858.

[39] Fogli A, Rodriguez D, Eymard-Pierre E, Bouhour F, Labauge P, Meaney BF, et al. Ovarian failure related to eukaryotic initiation factor 2B mutations. Am J Hum Genet 2003;72:1544–50.

[40] Crisponi L, Deiana M, Loi A, Chiappe F, Uda M, Amati P, et al. The putative forkhead transcription factor FOXL2 is mutated in blepharophimosis/ptosis/epicanthus inversus syndrome. Nat Genet 2001;27:159–66.

[41] Kagey MH, Newman JJ, Bilodeau S, Zhan Y, Orlando DA, van Berkum NL, et al. Mediator and cohesin connect gene expression and chromatin architecture. Nature 2010;467:430–5.

[42] Raimundo N, Song L, Shutt TE, McKay SE, Cotney J, Guan MX, et al. Mitochondrial stress engages E2F1 apoptotic signaling to cause deafness. Cell 2012;148(4):716–26.

[43] Tanackovic G, Ransijn A, Thibault P, Abou Elela S, Klinck R, Berson EL, et al. PRPF mutations are associated with generalized defects in spliceosome formation and pre-mRNA splicing in patients with retinitis pigmentosa. Hum Mol Genet 2011;20:2116–30.

[44] Simpson GG. The principles of classification and a classification of mammals. Bull Am Mus Nat Hist 1945;85:.

[45] Manolio TA, Collins FS, Cox NJ, Goldstein DB, Hindorff LA, Hunter DJ, et al. Finding the missing heritability of complex diseases. Nature 2009;461:747–53.

[46] Berman JJ. Rare diseases and orphan drugs: keys to understanding and treating common diseases. Cambridge, MD: Academic Press; 2014.

[47] Jonkman MF, Pasmooij AMG, Pasmans SGMA, van den Berg MP, ter Horst HJ, Timmer A, et al. Loss of desmoplakin tail causes lethal acantholytic epidermolysis bullosa. Am J Hum Genet 2005;77:653–60.

[48] Nagel ZD, Chaim IA, Samson LD. Inter-individual variation in DNA repair capacity: a need for multi-pathway functional assays to promote translational DNA repair research. DNA Repair (Amst) 2014;19:199–213.

[49] Cohen A, Thompson E. DNA repair in nondividing human lymphocytes: inhibition by deoxyadenosine. Cancer Res 1986;46:1585–8.

[50] Herbst AL, Ulfelder H, Poskanzer DC. Association of maternal stilbestrol therapy and tumor appearance in young women. N Engl J Med 1971;284:878–81.

[51] Herbst AL, Scully RE, Robboy SJ. The significance of adenosis and clear-cell adenocarcinoma of the genital tract in young females. J Reprod Med 1975;15:5–11.

[52] Anderson LM, Bhalchandra A, Diwan BA, Nicola T, Fear NT, Roman E. Critical windows of exposure for children's health: cancer in human epidemiological studies and neoplasms in experimental animal models. Environ Health Perspect 2000;108(Suppl. 3):573–94.

[53] Berman JJ. Neoplasms: principles of development and diversity. Sudbury: Jones & Bartlett; 2009.

[54] Berman JJ. Modern classification of neoplasms: reconciling differences between morphologic and molecular approaches. BMC Cancer 2005;5:100.

[55] Berman JJ. Tumor taxonomy for the developmental lineage classification of neoplasms. BMC Cancer 2004;4:88.

[56] Berman JJ. Tumor classification: molecular analysis meets Aristotle. BMC Cancer 2004;4:10.

[57] Saif MW. Pancreatoblastoma. JOP J Pancreas (Online) 2007;8:55–63.

[58] Cuilliere P, Lazure T, Bui M, Fabre M, Buffet C, Gayral F, et al. Solid adenoma with exclusive hepatocellular differentiation: a new variant among pancreatic benign neoplasms? Virchows Arch 2002;441:519–22.

[59] Wells JM, Esni F, Boivin GP, Aronow BJ, Stuart W, Combs C, et al. Wnt/beta-catenin signaling is required for development of the exocrine pancreas. BMC Dev Biol 2007;7:4.

[60] Ritter MR, Reinisch J, Friedlander SF, Friedlander M. Myeloid cells in infantile hemangioma. Am J Pathol 2006;168:621–8.

[61] Nomina Anatomica. Prepared by the international anatomical nomenclature committee. 5th ed. Baltimore: Williams and Wilkins; 1980.

[62] Fujii H, Yoshida M, Gong ZX, Matsumoto T, Hamano Y, Fukunaga M, et al. Frequent genetic heterogeneity in the clonal evolution of gynecological carcinosarcoma and its influence on phenotypic diversity. Cancer Res 2000;60:114–20.

[63] Calabrese P, Tavare S, Shibata D. Pretumor progression: clonal evolution of human stem cell populations. Am J Pathol 2004;164:1337–46.

[64] Antonescu CR, Elahi A, Healey JH, et al. Monoclonality of multifocal myxoid liposarcoma: confirmation by analysis of TLS-CHOP or EWS-CHOP rearrangements. Clin Cancer Res 2000;6:2788–93.

[65] Bacher U, Haferlach T, Hiddemann W, Schnittger S, Kern W, Schoch C. Additional clonal abnormalities in Philadelphia-positive ALL and CML demonstrate a different cytogenetic pattern at diagnosis and follow different pathways at progression. Cancer Genet Cytogenet 2005;157:53–61.

[66] Bennett M, Stroncek DF. Recent advances in the bcr-abl negative chronic myeloproliferative diseases. J Transl Med 2006;4:41.

[67] Chen L, Shern JF, Wei JS, Yohe ME, Song YK, Hurd L, et al. Clonality and evolutionary history of rhabdomyosarcoma. PLoS Genet 2015;11:e100507.

[68] Hacein-Bey-Abina S, Von Kalle C, Schmidt M, McCormack MP, Wulffraat N, Leboulch P, et al. LMO2-associated clonal T cell proliferation in two patients after gene therapy for SCID-X1. Science 2003;302(5644):415–9.

[69] Landgren O, Kyle RA, Pfeiffer RM. Monoclonal gammopathy of undetermined significance (MGUS) consistently precedes multiple myeloma: a prospective study. Blood 2009;113:5412–7.

[70] Hubalek M, Ramoni A, Mueller-Holzner E, Marth C. Malignant mixed mesodermal tumor after tamoxifen therapy for breast cancer. Gynecol Oncol 2004;95:264–6.

[71] Faustino NA, Cooper TA. Pre-mRNA splicing and human disease. Genes Dev 2003;17:419–37.

[72] Horike S, Cai S, Miyano M, Chen J, Kohwi-Shigematsu T. Loss of silent chromatin looping and impaired imprinting of DLX5 in Rett syndrome. Nat Genet 2005;32:31–40.

[73] Preuss P. Solving the Mechanism of Rett Syndrome: how the first identified epigenetic disease turns on the genes that produce its symptoms. Research News Berkeley Lab; 2004.

[74] Soejima H, Higashimoto K. Epigenetic and genetic alterations of the imprinting disorder Beckwith-Wiedemann syndrome and related disorders. J Hum Genet 2013;58:402–9.

[75] Agrelo R, Setien F, Espada J, Artiga MJ, Rodriguez M, Pérez-Rosado A, et al. Inactivation of the lamin A/C gene by CpG island promoter hypermethylation in hematologic malignancies, and its association with poor survival in nodal diffuse large B-cell lymphoma. J Clin Oncol 2005;23:3940–7.

[76] Bartholdi D, Krajewska-Walasek M, Ounap K, Gaspar H, Chrzanowska KH, Ilyana H, et al. Epigenetic mutations of the imprinted IGF2-H19 domain in Silver-Russell syndrome (SRS): results from a large cohort of patients with SRS and SRS-like phenotypes. J Med Genet 2009;46:192–7.

[77] Chen J, Odenike O, Rowley JD. Leukemogenesis: more than mutant genes. Nat Rev Cancer 2010;10:23–36.

[78] McKenna ES, Sansam CG, Cho YJ, Greulich H, Evans JA, Thom CS, et al. Loss of the epigenetic tumor suppressor SNF5 leads to cancer without genomic instability. Mol Cell Biol 2008;28:6223–33.

[79] Feinberg AP. The epigenetics of cancer etiology. Semin Cancer Biol 2004;14:427–32.

[80] Derissen EJB, Beijnen JH, Schellens JHM. Concise drug review: azacitidine and decitabine. Oncologist 2013;18:619–24.

[81] Cruijsen M, Lubbert M, Wijermans P, Huls G. Clinical results of hypomethylating agents in AML treatment. J Clin Med 2015;4(1):1–17.

[82] Kim H, Bae S. Histone deacetylase inhibitors: molecular mechanisms of action and clinical trials as anti-cancer drugs. Am J Transl Res 2011;3:166–79.

[83] Bentwich I, Avniel A, Karov Y, Aharonov R, Gilad S, Barad O, et al. Identification of hundreds of conserved and nonconserved human microRNAs. Nat Genet 2005;37:766–70.

[84] Wei Q, Bhatt K, He HZ, Mi QS, Haase VH, Dong Z. Targeted deletion of Dicer from proximal tubules protects against renal ischemia-reperfusion injury. J Am Soc Nephrol 2010;21:756–61.

[85] Zheng Y, Cai X, Bradley JE. microRNAs in parasites and parasite infection. RNA Biol 2013;10:371–9.

[86] Miska EA, Alvarez-Saavedra E, Abbott AL, Lau NC, Hellman AB, McGonagle SM, et al. Most *Caenorhabditis elegans* microRNAs are individually not essential for development or viability. PLoS Genet 2007;3:e215.

[87] Baek D, Villen J, Shin C, Camargo FD, Gygi SP, Bartel DP. The impact of microRNAs on protein output. Nature 2008;455:64–71.

[88] Musilova K, Mraz M. MicroRNAs in B-cell lymphomas: how a complex biology gets more complex. Leukemia 2015;29:1004–17.

[89] Quinonez S, Innis JW. Human HOX gene disorders. Mol Genet Metab 2014;111:4–15.

[90] Seidman JG, Seidman C. Transcription factor haploinsufficiency: when half a loaf is not enough. J Clin Invest 2002;109:451–5.

[91] Pagani F, Baralle FE. Genomic variants in exons and introns: identifying the splicing spoilers. Nat Rev Genet 2004;5:389–96.

[92] Fraser HB, Xie X. Common polymorphic transcript variation in human disease. Genome Res 2009;19 (4):567–75.

[93] Balaci L, Spada MC, Olla N, Sole G, Loddo L, Anedda F, et al. IRAK-M is involved in the pathogenesis of early-onset persistent asthma. Am J Hum Genet 2007;80:1103–14.

[94] International Consortium for Blood Pressure Genome-Wide Association Studies. Genetic variants in novel pathways influence blood pressure and cardiovascular disease risk. Nature 2011;478:103–9.

[95] Billings LK, Florez JC. The genetics of type 2 diabetes: what have we learned from GWAS? Ann N Y Acad Sci 2010;1212:59–77.

[96] Couzin-Frankel J. Major heart disease genes prove elusive. Science 2010;328:1220–1.

[97] Saey TH. Rare genetic tweaks may not be behind common diseases: variants thought to be behind inherited conditions prove difficult to pin down. Sci News 2013;183(11).

[98] Zuk O, Hechter E, Sunyaev SR, Lander ES. The mystery of missing heritability: genetic interactions create phantom heritability. Proc Natl Acad Sci U S A 2012;109:1193–8.

[99] Alfoldi J, Lindblad-Toh K. Comparative genomics as a tool to understand evolution and disease. Genome Res 2013;23:1063–8.

[100] Costa V, Esposito R, Aprile M, Ciccodicola A. Non-coding RNA and pseudogenes in neurodegenerative diseases: "The (un)Usual Suspects" Front Genet 2012;3:231.

[101] Kleaveland B, Shi CY, Stefano J, Bartel DP. A network of noncoding regulatory RNAs acts in the mammalian brain. Cell 2018;174:350–62.

[102] Shen E, Shulha H, Weng Z, Akbarian S. Regulation of histone H3K4 methylation in brain development and disease. Philos Trans R Soc Lond Ser B Biol Sci 2014;369:20130514.

[103] Weissman J, Naidu S, Bjornsson HT. Abnormalities of the DNA methylation mark and its machinery: an emerging cause of neurologic dysfunction. Semin Neurol 2014;34:249–57.

[104] Sorek R, Dror G, Shamir R. Assessing the number of ancestral alternatively spliced exons in the human genome. BMC Genomics 2006;7:273.

[105] Venables JP. Aberrant and alternative splicing in cancer. Cancer Res 2004;64:7647–54.

[106] Srebrow A, Kornblihtt AR. The connection between splicing and cancer. J Cell Sci 2006;119:2635–41.

[107] Wiestner A, Schlemper RJ, van der Maas AP, Skoda RC. An activating splice donor mutation in the thrombopoietin gene causes hereditary thrombocythaemia. Nat Genet 1998;18:49–52.

[108] Bose P, Holter JL, Selby GB. Bevacizumab in hereditary hemorrhagic telangiectasia. N Engl J Med 2009;360:2143–4.

[109] Plotkin SR, Merker VL, Halpin C, Jennings D, McKenna MJ, Harris GJ, et al. Bevacizumab for progressive vestibular schwannoma in neurofibromatosis type 2: a retrospective review of 31 patients. Otol Neurotol 2012;33:1046–52.

[110] Eyetech Study Group. Anti-vascular endothelial growth factor therapy for subfoveal choroidal neovascularization secondary to age-related macular degeneration: phase II study results. Ophthalmology 2003;110:979–86.

[111] Leung E, Landa G. Update on current and future novel therapies for dry age-related macular degeneration. Expert Rev Clin Pharmacol 2013;6:565–79.

[112] Zhang B, McGee B, Yamaoka JS, Guglielmone H, Downes KA, Minoldo S, et al. Combined deficiency of factor V and factor VIII is due to mutations in either LMAN1 or MCFD2. Blood 2006;107:1903–7.

[113] Amir RE, Van den Veyver IB, Wan M, Tran CQ, Francke U, Zoghbi HY. Rett syndrome is caused by mutations in X-linked MECP2, encoding methyl-CpG-binding protein 2. Nat Genet 1999;23:185–8.

[114] Pitteloud N, Quinton R, Pearce S, Raivio T, Acierno J, Dwyer A, et al. Digenic mutations account for variable phenotypes in idiopathic hypogonadotropic hypogonadism. J Clin Invest 2007;117:457–63.

[115] Bruning JC, Winnay J, Bonner-Weir S, Taylor SI, Accili D, Kahn CR. Development of a novel polygenic model of NIDDM in mice heterozygous for IR and IRS-1 null alleles. Cell 1997;88:561–72.

[116] Yu J, Vodyanik MA, Smuga-Otto K, Antosiewicz-Bourget J, Frane JL, Tian S, et al. Induced pluripotent stem cell lines derived from human somatic cells. Science 2007;318:1917–20.

[117] Agrelo R, Cheng WH, Setien F, Ropero S, Espada J, Fraga MF, et al. Epigenetic inactivation of the premature aging Werner syndrome gene in human cancer. Proc Natl Acad Sci U S A 2006;103:8822–7.

[118] Malhotra R, Mason PK. Lamin A/C deficiency as a cause of familial dilated cardiomyopathy. Curr Opin Cardiol 2009;24:203–8.

[119] Smulders YM, Eussen BHJ, Verhoef S, Wouters CH. Large deletion causing the TSC2-PKD1 contiguous gene syndrome without infantile polycystic disease. J Med Genet 2003;40:e17.

[120] Narod SA, Stiller C, Lenoir GM. An estimate of the heritable fraction of childhood cancer. Br J Cancer 1991;63:993–9.

[121] Jain M, Verma D, Thomas S, Chauhan R. Mixed medullary—papillary carcinoma thyroid: an uncommon variant of thyroid carcinoma. J Lab Physicians 2014;6:133–5.

[122] Kameda Y. Cellular and molecular events on the development of mammalian thyroid C cells. Dev Dyn 2016;245:323–41.

[123] Bierig JR. Actions for damages against medical examiners and the defense of sovereign immunity. Clin Lab Med 1998;18:139–50.

[124] Loeb LA. Mutator phenotype may be required for multistage carcinogenesis. Cancer Res 1991;51:3075–9.

[125] Tomlinson IP, Novelli MR, Bodmer WF. The mutation rate and cancer. Proc Natl Acad Sci U S A 1996;93:14800–3.

[126] Lotery A, Trump D. Progress in defining the molecular biology of age related macular degeneration. Hum Genet 2007;122:219–36.

[127] Maller J, George S, Purcell S, Fagerness J, Altshuler D, Daly MJ, et al. Common variation in three genes, including a noncoding variant in CFH, strongly influences risk of age-related macular degeneration. Nat Genet 2006;38:1055–9.

[128] Katta S, Kaur I, Chakrabarti S. The molecular genetic basis of age-related macular degeneration: an overview. J Genet 2009;88:425–49.

[129] Eichers ER, Lewis RA, Katsanis N, Lupski JR. Triallelic inheritance: a bridge between Mendelian and multifactorial traits. Ann Med 2004;36:262–72.

[130] Pober BR. Williams-Beuren syndrome. N Engl J Med 2010;362:239–52.

[131] Ng BG, Kranz C, Hagebeuk EE, Duran M, Abeling NG, Wuyts B, et al. Molecular and clinical characterization of a Moroccan Cog7 deficient patient. Mol Genet Metab 2007;91:201–4.

[132] Thomashow AI, Angle WD, Morrione TG. Primary cardiac amyloidosis. Am Heart J 1953;46:895–905.

[133] Oliveira AM, Perez-Atayde AR, Inwards CY, Medeiros F, Derr V, Hsi BL, et al. USP6 and CDH11 oncogenes identify the neoplastic cell in primary aneurysmal bone cysts and are absent in so-called secondary aneurysmal bone cysts. Am J Pathol 2004;165:1773–80.

[134] Tang Z, Chen Z, Tang B, Jiang H. Primary erythromelalgia: a review. Orphanet J Rare Dis 2015;10:127.

[135] de Vooght KMK, van Wijk R, van Solingel WE. Management of gene promoter mutations in molecular diagnostics. Clin Chem 2009;55:698–708.

[136] Ng SY, Gunning P, Eddy R, Ponte P, Leavitt J, Shows T, et al. Evolution of the functional human beta-actin gene and its multi-pseudogene family: conservation of noncoding regions and chromosomal dispersion of pseudogenes. Mol Cell Biol 1985;5:2720–32.

[137] Zhang Z, Harrison P, Gerstein M. Identification and analysis of over 2000 ribosomal protein pseudogenes in the human genome. Genome Res 2002;12:1466–82.

[138] Poliseno L. Pseudogenes: newly discovered players in human cancer. Sci Signal 2012;5:5.

[139] Mencia A, Modamio-Hoybjor S, Redshaw N, Morín M, Mayo-Merino F, Olavarrieta L, et al. Mutations in the seed region of human miR-96 are responsible for nonsyndromic progressive hearing loss. Nat Genet 2009;41:609–13.

[140] Baldauf SL. An overview of the phylogeny and diversity of eukaryotes. J Syst Evol 2008;46:263–73.

[141] Pavitt GD. EIF2B, a mediator of general and gene-specific translational control. Biochem Soc Trans 2005;33:1487–92.

[142] Gissen P, Maher ER. Cargos and genes: insights into vesicular transport from inherited human disease. J Med Genet 2007;44:545–55.

[143] Altman LK. For 3 Nobel winners, a molecular mystery solved. The New York Times; 2013. October 7.

4

Speciation

Section 4.1 A Species is a Biological Entity

The purpose of narrative is to present us with complexity and ambiguity.

Scott Turow

It has been argued that nature produces individuals, not species; the concept of species being a mere figment of the human imagination, created for the convenience of taxonomists who need to group similar organisms. This view is anathema to classic taxonomists, who have long held that a species is a natural unit of biological life and that the nature of each species is revealed through the intellectual process of building a consistent taxonomy [1].

In point of fact, there are many excellent reasons to believe that species are biological entities on equal or better scientific footing than individual organisms. Let's look at some of the biological properties of species.

–1.   Species are defined based on a fundamental biological attribute and membership within a species is immutable.

Early definitions of species were fashioned to exclude most organisms, including all bacteria, all unicellular eukaryotes, and all fungi. One long-held definition for a species was that it was a class of animals that shared important characteristics and that the members of a species could successfully procreate with one another but not with the members of other species. This long-standing definition did not help us to understand how species come into existence and did not inform us how to choose the important characteristics that determine membership in a species.

Evolution's Clinical Guidebook. https://doi.org/10.1016/B978-0-12-817126-4.00004-7

The modern definition of species can be expressed in three words: "evolving gene pool" [2]. This elegant definition is easy to comprehend and as we shall soon learn, serves to explain how new species come into existence [1, 3]. Because each member of a species has a genome constructed from the species gene pool, it is clear that membership within a species is immutable (e.g., a fish cannot become a cat and a cat cannot become a goat).

–2.   Membership within a species is biologically determined for every living organism.

Every organism came into existence by drawing its genome from the community gene pool and must therefore be a member of the species associated with its gene pool.

–3.   Species respond biologically to natural selection.

Natural selection operates on the gene pool of a species, changing the balance of available genes. Gene variants that enhance the fitness of the species, are preserved at a high prevalence (i.e., found in a high percentage of the species population). Unfavorable genes are relegated to a low prevalence (i.e., found in a small percentage of the population). Hence, natural selection operates primarily on the gene pool; effects on individuals are secondary.

–4.   Species live and die and have a primary biological purpose: speciation.

Nobody really knows how many species have existed and perished on this planet, but the number that keeps popping up is that there have been about 5–50 billion species of organisms on earth and more than 99% of them have met with extinction, leaving a relatively scant 10–100 million living species [4]. The wide variation in the estimated number of living species is accounted for by what we include (e.g., bacteria and viruses), how we subdivide (i.e., subtypes, strains, and breeds), and where we look (e.g., deep in the oceans, ponds, ditches, stratosphere, and heat vents). Regardless, we should probably assume that the overwhelmingly likely destiny of every species is extinction.

If the purpose of every species is to ensure its own survival, then they are all doing a very bad job of it, insofar as more than 99% of species are extinct. The observation of the remarkable diversity of life on earth suggests that the purpose and destiny of a species is not to endure, but to speciate; to produce new offspring species. Hence, the success of a species is not determined by whether it has produced lots of individuals of the species, or whether it has persisted for a long time, but whether it has produced a lineage of new species. In this sense, if we look at the mammals (i.e., monotremes, eutherians, and metatherians), the monotremes are the least successful (just a few extant species) and the eutherians are the most successful (many thousands of extant species). We will return to this topic in Section 6.3, "Mammals to Therians."

In summary, species have the biological properties associated with any living entity: uniqueness, life, death, the issuance of progeny, and the benefit of evolution through natural selection. Hence, we should think of a species as a biological entity; not as an abstraction.

Section 4.2 The Biological Process of Speciation

One of the most fundamental goals of modern biological research is comprehension of the way in which species arise.

George Gaylord Simpson (1902–84), in 1945 [5]

George Gaylord Simpson, a mid-20th century evolutionary biologist and taxonomist, had a gift for posing some of the best and most challenging questions in his field. It is intriguing that one year after Simpson explained the importance of asking how species arise, the science fiction author, Ray Bradbury, inadvertently found the answer. In "The Million-Year Picnic," one of the stories in "*The Martian Chronicles*," Bradbury relates how a man takes his family from earth to live permanently on Mars. Once there, they burn the rocket that transported them, so that they can never return to earth. One evening, the father tells his children that he is going to take them to see the martians. They walk together to a canal, and the father shows them their own reflections in the water. The family had become, by virtue of leaving the earth behind, a new species; a species of martians.

The family depicted by Ray Bradbury had separated itself from the gene pool of earthlings, and had established, along with the other martian colonists, a gene pool that would evolve in its own manner, to produce a species that would, over time, become distinguishable from its parent species: *Homo sapiens*. The martians might develop a horny coat to protect against cosmic radiation, or any number of modifications to cope with the toxic martian atmosphere. Regardless, we can be certain that the gene pool available to the colonizing martians would evolve differently than the gene pool of earthlings.

1. Species speciate. Individuals do not.

If you remember "*The Martian Chronicles*," then surely you recall the 25th episode of the third season of Star Trek: The Next Generation, titled, "Transfigurations." In this episode, a Zalkonian named "John" takes refuge aboard the Enterprise to escape pursuit by Captain Sunad, a zealous Zalkonian determined to capture John and return him to Zalkon, to face punishment for crimes unspecified.

During the ensuing drama, and between commercial breaks, John acquires strange powers, including the power to heal. In the last moment, John evolves into a being of pure energy, and flies under his own power through space. Presumably, he is headed back to Zalkon, where he will be the father of a new species of energetic Zalkonians.

Of course, this book is focused on non-terrestrial entities, but it should be apparent that evolution on the planet Zalkon is very different from evolution here on earth. For starters, no earthling organism has ever transformed into a member of a new species with a new set of physical properties. The most that a single organism can hope to accomplish is to participate in a new evolving gene pool that is different from that of the parental species.

Can a single organism account for a diverse collection of subspecies? Yes, and we suspect that it happens all the time on islands or in geographic areas isolated by mountains, or rivers, or habitat barriers. The key point to remember is that the founder of a new species is not itself a new species; it is simply the first contributor to a gene pool that will enlarge and evolve.

To understand how individual gene pools, not organisms, speciate, imagine a female African primate of Class Lemuriformes desperately clinging to a tree limb as she is carried to sea, only to wash up on the shore of Madagascar. After a few days, she assesses her situation and determines that she is the only strepsirrhine (a subclass of Class Lemuriformes) on the island. Adding to the drama, she discovers that she is pregnant.

If she and her brood survives, and if the children successfully breed, they will establish a new gene pool from which a new species will evolve. In fact, a scenario something like the one described has been suggested for the origin of all the species of lemur on Madagascar. The reasoning is simple: lemurs are found only on the island of Madagascar and on the Comoros Islands, just northwest of Madagascar. Hence, they must have arisen on the island from an ancestral species that lived only on the island. The non-lemur members of Class Lemuriformes, the galagos and lorisids, live in Africa and Asia. Hence, the founding parent of the lemurs of Madagascar most likely came from Africa.

The important point here is that the member of Class Lemuriformes who landed in Madagascar was nothing special. Unlike John the Zalkonian, she didn't evolve into another species. Her only claim to fame was that she contributed to a gene pool. The gene pool evolved over time, and the process of evolution required nothing more than natural selection of pooled genes. Eventually, all the different species of lemur now living in Madagascar came to be. Should we give credit to the original Madagascar primate for all the species of lemur living on earth today? Technically, no. We owe credit to the gene pool that she helped fill. Had there been a different pregnant primate, arriving under the same set of circumstances, we would have no lemurs as such, but we would have a different set of species in their place.

Evolutionary biologists suspect that just as one strepsirrhine primate may account for all the different types of lemur on Madagascar, one ancient rodent, of Class Hystrichomoropha, may have been responsible for all the extant rodents of South America (e.g., capybaras, coypus, chinchillas, pacas, cavies).

2. Species can only speciate into something that they already are

Throughout this book, we have been stressing the point that we, as humans, are members of every class of organisms in our ancestral lineage. Hence, we are eukaryotes, and we are deuterstomes, and we are craniates, and so on. Furthermore, the assertion that humans are eukaryotes must not be interpreted as a didactic device, intended to remind us of our early roots. In point of fact, we actually are eukaryotes; the reason being that the gametes in every organism represent the true progeny of the gametes in the animal's ancestry. Skeptics are counseled to follow the history of their own gametes backwards through evolution. The gametes in our bodies are the progeny of the fusion of two parental gametes. This process can be followed iteratively up through the chain of ancestors within a species, and up through the gametes of the parent species, and on and on, until reaching the single-cell eukaryotes, organisms that are, for all practical purposes, "all gamete." This explains why we have hundreds of the same core genes found in eukaryotic organisms, and why we are dues-paying members of Class Eukaryota, and of Class Deuterotomia, and of Class Craniata, down the line to Class *Homo sapiens.*

3. Evolution is inevitable (the horseshoe crab is no exception)

All species that survive will speciate, given time. Eventually, something will happen that isolates a subpopulation (e.g., migration to an island, change in the habitat in a geographic region, even a new highway that divides a population). New species begin at the moment when a population is split, even if there are no outward signs of change. The horseshoe crab is often touted as a species that never speciates, insofar as the horseshoe crabs living today look much like the horseshoe crabs living 100 million years ago. Nevertheless, the living horseshoe crabs belong to at least four modern species, with their own phylogenies determined by molecular analysis [6] (Fig. 4.1).

Speciation begins the moment that populations split apart from one another. Over time, small variations in the separate gene pools will produce subtle differences in the respective members of each population [7]. We can guess that several million years ago, some horseshoe crabs split from the herd and started their own species.

4. We can learn about class properties by studying divergent sister classes

Robert Frost, in his poem "The Road Not Taken," wrote: "Somewhere ages and ages hence: Two roads diverged in a wood, and I—I took the one less traveled by, And that has made all the difference."

How could Robert Frost be certain that choosing the road less traveled "made all the difference" in his life? Without putting too fine a point on it, wouldn't Frost need to have somehow duplicated himself at the divergence of roads, and watched how each of the

FIG. 4.1 Horseshoe crabs (*Limulus polyphemus*), an example of an animal that has evolved, but slowly. *Source: Wikipedia, and released into the public domain by its author, Breese Greg, of the US Fish and Wildlife Service.*

duplicate Frosts proceeded down their separate paths? It is conceivable that the two Frosts may have ended at the same destination, with nothing to indicate that their lives followed divergent pathways. Frost, you see, was operating at a disadvantage that modern-day biologists have overcome. In essence, we can watch the evolutionary pathways followed by species that diverge at a point in time. We do this by studying sister classes and their descendant subclasses.

In many cases, we have lost the living representatives of ancestral sister classes through extinction. In some cases, we risk losing sister classes through inattention and negligence. Currently, there is one living species that represents the sister class to all other angiosperms (i.e., flowering plants). This species, *Amborella trichopoda*, is an unassuming shrub found only on the small Pacific island of New Caledonia. It is feared that *A. trichopoda* is on the brink of extinction. A botanist wrote, in a 2008 PhD thesis (translated from the French), "The disappearance of *A. trichopoda* would imply the disappearance of a genus, a family and an entire order, as well as the only witness to at least 140 million years of evolutionary history," [8]. [Glossary Sister class and cousin class]

Having a verifiable member of a sister class has a number of uses. By studying sister species, we can sometimes:

— Verify phylogenetic relationships and taxonomic trees

If a class and its sister both share the same trait, then it is reasonable to infer that the trait was inherited from the parent class.

— Distinguish acquired traits from inherited ancestral traits

If a class and its sister do not share the same trait, and if the trait is present in a class that is not closely related to the sibling classes, then it is reasonable to infer that the trait is most likely convergent (i.e., acquired independently, and not inherited by shared ancestry) in the classes that have the common trait. [Glossary Convergence, Homoplasy, Mimicry versus convergence]

— Improve our understanding of embryologic development

As an example, amphioxus, also known as lancelet, is a cephalochordate, the sister class to Class Craniata. Amphioxus lacks a neural crest. Therefore, structures present in amphioxus that are homologous to structures in craniates could not have arisen purely through the action of the neural crest. By comparing amphioxus development with the development of craniates, we can learn a great deal about the specific role of the neural crest in craniate development [9].

— Serve as the comparison species for a molecular clock analysis determining the time at which the sister classes diverged. [Glossary Molecular clock]

When two classes split, the molecular clock is essentially reset, with new mutations arising in either class at a rate that is presumed to be fairly constant for homologous genes. By comparing the sequences in homologous genes in sister classes, we can determine whether a gene has been tightly conserved.

 — Determine the pathways operative in diseases

For example, it is common to find viruses that are present in the genome of a species or a class of animals, and absent from its sister class. To name a few, the PtERV retrovirus is not present in the human genome, but is present in Class Pan (chimpanzees and bonobos) and Class Gorillini (gorillas). Similarly, a lineage of rhadinovirus as well as a species of foamy virus are absent from humans but are found in closely related primates [10]. Contrariwise, HIV infections are devastating for humans but less so for several closely related primates [10]. In some cases, it may be shown that an effective antiviral gene was present in the species that is currently free of virus infection, and that the virus-infected sister species experienced degeneration of the antiviral gene, over time. In such cases, it may be possible to identify the effective antiviral gene in the resistant species, and to develop an effective antiviral treatment for every member of the nonresistant sister class [10].

 We tend to think of life in terms of individuals, not species. We should remember that species are living entities, and that species bear within them many of the deepest secrets of biology and medicine.

Speciation does not require newly acquired genetic mutations

 As remarked previously, when a group of individuals, each representing a sampling from the species' gene pool, wander off somewhere and mate exclusively with one another, they create their own evolving gene pool; hence their own species. The initial gene pool of the new species is a subset of the gene pool of the parent species, and the members of the new species are indistinguishable as a group from the members of the parent species. Hence, the creation of the new species does not require the acquisition of new genetic mutations.

 It's reasonable to ask, "If the new species is indistinguishable from the parent species, and has a gene pool that was present in the parent species, then how can we possibly assert that a new species was created?" Over time the gene pool of the new species will evolve, accumulating new variants of genes, and serving as the genetic material for individuals who will look less and less than the parent species. The greatest threat to the survival of a new species is cross-mating between members of the parent species and the child species; this results in a mixing of the genes, and we would no longer have two separately evolving gene pools.

 Just for fun, let's assume that we are wrong, and that each new species arises from a new gene that evolved from the parent species, rendering the offspring sufficiently different to earn themselves a place in the list of terrestrial species. If this were the case, then we should be able to identify the "squirrel" gene and the "tulip" gene and the "mosquito" gene that distinguishes each of these species from every other species on earth. Of course, this is an impossibility. We cannot identify the defining "species gene" or the set of genes that is characteristic for any species on earth. We cannot even find the "human" gene that separates us from gorillas. If our species were defined by the acquisition of one new gene, then a loss-of-function mutation in our "human gene" would cause affected individuals to regress back to our ancestral species. We would have proto-humans walking among us. This never happens. There simply is no such thing as the "human gene." What, then, distinguishes one species from another? The answer, is "the gene pool." Humans have their own gene pool. Squirrels have their gene pool. Tulips have their gene pool. It is as simple as that.

Section 4.3 The Diversity of Living Organisms

There can be only one.

Motto of the immortals in the fictional Highlander epic

Each organism's environment, for the most part, consists of other organisms.

Kevin Kelly

When we think about biological diversity, there is a tendency for each of us to contemplate the issue in relation to their chosen field of study. A zoologist is likely to think of diversity in the number and behavior of living animal species. A geneticist will consider the totality of different functional genes available to the biosphere. A chemist might think in terms of all the different molecular species synthesized by living organisms. Of course, the different modes of diversity are biologically related. For example, a species is basically an evolving gene pool, and each species will eventually contribute its gene pool to the number of genes available to bioinformaticians who search large genomic databases. Due to the overwhelming complexity of the various types of biological diversity, as they relate to one another, it is probably best to focus on the most familiar concepts, one by one: species diversity, genetic and proteomic diversity, regulatory diversity, structural (i.e., anatomic and cytologic) diversity, and chemical diversity [11].

Diversity of species

As noted previously, it is estimated that from 5 to 50 billion species have lived on earth [4], with somewhere between 10 and 100 million now inhabiting the planet. Perhaps the greatest number of species comes from the prokaryotes (bacteria plus archaea), which are estimated to have between 100 thousand and 10 million species. These estimates almost certainly underestimate the true number of prokaryotic species, as they are based on molecular techniques that would exclude valid species that happen to have sequence similarities to other species [12]. As an example of how methodology impacts numbers, samples of soil yield a few hundred different species per gram, based on culturing. If the species are counted on the basis of 16 s RNA gene sequencing, we find a few thousand different species of bacteria in each gram of soil. If we base the count on DNA-DNA reassociation kinetics, the number of different bacterial species, per gram of soil, rises to several million [13].

Let's not forget the viruses. Later in this chapter we'll be considering whether viruses qualify as living organisms. For now, let's agree that there are distinct species of viruses, with each species conforming to the aforementioned species definition as "an evolving gene pool." Viruses are plentiful in the oceans. It is estimated that there may be several hundred thousand marine species of viruses and phages [14]. The number of virus species that live in or infect mammals is estimated to be about 320,000 [15]. These are only estimates, but they indicate that viruses, despite their minimalist physicality, have accounted for themselves admirably [16].

The eukaryotes are estimated to have about 9 million species [17]. Adding up the estimates for prokaryotes, eukaryotes and viruses, we get a rough and conservative 10–20 million living species. To get an idea of the wide range of estimates, we must not overlook a 2016 study that estimates at least a trillion species of organisms on earth [18].

From this staggeringly large number of terrestrial species, we can infer that speciation is a relatively easy, almost inevitable, process. How does this large number of past and present species reconcile with Darwin's theory of natural selection through the survival of the fittest? Intuitively, wouldn't we anticipate that as numerous species compete for survival, there would be one species that attains maximum survivability, and that this species would displace all of the other species? Hence, we would expect that if we started with a very large number of species, then that number would diminish over time, until only one species would inhabit the planet. Anyone who has watched any of the Highlander movies or television shows is familiar with the concept.

Of course, this is utter nonsense, and we can see several reasons why species tend to diversify (i.e., increase in number) over time:

- –1. Species need other species. Aside from the many carnivorous organisms that prey on other living organisms, there are many organisms that live off of dead, or decaying, or fully decayed organisms. Even plants, that live off of sunlight and water and carbon dioxide, rely on soil nutrients containing the decomposed detritus of formerly living species.
- –2. The term "fittest" has no absolute meaning. An organism that is more fit for survival under one set of conditions may be totally unfit under another set of circumstances (i.e., extremes of weather, susceptibility to infection, diminished food supplies).
- –3. Third, and most importantly, species speciate. A species will always produce another species if the genetic and environmental conditions permit (as discussed in the earlier sections of this chapter).

This tendency to speciate is something that some classes of organisms do much better than others. Evolvability seems to be a trait of some classes of organisms. For example, there are over 350,000 species of beetle, exceeding the combined number of plant species (250,000) plus roundworm species (12,000) plus mammals (4000). Vertebrates are biological underachievers, compared with beetles, when it comes to speciation. It would seem that the ability to produce other species is itself a trait held by species, and this trait is referred to by the term "evolvability."

One of the most evolvable vertebrates is the cichlid, with about 2000 known species, and with hundreds of different species occupying some of the same African lakes. The hundreds of different species found in Lake Victoria took an estimated 15,000–100,000 years to radiate (i.e., to diversify from a single founder species) [19]. It is impossible to determine

FIG. 4.2 Long-finned Oscar, one of about 2000 known species of cichlid fish. *Source: Wikipedia, and entered into the public domain by Auerdoan.*

any specific factor that renders a species capable of diversifying. In the case of cichlids, nearly every variable examined in the geographic and ecological history of this fish would seem to encourage diversity (populations that expand and contract, lakes that swell up and dry out, changes in flora and fauna cohabiting the lake, etc.). On a molecular level, cichlids are endowed with a genome containing a large number of gene duplications, an abundance of noncoding elements, and many novel miRNAs [19]. Any of these factors may have played an important role in the adaptive radiation of cichlid species. This serves as an example of the relationship between genetic diversity and species diversity (Fig. 4.2).

Gene diversity

The earth's proteome consists of all the different protein-coding genes on the planet. The estimates of the planetary proteome vary widely. The lowest number seems to be 5 million [20]. Elsewhere, we read that the human intestine, alone, contains about 40,000 species of bacteria producing a whopping 9 million unique bacterial genes [21–23]. It seems plausible that before all the counts come in, we'll find that there are billions of genes in the total collection.

The human genome contributes a meager 20–25 thousand genes to the proteome. Other, seemingly less complex, animals have a larger genetic repertoire than humans. For example, the nearly microscopic crustacean *Daphnia pulex* (the water flea) has 31,000 genes. Plants tend to have way more genes than animals. For example, an unassuming grain of rice, contributes an estimated 46–56 thousand genes to earth's proteome [24].

Chemical diversity

We must begin by confessing that the metazoans, as a class, are metabolic underachievers in terms of chemical and metabolic diversity [25]. The prokaryotes are far more advanced [26]. Among the eukaryotes, only Class Archiplastida (i.e., plants) and Class

Fungi seem to be making any effort to impress [27, 28]. The eukaryotes largely rely on endosymbiotic relationships with current or former prokaryotes to perform complex biosyntheses (e.g., mitochondria and chloroplasts captured from former bacteria). Otherwise, eukaryotes are saddled with the rather humdrum tasks of synthesizing organelles and membranes and manipulating the various ingredients for cellular life. An awful lot of the metabolic effort in eukaryotes is devoted to the eternal activities of eating and digesting food. The fundamental chemical constituents of living organisms were established about 3 or 4 billion years ago, and have not changed much since. DNA, RNA, proteins, carbohydrates, lipids, and the biological machinery for their manufacture and modification were popular molecules 2 billion years before the Cambrian explosion.

We do not know when the first fungi and plants evolved. Some estimates suggest that the first fungi appeared as early as 1.3 billion years ago, while the first land plants may have evolved 700 million years ago. The first fossils of modern vascular land plants are dated to about 480 million years ago, just after the end of the Cambrian explosion (about 500 million years ago). Regardless of the timing, we can surmise that following the Cambrian explosion, plants, fungi, and metazoans were obliged to cohabitate with diverse classes of organisms.

Whereas animals rely on their body structure for both aggressive and defensive activities (e.g., running after prey and running away from predators), plants and fungi rely on their ability to synthesize bioactive chemicals that act as respiratory poisons (e.g., cyanide), neuromuscular agents (e.g., nicotine), irritants (e.g., capsaicin), and a host of other chemical warfare tactics [28]. Plants devote 15%–25% of their genes to producing enzymes involved in the synthesis of secondary metabolites (i.e., synthesized molecules that play no role in the primary functions of plant cells). Several hundred thousand secondary plant metabolites have been reported [29]. We presume that every secondary metabolite is bioactive under some set of circumstances; otherwise, the synthetic method for creating the chemical could not have evolved. In point of fact, a good bit of medicinal chemistry, as it was pursued in the 20th century, consisted of finding appropriate secondary metabolites from bacteria, fungi or plants that would have some utility in the prevention or treatment of human diseases.

When we eat plants and mushrooms, we can expect to ingest a bit of poison. For example, cycasin is a toxin and carcinogen found in the seeds and the pollen of every class of cycad tree [30]. Among the fungi, *Aspergillus flavus*, a ubiquitous fungus found growing on peanuts and other crop plants, synthesizes aflatoxin, one of the most powerful liver carcinogens known [31]. Peanut butter manufacturers take great pains to insure that peanuts are harvested under conditions that minimize their contamination with aflatoxin, and they measure the amounts of aflatoxin in manufactured peanut butter to ensure that batches exceeding a legal limit will never reach the market. Fungi have evolved their own chemical arsenals. For example, alpha-amanitin, a strong, often fatal hepatotoxin, is produced by the mushroom *Amanita phalloides*. Another fungal product is gyromitrin, a hydrazine compound present in edible mushrooms, including most members of the common False Morel genus. Nobody is quite sure what effects gyromitrin and other related hydrazine molecules may have on human consumers, but we eat mushrooms just the same.

Structural diversity

The one area of diversity wherein animals take the lead is structural diversity. Structural diversity took off in the Cambrian explosion, when nearly all the extant metazoan body plans were established. Although there were metazoans living prior to the Cambrian, it would seem that the dominant eukaryotes at that time were standard issue unicellular organisms. These organisms varied in terms of size and shape and external structures (e.g., wavy membranes, pseudopods, undulipodia, and cilia), but they couldn't compete with the diversity of structures that arose in the Cambrian explosion.

The key acquisition that propelled the attainment of structural diversity in metazoans was almost certainly the evolution of specialized junctions, particularly the desmosome that uniquely characterize animal cells. As previously discussed, desmosomes act like rivets, and soft animal cells serve as somewhat modular building blocks that can be fastened together into almost any shape and size imaginable (e.g., ducts, glands, acini, and layers). The application of cuticles from keratin, or chitin, and bone from hydroxyapetite deposited into collagenous matrices, allowed the animal kingdom to produce a multitude of species with variably shaped internal and external structures.

Among the animals, the holometabolous insects probably take the prize when it comes to structural diversity, insofar as a single organism may pass through multiple stages of life, each having a its own structural morphology. [Glossary Holometabolism]

Fungi and plants display a fair amount of structural diversity; usually simple variations on common themes (e.g., stalk, leaves, and flowers). With few exceptions, the plant kingdom does a much better job with colors than does the animal kingdom. Plants rely on flavonoids (particularly anthocyanins) and carotenoids to produce their vivid colors. For the most part, animals have a single pigment molecule, melanin, as their primary source of coloration. An assortment of colors can be coaxed from the melanin molecule by controlling the concentration and spatial distribution of melanin within cells, and by making small modifications to the base molecule. Other colors, such as the red of hemoglobin, are produced with iron and other metal cofactors bound to proteins. [Glossary Cofactor]

Regulatory diversity

Relatively early in eukaryotic phylogeny, cells evolved a diverse methodology for regulating their genomes. This would include the evolution of the epigenome, wherein the DNA sequence of genetic material is modified by base methylations, and these modifications are themselves remodified at every step of cell-type development.

Chromatin, the structural backbone of the genome is modified by the attachment of proteins (histones and nonhistone varieties), and by the wrapping of units of DNA into tight nucleosomes. There are numerous ways in which chromatin is modified [32]. [Glossary Nucleosome]

Aside from the complexities of the epigenome, there are a host of genome modifiers that were discussed in Section 3.5, "Pathologic Conditions of the Genomic Regulatory Systems." These genomic regulators will not be discussed further here except to say that some level of gene regulation is found in every class of organisms (i.e., prokaryotes, single-celled eukaryotes, animals, plants, fungi, and viruses). Many of these regulatory systems are common to all eukaryotes, while others seem to be specific for particular subclasses. For example, imprinting among animals seems to be confined to eutherians. Similarly, chaperone proteins seems to be somewhat exclusive to mammals.

What does all this mean?

We are taught in school that evolution brings us greater and greater fitness, producing organisms that are more complex and more functional over time. This is why there are humans rocketing into space today, and why there is no (convincing) evidence of rocket scientists living in the Jurassic period. Actually, though, evolution is more of a random, almost Brownian movement, lurching toward organisms that are more fit for some immediate threat to the species, but possibly less fit for some future challenge.

The natural process that accounts for the vitality and resilience of our biosphere, more so than evolution, is diversification. Because successful species always speciate, we can expect the number of species on the planet to continually increase, until such time as we face the next global extinction event. When that day comes, and it may be quite soon, the factor that will preserve life on earth will not be evolution. We cannot evolve our way out of the consequences of a nuclear catastrophe or a moon-sized meteoric collision, or a 4° rise in temperature occurring in the span of a human lifetime. The process that will preserve life on earth will be the diversification of species. Of the hundreds of millions of species on earth, the odds are that some of them will have the genetic wherewithal to survive, when the other species die.

Of greater immediate practicality, species diversity affords us the opportunity to find new antibiotics, new anticancer agents, and new methods to control and modify just about any metabolic pathway or regulatory mechanism we choose to study. It is due to genetic diversity among many species that we have found the thermophilic taq polymerase (from *Thermus aquaticus* bacteria) used in PCR (polymerase chain reactions) and the gene-editing enzymes used in CRISPR/Cas9 (prokaryotic species) and Cre-LoxP (bacteriophage P1), and CAR-T (lentiviral and gammaretroviral vectors) [33–36]. [Glossary CAR T-cell therapy]

Section 4.4 The Species Paradox

It is once again the vexing problem of identity within variety; without a solution to this disturbing problem there can be no system, no classification.

Roman Jakobson

A horse is a horse of course of course

Mr. Ed television show, theme music, 1961–66

In the introductory section of this chapter, we defined species as an evolving gene pool. Does this simple definition account for what we observe whenever we try to identify a species? Not really. When we study a species, we never look at the gene pool. Instead, we collect and inspect a bunch of individual organisms that are members of the species. From these organisms, we try to find the features that characterize all the members of the species and that distinguishes the species from every other species. When doing so, we always make the same observation: that every individual member of a species is a unique organism and that every offspring of every organism is uniquely different from either of its parents.

Here is the apparent paradox. We observe that every species is a collection of organisms that are all different from one another; and the differences among the organisms are constantly reassorted into new and unique individuals, with every generation. With its members constantly changing, how can we ever have a species that has stable characteristics that distinguish the species from other species? How can we uphold the intransitive property of species that forbids individuals to change their membership from one species to another, when new species are constantly evolving from existing species? It doesn't seem to make any sense! [Glossary Intransitive property]

The solution to the paradox is very simple. Every species is defined by its ancestral lineage; not by the collective diversity of the individuals within the species. Two individuals belong to the same species if they share the same ancestry; regardless of differences in their genomes.

For example, anyone who visits the produce department of a well-stocked grocery store will encounter varieties of *Brassica oleracea*, each producing a different menu item:

- *Brassica oleracea* Acephala Group—kale and collard greens.
- *Brassica oleracea* Alboglabra Group—kai-lan (Chinese broccoli).
- *Brassica oleracea* Botrytis Group—cauliflower, Romanesco broccoli, and broccoflower.
- *Brassica oleracea* Capitata Group—cabbage.
- *Brassica oleracea* Gemmifera Group—brussels sprouts.
- *Brassica oleracea* Gongylodes Group—kohlrabi.
- *Brassica oleracea* Italica Group—broccoli.

The Acephala group (from the root meaning without a head) represented by kale and collard greens is phenotypically most like the wild cabbage. Cauliflower differs from wild cabbage because of a mutation in a single gene (the CAL gene) which produces an inflorescence. This means that the stem cells (of the meristem) grow into a mass of undifferentiated cells; basically a Brassica hamartoma. [Glossary Hamartoma, Wild-type]

One of the many cultivars of *Brassica oleracea*, known as Jersey cabbage, grows up to three meters tall. These giant-sized cabbages have woody stalks that look like tree limbs. Hence, a single species of plant can provide nearly every common green vegetable that appears on American dinner plates, as well as stalks suitable as walking canes (Fig. 4.3).

The different cultivars of *Brassica oleracea* are analogous to serotypes of bacteria or breeds of animals. They all represent variants of the same species. Although the cultivars may be distinguished from one another by simple genetic variations, sometimes involving a single gene, they all belong to the same species because they all share the same ancestry and can be interbred. [Glossary Serotype]

We can glimpse some of the enormous genetic variation within the members of a species by focusing our attention on single nucleotide polymorphisms (SNPs), which are easy

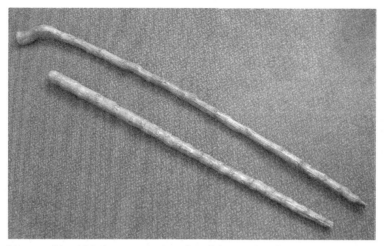

FIG. 4.3 Jersey cabbage walking sticks from another member of the *Brassica oleracea* species. *Source: Wikipedia, from a photography by Man Vyi and entered into the public domain.*

to detect and for which much data has been collected. A SNP is a nucleotide variation between members of a species that occurs in at least 1% of the population. To get an accurate determination of all the SNPs in the human population, we would need to sequence everyone's genome. Sequencing everyone's genome is impossible, at present, but we can do our best to get a fair sampling of the human population. The current rough estimate of the number of SNPs in the human population is 10–30 million. This number may increase substantially, as we improve the gene sampling process, and it is fair to assume that the number would be much higher if we included rare polymorphisms occurring in less than 1% of the population. **It is startling to acknowledge that despite the enormous genetic variations among individuals, we all belong to the same species, we all descended from the same parent species, and we all share the same evolving gene pool.**

Section 4.5 Viruses and the Meaning of Life

The question whether or not "viruses are alive" has caused considerable debate over many years. Yet, the question is effectively without substance because the answer depends entirely on the definition of life or the state of "being alive" that is bound to be arbitrary.

Eugene V. Koonin and Petro Starokadomskyy [37]

The origin of retroviruses is lost in a prebiotic mist.

Patric Jern, Goran Sperber, Jonas Blomberg [38]

In my prior work, "Taxonomic Guide to Infectious Diseases: Understanding the biologic classes of pathogenic organisms," published in 2012, we find the following text: "In this book, viruses and prions are referred to as "biological agents"; not as living organisms.

Viruses lack key features that distinguish life from nonlife. They depend entirely on host cells for replication; they do not partake in metabolism and do not yield energy; they neither adjust to changes in their environment (i.e., no homeostasis), nor can they respond to stimuli. Most scientists consider viruses to be mobile genetic elements that travel between cells (much as transposons are mobile genetic elements that travel within a cell)" [39]. [Glossary Prion disease]

For nonliving organisms dependent entirely on host cells for their continued existence, viruses have done extremely well for themselves. Every class of living organism hosts viruses. Viruses are literally everywhere in our environment and are the most abundant life form in the oceans, in terms of numbers of organisms [14]. At least 8% of the human genome is composed of fragments of RNA viruses, acquired in our genetic past [10, 40, 41]. As far as anyone knows, viruses are as ancient as the earliest forms of terrestrial life.

Over the past 7 years, the evidence and the arguments supporting viruses as living organisms have won my support. Consider the following points:

Viruses replicate and create virocells

Viruses employ the replicative machinery that they find in host cells, manufacturing viral assembly plants in the process. After infection, host cells forego many of their normal functions, and the newly synthesized viral products may cause cell death (so-called viral cytolytic effect) or may produce disturbances in cellular physiology (so-called cytopathic effects). In either case, the infected host cells become something more akin to viral factories than to eukaryotic organisms, and these new living entities are referred to as virocells, indicating that the virus has created its own form of cellular life. [Glossary Host]

Viruses exhibit chemical, structural, and physiological diversity

Viruses display great diversity in their range of hosts, habitat, size, genes, mechanisms of infection, and capsids. Unlike eukaryotes, viruses have a range of genomic types, which include: double-stranded DNA; single-stranded DNA; double-stranded RNA; positive sense single-stranded RNA; negative sense single-stranded RNA, single-stranded RNA with a reverse transcriptase, and double-stranded DNA with a reverse transcriptase. Viruses of each type breed true, meaning that a viruses with a double-stranded DNA genome will replicate to produce another virus with a double-stranded DNA genome. Thus, the types of viruses, determined by their genomic structure and replicative modes, represent biological classes.

Viruses do not lead a totally intracellular life

Viruses are accused of having no life outside of their host cells. It has long been assumed that extracellular viral existence is relegated to long-term storage: a lifeless viral genome double-wrapped in a protein capsid, and an envelope that is mostly purloined from the cell membranes of its former host.

It has been recently demonstrated that some viruses enjoy life outside of their hosts. The Acidianus Tailed Virus can be cold-stored at room temperature for long periods, without changing their morphology. When the temperature rises they undergo a structural transformation, forming bipolar tails [42]. This tells us that viruses can react to external stimuli and perform biological activities outside their hosts, much as any free-living organisms does.

Viruses evolve and speciate

Viruses lack the elaborate set of DNA repair mechanisms observed in eukaryotes and prokaryotes. Consequently, the rates of viral mutation are very high, and the genetic variants of viruses occur frequently. We have noted that viruses can be classified by their genomic type and that these genomic types replicate true. If the individual members of classes of viruses are mutating, and if viruses replicate to produce additional viruses of the same type (e.g., same class and same morphologic traits), then the resulting generations of viruses seem to meet the basic definition of a species; that is to say, they are evolving gene pools.

Aside from the basic definition of a species as an evolving gene pool, we have come to expect some degree of stability in a species. In the case of viruses, we can easily imagine that the rate of evolution and speciation might exceed the rate at which we can isolate, study, and describe any new species. As it happens, viral species have stability, and much of the stability of viral species is imposed by host organisms. How so? All viruses must replicate within a host, and the process by which a eukaryotic cell is infected and transformed into a virocell is very complex. Viruses are forced to adapt some level of host preference, and must fine-tune all their functions to the available conditions within the preferred host [43]. Thus, for viruses to successfully replicate, they must evolve into relatively stable species. And this is what we observe when we study viruses. [Glossary Evolvability]

Viruses participate in the evolution of eukaryotic organisms

As previously noted, some of the greatest evolutionary leaps among eukaryotic organisms can be attributed to genes taken from non-eukaryotes (e.g., mitochondria and chloroplasts). The viruses have also contributed to eukaryotic evolution, and it's fair to say that if it were not for viral genes, you would not be reading this book (it would be left to some octopus or insect to finish out the chapter in your stead).

Specifically, at least three major metazoan advancements can be attributed to retroviruses:

1. The acquisition of adaptive immunity, beginning in Class Gnathostomata (to be described in detail in Section 6.2, "Vertebrates to Synapsids" [44]. [Glossary Adaptive immunity]

2. The acquisition of the placenta in Class Theria (to be described in detail in Section 6.3, "Mammals to Therians").

3. The maintenance of pluripotent stem cells, important in embryogenesis, is achieved, in part, with transcripts of stem cell specific long terminal repeat retrotransposons, themselves derived from retroviral sequences [45, 46]. [Glossary Transposon]

4. In Class Mus (containing the common house mouse), an MuERV-L retrovirus capsid gene was domesticated to code an antiviral factor [47, 48].

We should note that the adaptation of retroviruses does not always have purely beneficial consequences for the individual members of a species. In Section 7.5, "Why Good People Get Bad Diseases," we will see that oncogenes, many of which are of retroviral origin, may have served a beneficial role for the species, while playing a deleterious role for some individuals of the species (i.e., by causing cancer). One particularly striking example occurs in a polycythemia strain of mice carrying the Friend murine leukemia virus. All of the mice of this particular strain have polycythemia, a condition characterized by a sustained increase in the number of circulating red blood cells. It seems that the resident virus codes for a nonfunctional Env (viral envelope) gene. The protein coded by the degenerate Env gene serves as a mimic for the normal erythropoietin gene of the mouse, producing an increase in red cell production (i.e., polycythemia) [49, 50].

Participating in eukaryotic evolution, and accounting in no small way for the generation of new metazoan species, as viruses certainly do, is itself a biological function and another indicator of viral life.

Virus species speciate, yielding phylogenetic lineages (i.e., descendant families)

We can infer that viruses speciate by observing that the mutation rate in viruses is very high (i.e., they have a mechanism whereby new genetic species may develop from existing species). We also know that the number of viral species is enormous, indicating that the process that produces new viral species must be efficient and prolific.

We do not know the number of different viral species on earth, but as of 2017, the ICTV (International Committee on Taxonomy of Viruses) has recognized 9 orders, 131 families, 46 subfamilies, 803 genera, and 4853 different viral species [51]. The number of known viral species (4853) is thought to represent a tiny fraction of the total number, there being an estimated 320,000 viral species in mammals [15]. The majority of viral species seem to reside in the oceans [14, 52]. The definition of "species" as used by ICTV, indicates that viral species emerge from a replicating lineage, thus inferring that species speciate.

Highly innovative work in the field of viral phylogeny is proceeding, using a variety of different approaches, including: inferring retroviral phylogeny by sequence divergences of nucleic acids and proteins in related viral species [38]; tracing the acquisition of genes in DNA viruses [53]; and dating viruses by the appearance of antiviral genes in ancient host cells [10]. As viruses evolve very rapidly, it is possible to trace the evolution of some viruses, with precision, over intervals as short as centuries or even decades [54–57].

Relatively recently we have seen the discovery of complex viruses known as nucleocytoplasmic large DNA viruses, popularly known as giant viruses (NCLDVs). The life of an NCLDV is not much different from that of obligate intracellular bacteria (e.g., Rickettsia). The NCLDVs, with their large genomes and complex sets of genes, have provided taxonomists with an opportunity to establish ancestral lineages among some of these viruses [58, 59].

Life is what we choose to make of it

Are viruses living organisms, or are they simply sequences of nucleic acid that have the wherewithal to move from cell to cell. Much of the debate focuses around the definition of life [37, 60]. How we define life is somewhat arbitrary, and we can certainly produce a definition of life that includes the viruses, if that is what we choose. There's no urgency to the matter. In a few years or decades, after our personal robots become self-replicating, we'll need to address the definition of life anew. When that time comes, perhaps robots will take the lead and create a definition of life that includes both organic viruses and software viruses.

Who is the better class of organism: virus or human?

For a moment, let's pretend that we are viruses [61]. As viruses, what would we think of humans and other eukaryotic organisms? Would we be willing to accept humans as fellow living organisms, or would we point to the following list of disqualifiers to conclude that humans just don't make the grade?

–1. Humans cannot replicate (they merely procreate).

In an effort to strengthen their species' gene pool, humans undergo a strange mating process in which the chromosomes of both parents are hopelessly jumbled together to produce an offspring that is unique, and unlike either the father or the mother. In doing so, humans miss out on the replicative process, performed with the greatest enthusiasm by every virus.

Self-replication is one of the fundamental features of life. As humans cannot replicate themselves, they barely qualify as living organisms. At least, this is what the viruses think.

–2. Humans do not react in a manner that preserves the survival of their species.

Humans love to attack other humans. Even now, they are busy developing new and more powerful weapons for the purpose of wiping out the human race. Viruses respect each other's right to live.

–3. Humans rely on viruses for their existence.

Viruses are constantly donating DNA to humans, and humans have used this DNA to evolve [10]. In point of fact, if there were no viruses, there would be no DNA replication, no adaptive immune system, no placentas, and no humans [62].

We also see that viral species acquire genes from their hosts, and that viruses retain these genes as they evolve. Presumably, the acquisition of host genes confers some survival advantage upon the virus. Nonetheless, viruses are a self-sufficient class of organism, and no specific instance comes to mind wherein a viral species depended on a human gene for its survival.

–4. Humans may have descended from viruses.

Nobody really knows much about the earliest forms of life, and there is plenty of room for conjecture. It's quite feasible that the earliest genetic material consisted of sequences of RNA, and that these RNA molecules moved between the earliest forms of cells (discussed in Section 1.2, "Bootstrapping Paradoxes") If this were the case, then the earliest genomes were essentially RNA viruses, and this would place humans as direct, but distant descendants of viruses.

As we will describe in Section 5.3, "Eukaryotes to Obazoans," there is a current theory, among many competing theories, that the first eukaryotic nucleus was a giant virus that was not totally successful in transforming its proto-eukaryotic host into a virocell

[61, 63, 64]. A hybrid giant virus/virocell/proto-eukaryote may have stabilized and replicated to form an early, nucleated cell; the first eukaryote.

  –5. Humans serve viruses; not vice versa.

We are taught to think of viruses as fragments of nucleic acids wrapped by a capsid. To a virus, extracellular existence must be akin to a state of suspended animation. Viruses come to life when they invade a eukaryotic cell and create a virocell; a living organism consisting of the hijacked eukaryotic cell whose nuclear machinery is redirected to synthesize viral progeny. If every eukaryotic cell is conceptualized as a potential virocell, then every eukaryotic species is a potential slave owned by the viral kingdom.

Glossary

Adaptive immunity Immunity in which the response adapts to the specific chemical properties of foreign antigens. Adaptive immunity is a system wherein somatic T cells and B cells are produced, each with a unique and characteristic immunoglobulin (in the case of B cells) or T-cell receptor (in the case of T cells). Through a complex presentation and selection system, a foreign antigen elicits the replication of a B cell that produces an antibody whose unique immunoglobulin attachment site matches the antigen. The antigen-antibody complexes may deactivate and clear circulating antibodies, or may lead to the destruction of the organism that carries the antigen (e.g., virus or bacteria).

   The process of producing unique proteins requires that recombination and hypermutation take place within a specific gene region. Recombinations yield on the order of about a billion unique somatic genes, starting with one germinal genome. This process requires the participation of recombination activating genes (RAGs). The acquisition of an immunologically active RAG, from a retrovirus, is presumed to be the key evolutionary event that led to the development of the adaptive immune system.

   This event, which occurred in one of the early species of gnathostomes, established the adaptive immune system in all jawed vertebrates and their descendants. As one might expect, inherited mutations in RAG genes cause immune deficiency syndromes [65, 66].

CAR T-cell therapy In August 2017, a gene therapy for the treatment of children and young adults with B-cell acute lymphoblastic leukemia was approved by the US Food and Drug Administration [67]. The therapy is centered on several stunning refinements of an old and nearly defunct approach, using a patient's own immune system to destroy cancer cells. The successful methodology that was developed is known as CAR-T (Chimeric Antigen Receptor for T cells). Here are the steps involved in deploying CAR-T [68]:

  –1. Choose some antigen in the disease-causing cells that is present in high concentration in those cells or that is unique to those cells. It is best if the antigen lies on the cell surface, where it can easily bind to T-cell receptors. In cancers, the target cells for this procedure will be the cancer cells themselves. In the case of CAR-T therapy for B-cell acute lymphoblastic leukemia, the target antigen is CD-19, a B-cell surface antigen. Researchers are hoping that this technique can be applied to diseases other than cancer [69]. Perhaps, in the future, the target cells may be the chief effector cells within many different types of lesions (e.g., parasitic infections).

  –2. Create the chimeric receptor to the antigen. This is the step that awaited several advances in genetic engineering [36]. Basically, the different units of a receptor designed to provide the greatest possible biological response, when bound to an antigen, are combined with a unique antigen-recognizing subunit. The final product is an artificial receptor molecule capable of arming T cells.

  –3. Determine the gene sequence that encodes the artificial receptor, and synthesize it in vast number.

—4.   Enclose copies of the synthesized gene in an appropriate vector (e.g., lentivirus).

—5.   Extract a sample of T cells from the patient.

—6.   Transfect those T cells with the vector.

—7.   Grow the transfected T cells, now expressing the chimeric receptor, in culture.

—8.   Deplete the patient of his or her own T cell population, to "make room" for the cultured T cells that it will soon receive.

—9.   Transfuse the cultured and transfected T cells into the patient.

—10.   Wait as the transfused T cell recognize and destroy the patient's cancer cells.

—11.   Watch to see if the patient develops any adverse events, and treat complications aggressively [70].

In the case of B-cell acute lymphoblastic leukemia, a rare form of cancer, the results have been nothing short of miraculous. In early trials, complete and lasting remissions have been achieved in more than two-thirds of treated patients [69]. As is so often the case, early successes come to the rare diseases. The common diseases come later, if at all. At present, CAR-T therapy is not particularly effective against common solid tumors, possibly due to the enormous phenotypic heterogeneity of epithelial tumors, and the emergence of cells that lack the target antigen, following an initial response to the treatment [69].

Cofactor When biochemists use the term "cofactors," they are referring to chemicals that bind to enzymes, to activate the enzyme or to enhance the activity of the enzyme. Some enzymes or enzyme complexes need several cofactors (e.g., the pyruvate dehydrogenase complex which has five organic cofactors and one metal ion). Vitamins are often cofactors for enzymes.

Convergence When two species independently acquire an identical or similar trait through adaptation; not through inheritance from a shared ancestor. Examples are: the wing of a bat and the wing of a bird; the opposable thumb of opossums and of primates; the beak of a platypus and the beak of a bird. As applied to diseases, convergence occurs when diverse pathogenetic sequences all lead to the same clinical phenotype. Convergence accounts for all of the common diseases, because a disease with many convergent pathways is apt to occur often.

In the case of systemic responses to injury, convergence clearly has an evolutionary origin. In this instance, the species has evolved to respond in an orchestrated way to a variety of pathologic stimuli (e.g., Systemic Inflammatory Response Syndrome [71]). Convergence is also observed in rare diseases that have genetic heterogeneity. In these instances, different mutated genes may lead to the same clinical phenotype (e.g., multiple causes for epidermolysis bullosa, retinitis pigmentosa, long QT syndrome).

The biologic basis for convergence is that there are a limited number of ways in which an organism can respond to a biological perturbation. By this reasoning, we would expect that there would be a limited number of clinical phenotypes. Furthermore, if the number of disease phenotypes is constant for any species, then this would imply that the human race cannot encounter any new diseases. No matter what may happen to a human (e.g., nibbled by 10,000 ducks, attacked by an alien virus), the unfortunate aftermath would be one of a finite number of known clinical phenotypes.

Evolvability Evolution by natural selection is not a physiological process, like respiration or replication. The theory of evolution is little more than a restatement of a probabilistic truism, in biological terms. Namely, organisms that are most likely to survive will be the organisms most likely to reproduce. Biologists speak in terms of evolvability, to indicate certain factors that may tip the evolutionary scales in an organism's favor.

For the most part, these features all involve mechanisms by which new or modified genetic material may serve as the source of new genes, is obtained. These would include:

—   Having mechanisms for horizontal gene transfer.

—   Having mechanisms for increasing the rate of mutation under environmentally stressful circumstances (e.g., radiation, heat, cold).

—   Tendency toward endoduplication of genes.

—   Having large, diverse gene pool.

—   Presence of pseudogenes and junk DNA.

Hamartoma   Hamartomas are benign growths that occupy a peculiar zone lying between neoplasia (i.e., a clonal expansion of an abnormal cell) and hyperplasia (i.e., the localized overgrowth of a tissue). Some hamartomas are composed of tissues derived from several embryonic lineages (e.g., ectodermal tissues mixed with mesenchymal tissue). This is almost never the case in cancers, which are clonally derived neoplasms wherein every cell is derived from a single cell type.

Hamartomas occasionally occur in abundance in inherited syndromes; as in tuberous sclerosis. The pathognomonic lesion in tuberous sclerosis it the brain tuber, the hamartoma from which the syndrome takes its name. Tubers of the brain consist of localized but poorly demarcated malformations of neuronal and glial cells. Like other hamartoma syndromes, the germline mutation in tuberous sclerosis produces benign hamartomas as well as carcinomas; indicating that hamartomas and cancers are biologically related. Hamartomas and cancers associated with tuberous sclerosis include cortical tubers of brain, retinal astrocytoma, cardiac rhabdomyoma, lymphangiomyomatosis (very rarely), facial angiofibroma, white ash leaf-shaped macules, subcutaneous nodules, cafe-au-lait spots, subungual fibromata, myocardial rhabdomyoma, multiple bilateral renal angiomyolipoma, ependymoma, renal carcinoma, and subependymal giant cell astrocytoma [72].

Another genetic condition associated with hamartomas is Cowden syndrome. Cowden syndrome is associated with a loss of function mutation in PTEN, a tumor suppressor gene [73]. Features that may be encountered are macrocephaly, intestinal hamartomatous polyps, benign hamartomatous skin tumors (multiple trichilemmomas, papillomatous papules, and acral keratoses), dysplastic gangliocytoma of the cerebellum, and a predisposition to cancers of the breast, thyroid and endometrium [74–76].

Holometabolism   Complete metamorphosis, as observed in all insect species of Class Endopterygota, involving four developmental stages: egg, larva, pupa, and imago or adult. The term "holometabolism" is reserved for insects, but we see complete multistage metamorphosis in other animals, such as the European eel, a type of fish.

Homoplasy   A feature found in a species that was not present in the ancestor. It is generally believed that species develop homoplasies when none of the features inherited from ancestors happens to fulfill their needs [77]. A homoplasy may appear in a skipwise manner in different subclasses of species belonging to the same class of animals, depending on their survival requirements. An example of a common homoplasy in various insect species is the loss of wings. Species of some subclasses of insects have lost their wings, while most other classes of insects have retained their ancestral wings.

Host   The animal in which an infectious organism resides.

Intransitive property   One of the criteria for a classification is that every member belongs to exactly one class. From this criteria comes the intransitive property of classifications; namely, an object cannot change its class. Otherwise, over time, an object would belong to more than one class. It is easy to apply the intransitive rule under most circumstances. A cat cannot become a dog and a horse cannot become a sheep. What do we do when a caterpillar becomes a butterfly? In this case, we must recognize that caterpillar and butterfly represent phases in the development of one particular member of a species, and we do not create separate caterpillar classes or butterfly classes.

Mimicry vs convergence   In nature, there are many instances when two different organisms may look alike. Often, the species that mimics another species gains a survival advantage. For example, a predator may ignore a butterfly if the butterfly is mistaken for a leaf, and a small insect may ignore a predatory larger insect that appears to be a twig.

Evolutionary convergence is different from mimicry. In convergence, organisms of different classes evolve the same morphologic or biochemical traits independently (i.e., through different evolutionary lineages). One example is found in new world vultures and old world vultures, which look similar to one another and which both feed on carrion, but the old world vultures are related to eagles and hawks, while the new world vultures are related to storks. An example of chemical convergence is found in syntheses by elephants and butterflies of a compound [(Z)-7-dodecen-1-yl acetate] which serves as a pheromone for both species.

Molecular clock The molecular clock is a metaphor describing an analytic method by which the age of phylogenetic divergence of two species can be estimated by comparing the differences in sequence between two homologous genes or proteins. The name "molecular clock" and the basic theory underlying the method were described in the early years of the 1960s, when the amino acid sequence of hemoglobin molecule was determined for humans and other hominids [78]. It seemed clear enough at the time that if the number of amino acid substitutions in the hemoglobin sequence, compared among two species, was large, then a very great time must have elapsed since the phyogenetic divergence of the two species. The reason being that sequence changes occur randomly over time, and as more time passes, more substitutions will occur. Conversely, if the differences in amino acid sequence between species is very small, then the time elapsed between the species divergence must have been small.

As with all simple and elegant theories in the biological sciences, the devil lies in the details. Today, we know that analyses must take into account the presence or absence of conserved regions (whose sequences will not change very much over time). Indeed analysts must apply a host of adjustments before they can claim to have a fairly calibrated molecular clock [79, 80]. At the end of the process, biologists have additional information related to the timing of species divergences, possibly corroborating the prior chronology, or tentatively establishing new timelines.

Nucleosome The basic unit of DNA packaging in eukaryotes. A nucleosome consists of a length of DNA wound around a histone protein core.

Prion disease The term prion was introduced in 1982, by Stanley Prusiner [81]. Prions are the only infectious agent that contain neither DNA or RNA. Though few scientists would consider prions to be organisms, living or otherwise, they are included here to ensure that readers are aware of these biological agents. There are five known prion diseases of humans, and all of them produce encephalopathies characterized by decreasing cognitive ability and impaired motor coordination.

At present, all of the human prion disease are progressive and fatal, and comprise the following: Kuru prion (Kuru); CJD prion (Creutzfeldt-Jakob disease, CJD); nvCJD prion, or vCJD prion, or Bovine Spongiform Encephalopathy prion (New variant Creutzfeldt-Jakob disease, vCJD, nvCJD); GSS prion (Gerstmann-Straussler-Scheinker syndrome, GSS); and FFI prion (Fatal familial insomnia, FFI) [39].

Prions are not confined to mammals. They have been observed in fungi, where their accumulation does not seem to produce any deleterious effect, and may even be advantageous to the organism [82].

Serotype Same as serovar. A subtype of a species of bacteria or virus that is distinguished by its surface antigens.

Sister class and cousin class Alternately known as sister clade, sister group, and sister taxon. Two classes are sisters to one another if they have the same parent class. Similarly, two classes are cousins to one another if they have the same grandparent class. By custom, any two classes are considered cousins if they have in common any relatively recent class above the parent class. For example, humans and lungfish are considered cousins because we have a relatively recent ancestral class, Dipnotetrapodomorpha (also known as Rhipidistia).

Transposon Also called transposable element, and informally known as jumping gene. The name "transposable element" would seem to imply that a fragment of the genome (i.e., the transposable element) physically moves from one point in the genome to another. This is not the case. What actually happens (in the case of Class II transposons) is that a copy of the DNA sequence of the transposon is inserted elsewhere in the genome, resulting in the sequence now occupying two different locations in the genome. In the case of Class II transposons, the DNA sequence of the transposon is translated into RNA, then reverse-transcribed as DNA, and reinserted at another location; likewise resulting in two of the same sequence in two locations in the genome [83]. You can see how transposable elements might bloat the genome with repeated elements. Some transposons are the ancient remnants of retroviruses and other horizontally transferred genes that insinuated their way into the eukaryotic genome. Because transposon DNA is not necessary for cell survival, the sequences of transposons are not conserved, and mutations occurring over time yield degenerate sequences that no longer function as retroviruses. As luck would have it, not all mutations to transposable elements are without benefit to the host cell. A transposon is credited with the acquisition of adaptive immunity in animals. The RAG1 gene was acquired as a transposon. This gene enabled the DNA that encodes a segment of the immunoglobulin molecule to rearrange, thus producing a vast array of protein variants [44]. A role for transposons in the altered expression of genes in cancer cells has been suggested [84].

Wild-type The common, nonmutated version of a gene, occurring naturally in a population.

References

[1] DeQueiroz K. Ernst Mayr and the modern concept of species. PNAS 2005;102(Suppl. 1):6600–7.

[2] DeQueiroz K. Species concepts and species delimitation. Syst Biol 2007;56:879–86.

[3] Mayden RL. Consilience and a hierarchy of species concepts: advances toward closure on the species puzzle. J Nematol 1999;31:95–116.

[4] Raup DM. A kill curve for Phanerozoic marine species. Paleobiology 1991;17:37–48.

[5] Simpson GG. The principles of classification and a classification of mammals. Bull Am Mus Nat Hist 1945;85:1–350.

[6] Kamaruzzaman BY, Akbar JB, Zaleha K, Jalal KCA. Molecular phylogeny of horseshoe crab. Asian J Biochem 2011;3:302–9.

[7] Cande J, Andolfatto P, Prud'homme B, Stern DL, Gompel N. Evolution of multiple additive loci caused divergence between *Drosophila yakuba* and *D. santomea* in wing rowing during male courtship. PLoS ONE 2012;7:e43888.

[8] Pillon Y. Biodiversite, origine et evolution des Cunoniaceae. In: Implications Pour la Conservation de la Flore de Nouvelle-Caledonie. University of New Caledonia; 2008 [Ph.D. thesis].

[9] Holland LZ, Laudet The chordate amphioxus: an emerging model organism for developmental biology. Cell Mol Life Sci 2004;61:2290–308.

[10] Emerman M, Malik HS. Paleovirology: modern consequences of ancient viruses. PLoS Biol 2010;8: e1000301.

[11] Eakin RE. An approach to the evolution of metabolism. Proc Natl Acad Sci U S A 1963;49:360–6.

[12] Whitman WB, Coleman DC, Wiebe WJ. Prokaryotes: the unseen majority. Proc Natl Acad Sci 1998;95:6578–83.

[13] Schloss PD, Handelsman J. Toward a census of bacteria in soil. PLoS Comput Biol 2006;2:e92.

[14] Angly FE, Felts B, Breitbart M, Salamon P, Edwards RA, Carlson C, et al. The marine viromes of four oceanic regions. PLoS Biol 2006;4:e368.

[15] Anthony SJ, Epstein JH, Murray KA, Navarrete-Macias I, Zambrana-Torrelio CM, Solovyov A, et al. A strategy to estimate unknown viral diversity in mammals. MBio 2013;4(5):e00598–613.

[16] Koonin EV, Wolf YI. Evolution of microbes and viruses: a paradigm shift in evolutionary biology? Front Cell Infect Microbiol 2012;2:119.

[17] Mora C, Tittensor DP, Adl S, Simpson AGB, Worm B. How many species are there on earth and in the ocean? PLoS Biol 2011;9:e1001127.

[18] Locey KJ, Lennon JT. Scaling laws predict global microbial diversity. Proc Natl Acad Sci U S A 2016;113:5970–5.

[19] Brawand D, Wagner CE, Li YI, Malinsky M, Keller I, Fan S, et al. The genomic substrate for adaptive radiation in African cichlid fish. Nature 2014;513:375–81.

[20] Perez-Iratxeta C, Palidwor G, Andrade-Navarro MA. Towards completion of the Earth's proteome. EMBO Rep 2007;8:1135–41.

[21] Frank DN, Pace NR. Gastrointestinal microbiology enters the metagenomics era. Curr Opin Gastro-enterol 2008;24:4–10.

[22] Yang X, Xie L, Li Y, Wei C. More than 9,000,000 unique genes in human gut bacterial community: estimating gene numbers inside a human body. PLoS ONE 2009;4:e6074.

[23] Banuls A, Thomas F, Renaud F. Of parasites and men. Infect Genet Evol 2013;20:61–70.

[24] Yu J, Hu S, Wang J, Wong GK, Li S, Liu B, et al. A draft sequence of the rice genome (*Oryza sativa* L. ssp. *indica*). Science 2002;296:79–92.

[25] Baldauf SL. An overview of the phylogeny and diversity of eukaryotes. J Syst Evol 2008;46:263–73.

[26] Frias-Lopez J, Shi Y, Tyson GW, Coleman ML, Schuster SC, Chisholm SW, et al. Microbial community gene expression in ocean surface waters. Proc Natl Acad Sci 2008;105:3805–10.

[27] Demain AL. Regulation of secondary metabolism in fungi. Pure Appl Chem 1986;58:219–26.

[28] Theis N, Lerdau M. The evolution of function in plant secondary metabolites. Int J Plant Sci 2003;164: S93–S102.

[29] Pichersky E, Gang D. Genetics and biochemistry of secondary metabolites in plants: an evolutionary perspective. Trends Plant Sci 2000;5:439–45.

[30] Matsushima T, Matsumoto H, Shirai A, Sawamura M, Sugimura T. Mutagenicity of the naturally occurring carcinogen cycasin and synthetic methylazoxymethanol conjugates in *Salmonella typhimurium*. Cancer Res 1979;39:3780–2.

[31] Wales JH, Sinnhuber RO, Hendricks JD, Nixon JE, Eisele TA. Aflatoxin B1 induction of hepatocellular carcinoma in the embryos of rainbow trout (*Salmo gairdneri*). J Natl Cancer Inst 1978;60:1133–9.

[32] Zheng J, Xia X, Ding H, Yan A, Hu S, Gong X, et al. Erasure of the paternal transcription program during spermiogenesis: the first step in the reprogramming of sperm chromatin for zygotic development. Dev Dyn 2008;237:1463–76.

[33] Gersbach CA, Perez-Pinera P. Activating human genes with zinc finger proteins, transcription activator-like effectors and CRISPR/Cas9 for gene therapy and regenerative medicine. Expert Opin Ther Targets 2014;18:835–9.

[34] Lv Q, Yuan L, Deng J, Chen M, Wang Y, Zeng J, et al. Efficient generation of myostatin gene mutated rabbit by CRISPR/Cas9. Sci Rep 2016;6:25029.

[35] Shah RR, Cholewa-Waclaw J, Davies FCJ, Paton KM, Chaligne R, Heard E, et al. Efficient and versatile CRISPR engineering of human neurons in culture to model neurological disorders. Wellcome Open Res 2016;1:13.

[36] Zhang C, Liu J, Zhong JF, Zhang X. Engineering CAR-T cells. Biomarker Res 2017;5:22.

[37] Koonin EV, Starokadomskyy P. Are viruses alive? The replicator paradigm sheds decisive light on an old but misguided question. Stud Hist Phil Biol Biomed Sci 2016;59:125–34.

[38] Jern P, Sperber GO, Blomberg J. Use of endogenous retroviral sequences (ERVs) and structural markers for retroviral phylogenetic inference and taxonomy. Retrovirology 2005;2:50.

[39] Berman JJ. Taxonomic guide to infectious diseases: understanding the biologic classes of pathogenic organisms. Cambridge, MA: Academic Press; 2012.

[40] Horie M, Honda T, Suzuki Y, Kobayashi Y, Daito T, Oshida T, et al. Endogenous non-retroviral RNA virus elements in mammalian genomes. Nature 2010;463:84–7.

[41] Griffiths DJ. Endogenous retroviruses in the human genome sequence. Genome Biol 2001;2: reviews 1017.1–reviews 1017.5.

[42] Haring M, Vestergaard G, Rachel R, Chen L, Garrett RA, Prangishvili D. Virology: independent virus development outside a host. Nature 2005;436:1101–2.

[43] Bandin I, Dopazo CP. Host range, host specificity and hypothesized host shift events among viruses of lower vertebrates. Vet Res 2011;42:67.

[44] Kapitonov VV, Jurka J. RAG1 core and V(D)J recombination signal sequences were derived from Transib transposons. PLoS Biol 2005;3:e181.

[45] Fort A, Hashimoto K, Yamada D, Salimullah M, Keya CA, Saxena A, et al. Deep transcriptome profiling of mammalian stem cells supports a regulatory role for retrotransposons in pluripotency maintenance. Nat Genet 2014;46:558–66.

[46] Durruthy-Durruthy J, Sebastiano V, Wossidlo M, Cepeda D, Cui J, Grow EJ, et al. The primate-specific noncoding RNA HPAT5 regulates pluripotency during human preimplantation development and nuclear reprogramming. Nat Genet 2016;48:44–52.

[47] Patel MR, Emerman M, Malik HS. Paleovirology: ghosts and gifts of viruses past. Curr Opin Virol 2011;1(4):304–9.

[48] Blanco-Melo D, Gifford RJ, Bieniasz PD. Reconstruction of a replication-competent ancestral murine endogenous retrovirus-L. Retrovirology 2018;15:34.

[49] Li JP, D'Andrea AD, Lodish HF, Baltimore D. Activation of cell growth by binding of friend spleen focus-forming virus gp55 glycoprotein to the erythropoietin receptor. Nature 1990;343:762–4.

[50] Ruscetti SK, Janesch NJ, Chakraborti A, Sawyer ST, Hankins WD. Friend spleen focus-forming virus induces factor independence in an erythropoietin-dependent erythroleukemia cell line. J Virol 1990;64:1057–62.

[51] Taxonomy. International Committee on Taxonomy of Viruses, https://talk.ictvonline.org; 2018 [viewed October 18].

[52] Suttle CA. Environmental microbiology: viral diversity on the global stage. Nat Microbiol 2016;1:16205.

[53] Hughes AL, Friedman R. Poxvirus genome evolution by gene gain and loss. Mol Phylogenet Evol 2005;35:186–95.

[54] Mohammed MA, Galbraith SE, Radford AD, Dove W, Takasaki T, Kurane I, et al. Molecular phylogenetic and evolutionary analyses of Muar strain of Japanese encephalitis virus reveal it is the missing fifth genotype. Infect Genet Evol 2011;11:855–62.

[55] Bannert N, Kurth R. The evolutionary dynamics of human endogenous retroviral families. Annu Rev Genomics Hum Genet 2006;7:149–73.

[56] Nasir A, Caetano-Anolles G. A phylogenomic data-driven exploration of viral origins and evolution. Sci Adv 2015;1:e1500527.

[57] Rasmussen MD, Kellis M. A Bayesian approach for fast and accurate gene tree reconstruction. Mol Biol Evol 2011;28:273–90.

[58] Arslan D, Legendre M, Seltzer V, Abergel C, Claverie J. Distant Mimivirus relative with a larger genome highlights the fundamental features of Megaviridae. Proc Natl Acad Sci U S A 2011;108:17486–91.

[59] Yoosuf N, Yutin N, Colson P, Shabalina SA, Pagnier I, Robert C, et al. Related giant viruses in distant locations and different habitats: *Acanthamoeba polyphaga* moumouvirus represents a third lineage of the Mimiviridae that is close to the megavirus lineage. Genome Biol Evol 2012;4:1324–30.

[60] Bianciardi G, Miller JD, Straat PA, Levin GV. Complexity analysis of the Viking labeled release experiments. Intl J Aeronautical Space Sci 2012;13:14–26.

[61] Forterre P. Defining life: the virus viewpoint. Orig Life Evol Biosph 2010;40:151–60.

[62] Valas RE, Bourne PE. The origin of a derived superkingdom: how a gram-positive bacterium crossed the desert to become an archaeon. Biol Direct 2011;6:16.

[63] Filee J. Multiple occurrences of giant virus core genes acquired by eukaryotic genomes: the visible part of the iceberg? Virology 2014;466–467:53–9.

[64] Forterre P, Gaia M. Giant viruses and the origin of modern eukaryotes. Curr Opin Microbiol 2016;31:44–9.

[65] Zhang J, Quintal L, Atkinson A, Williams B, Grunebaum E, Roifman CM. Novel RAG1 mutation in a case of severe combined immunodeficiency. Pediatrics 2005;116:445–9.

[66] de Villartay JP, Lim A, Al-Mousa H, Dupont S, D chanet-Merville J, Coumau-Gatbois E, et al. A novel immunodeficiency associated with hypomorphic RAG1 mutations and CMV infection. J Clin Invest 2005;115:3291–9.

[67] CAR. FDA approval brings first gene therapy to the United States: CAR T-cell therapy approved to treat certain children and young adults with B-cell acute lymphoblastic leukemia. U.S. Food and Drug Administration; 2017 [August 30].

[68] Ren J, Zhang X, Liu X, Fang C, Jiang S, June CH, et al. A versatile system for rapid multiplex genome-edited CAR T cell generation. Oncotarget 2017;8:17002–11.

[69] Chmielewski M, Abken H. TRUCKs: the fourth generation of CARs. Expert Opin Biol Ther 2015;15:1145–54.

[70] Morgan RA, Yang JC, Kitano M, Dudley ME, Laurencot CM, Rosenberg SA. Case report of a serious adverse event following the administration of T cells transduced with a chimeric antigen receptor recognizing ERBB2. Mol Ther 2010;18:843–51.

[71] Seok J, Warren HS, Cuenca AG, Mindrinos MN, Baker HV, Xu W, et al. Genomic responses in mouse models poorly mimic human inflammatory diseases. Proc Natl Acad Sci U S A 2013;110:3507–12.

[72] Omim. Online Mendelian Inheritance in Man. Available from: http://omim.org/downloads [viewed June 20], 2013.

[73] Salmena L, Carracedo A, Pandolfi PP. Tenets of PTEN tumor suppression. Cell 2008;133:403–14.

[74] Brownstein MH, Mehregan AH, Bikowski JBB, Lupulescu A, Patterson JC. The dermatopathology of Cowden's syndrome. Br J Dermatol 1979;100:667–73.

[75] Haibach H, Burns TW, Carlson HE, Burman KD, Deftos LJ. Multiple hamartoma syndrome (Cowden's disease) associated with renal cell carcinoma and primary neuroendocrine carcinoma of the skin (Merkel cell carcinoma). Am J Clin Pathol 1992;97:705–12.

[76] Schrager CA, Schneider D, Gruener AC, Tsou HC, Peacocke M. Clinical and pathological features of breast disease in Cowden's syndrome: an underrecognized syndrome with an increased risk of breast cancer. Hum Pathol 1998;29:47–53.

[77] Wagner PJ, Ruta M, Coates MI. Evolutionary patterns in early tetrapods. II. Differing constraints on available character space among clades. Proc R Soc B Biol Sci 2006;273:2113–8.

[78] Zuk O, Hechter E, Sunyaev SR, Lander ES. The mystery of missing heritability: genetic interactions create phantom heritability. Proc Natl Acad Sci U S A 2012;109:1193–8.

[79] Schwartz JH, Maresca B. Do molecular clocks run at all? A critique of molecular systematics. Biol Theory 2006;1:357–71.

[80] Drummond AJ, Ho SYW, Phillips MJ, Rambaut A. Relaxed phylogenetics and dating with confidence. PLoS Biol 2006;4:e88.

[81] Prusiner SB. Novel proteinaceous infectious particles cause scrapie. Science 1982;216:136–44.

[82] Michelitsch MD, Weissman JS. A census of glutamine/asparagine-rich regions: implications for their conserved function and the prediction of novel prions. PNAS 2000;97:11910–5.

[83] Holmes I. Transcendent elements: whole-genome transposon screens and open evolutionary questions. Genome Res 2002;12:1152–5.

[84] Lerat E, Semon M. Influence of the transposable element neighborhood on human gene expression in normal and tumor tissues. Gene 2007;396:303–11.

5

Phylogeny: Eukaryotes to Chordates

OUTLINE

Section 5.1 On Classification

Individuals do not belong in the same taxon because they are similar, but they are similar because they belong to the same taxon.

George Gaylord Simpson [1]

Order and simplification are the first steps toward the mastery of a subject.

Thomas Mann

Classification is a discipline that nobody wants to learn, because everybody is certain that they have an adequate, intuitive understanding of the subject. Please perish this thought! In practice, classifications are incredibly tricky to build, and even the most brilliant taxonomists make mistakes. The 2000+ year history of the classification of animals is rife with errors, and modern taxonomists expend an inordinate amount of effort correcting the mistakes created by their mentors. The purpose of this section is to define classifications, explain their purposes, and describe how classifications are created, with special emphasis on the phylogenetic classification of living and extinct terrestrial organisms.

Let's begin with Aristotle, a Greek philosopher who lived c. 384 to 322 BCE. Aristotle amused his contemporaries by classifying dolphins as a type of mammal; not as a type of fish. It seemed obvious to nearly everyone, at the time, that dolphins were much more similar to fish than to mammals. Dolphins looked like fish, they lived in the ocean like fish, they swam like fish, and could be hooked or netted along with the other fish. If it looks like a fish, and acts like a fish, then it must be a fish. So thought Aristotle's many detractors (Fig. 5.1).

Aristotle reasoned otherwise. For Aristotle, classifications were built on relationships among animals, not superficial similarities. Aristotle found that one group of animals,

Evolution's Clinical Guidebook. https://doi.org/10.1016/B978-0-12-817126-4.00005-9

FIG. 5.1 Bottlenose dolphin. Aristotle know that a dolphin is not a fish, but his contemporaries did not. *Source: The US National Aeronautics and Space Administration, public domain work.*

the mammals, nourished their developing embryos with a placenta. At birth, the offspring are delivered into the world as formed, but small versions of the adult animals (i.e., not as eggs or as larvae), and newborn dolphin feed from milk excreted by specialized glandular organs (mammae). Aristotle knew that these birthing features characterized mammals, and distinguished mammals from all the other groups of animals. He correctly reasoned that dolphins must be a type of mammal, not a type of fish. For nearly 2000 years, Aristotle's classification was ridiculed by biologists. Belatedly, we now know that Aristotle was right all along.

Likewise, we must resist the urge to assign a terrier to the same class as a house cat, just because both animals have many phenotypic features in common (e.g., similar size and weight, presence of a furry tail, four legs, tendency to snuggle in a lap). A terrier is dissimilar to a wolf, and a house cat is dissimilar to a lion, but the terrier and the wolf are closely related to one another; as are the house cat and the lion. For the purposes of creating a classification, relationships are all that are important. Similarities, when they occur, arise as a consequence of relationships; not the other way around.

Aristotle, and legions of taxonomists (i.e., experts in classification) that followed him, understood that taxonomy is all about finding the relationships among species, not their similarities. But isn't a similarity a type of relationship? Actually, no. To better understand the difference, imagine the following scenario. You look up at the clouds, and you begin to see the shape of a lion emerging. The cloud has a tail, like a lion's tale, and a fluffy head, like a lion's mane. With a little imagination, the mouth of the lion seems to roar down from the sky. You have succeeded in finding similarities between the cloud and a lion. When you look at a cloud and you imagine a tea kettle producing a head of steam, you may recognize that the physical forces that create a cloud from the ocean's water vapor and the physical forces that produce steam from the water in a heated kettle are the same. At this moment, you have found a relationship. The act of searching for and finding relationships lies at the heart of science; this is how we make sense of reality. Finding similarities is an aesthetic joy, but it is not science.

The first crude classifications of living organisms were similarity-based, producing classes such as flying animals (e.g., birds, bats, and bumblebees) or swimming animals (e.g., fish, cephalopods, and lampreys), or walking animals (e.g., man, bears, and penguins). Among many and various shortcomings, similarity-based approaches to classification cannot take into account the dissimilarities within a single entity during various stages of its life. For example, butterflies fly, but caterpillars do not; yet they are the same organism and hence both forms would be contained within the class of flying animals. Similarity-based approaches to not take into account the disparate ways by which species of animals achieve a similar property. For example, flying squirrels, birds, and "helicopter seeds" of maple trees all fly through the air, but they all achieve flight through different methods. The earliest animal classifications did not add to our understanding of organisms and had value only as aids in animal identification. [Glossary Classification vs ontology, Diagnosis vs classification]

Modern classifications are all based on grouping organisms by their relationships to one another [1–5]. Nonetheless, it is exceedingly easy to fall back on similarity-based classifiers. Realistically, it is much easier to find similarities than to find relationships, and it is human nature to follow the path of least resistance.

Let's look at a somewhat formal definition of a classification, wherein we group related objects into classes according to the following rules [1, 3, 6]:

1. Each class of related objects can be described by a set of defining properties that apply to every member of the class and that establish the class.

2. Every class must have a parent class; with the exception of the top class, sometimes referred to as the root class. The defining properties of the parent class are inherited by the child class. [Glossary Child class, Parent class, Superclass, Subclass]

3. Any class may have one or more child classes of its own.

4. Every object (sometimes referred to as class member or as a class instance) belongs to exactly one class and cannot change its class.

Technically, that's all there is to a classification. What purposes do classifications fulfill?

– Simplification

You can begin to see how a classification can drive down the complexity of its contained objects. For example, classifications of organisms vastly simplify the field of clinical microbiology. With over 1400 described infectious diseases to deal with, it is important to have a way of grouping organisms into a manageable collection of biological classes. As it happens, every known disease-causing organisms has been assigned to one of about 40 well-defined classes of organisms, and each class fits within a simple ancestral lineage. This means that every known pathogenic organism inherits certain properties from its ancestral classes and shares these properties with the other members of its own class.

When you learn the class properties, along with some basic information about the infectious members of the classes, you gain a comprehensive understanding of relationships and properties of every included infectious organism.

Likewise, the entire lineage of humans can be reduced to a few dozen classes that intervene between the root class (single-celled eukaryotes) and the species class (*Homo sapiens*). In this chapter, and the next (Chapter 6, "Phylogeny: Craniates to Humans"), we'll be using the classification of living organisms to condense biological history to a few dozen pages, beginning with the first eukaryote and ending with the appearance of humans.

– Data retrieval

Classifications can function as the master index to the domain of included objects, supporting search and retrieval methods, and facilitating data analysis and data mining projects.

Because every class has only one parent class, it is easy to construct the full ancestry of every member of a classification. For example, if you have an instance of spider (e.g., "inky-dinky spider"), you need only determine the binomial name of the species (i.e., *Tegenaria domistica*), to determine the name of its parent class (i.e., its genus, *Tegenaria*). You can look up the parent class of Tegenaria (i.e., Agelenidae) to determine the grandfather class. By iteratively determining the parent class of each ancestor, you can reconstruct the complete lineage of "inky dinky spider," namely: *Tegenaria domestica*, Tegenaria, Agelenidae, Araneomorphae, Araneae, Arichnida, Chelicerata, Arthropoda, Panarthropoda, Protostomia, Coelomata, Bilateria, Eumetazoa, Metazoa, Fungi/Metazoa group, Eukaryota, and Cellular Organism (root class). Computers handle this kind of search and retrieval algorithm at incredibly fast speeds [7, 8]. [Glossary Metazoa]

– Inferencing

Every true classification should have something that information scientists refer to as competence. Competence refers to the ability to make inferences based on knowledge of class properties and of the relationships among the classes. For example, you can infer that every member of a class will inherit all of the class properties and class methods available to the parent class. In addition, we can infer that human organisms may contain some cells with a posterior undulipodium, because our class system informs us that humans are members of Class Unikonta (eukaryotes with a posterior undulipodium). As it happens, spermatocytes have a straight, posterior tail, as we would expect to find in any self-respecting unikont.

A competent classification allows us to make predictions. For example, if it is known that a particular pathway is a class property, then we can predict that a drug that blocks the pathway in members of the class will also block the pathway in members of the descendant classes. [Glossary Rank]

For the modern scientist, the classification of living organisms is the inference engine around which all biological science is unified. In the case of infectious diseases, when scientists find a trait that informs us that what we thought was a single species is actually two species, it permits us to develop treatments optimized for each species, and to develop new methods to monitor and control the spread of both organisms. When we correctly group organisms within a common class, we can test and develop new drugs that are effective against all of the organisms within the class, particularly if those organisms are characterized by a molecule, pathway or trait that can be targeted by a drug [9].

  – Self-correction

We use classification to create new hypotheses about the world and about the classification itself. The process of testing such hypotheses may sometimes reveal that the classification is flawed and that our early assumptions were incorrect. More often, testing hypotheses will reassure us that our assumptions were consistent with new observations, adding to our understanding of the relations between the classes and instances within the classification.

When an inference is demonstrated to be false, then we must revisit the classification and determine the reason for the error. Sometimes, we may find that the classification is valid, but that we made certain assumptions about the properties of classes that are untrue. At other times, we learn that our classification is imperfect, and must be revised. All classifications must be continuously and endlessly tested. Otherwise, the classification is simply a pseudo-scientific assertion. This testing process, which is based on hypotheses generated by the classification itself, is a self-correcting mechanism, and is perhaps the most scientifically valuable aspect of taxonomy. [Glossary Pseudoscience, Results, Science, Validation, Verification]

Aside from structural errors in classification, due to misinformation or poor guessing, taxonomists routinely encounter logic errors. It happens that it is very difficult to create self-consistent classifications, and the pitfalls are often due, in one way or another, to paradoxes of self-reference. Motivated readers are directed to the Glossary, for a detailed discussion of taxonomic paradoxes. [Glossary Paradoxes of classification]

Of late, the eminence of classifications has been challenged by a relatively new invention in the field of taxonomy: the ontology. Ontologies are classifications that have abandoned the one class one parent rule (i.e., an ontology class is permitted to have more than one parent class). The relative strengths and weakness of ontologies, as compared with classifications, have been thoroughly examined in several of my previously published works and won't be discussed here [7, 10–13]. Suffice it to say that a classification is a simple type of ontology, and that the current taxonomy of all living organisms on earth is a classification. [Glossary Cladistics, Monophyletic class, Apomorphy, Ontology, Multiclass classification, Multiclass inheritance, Synapomorphy]

Section 5.2 The Complete Human Phylogenetic Lineage

I always wanted to be somebody, but now I realize I should have been more specific.
Lily Tomlin

In a sense, all the preceding chapters of this book served to prepare us for the next few sections, wherein we finally specify the particular evolutionary milestones that determined, millions and billions of years ago, the classes of diseases that occur in modern humans.

Here is the ancestral lineage of human beings, beginning with the earliest indication of life on earth. Wherever possible, the major classes of organisms are annotated with an approximate chronology. It is useful to have some notion of the time interval between classes of ancestral organisms, even if the dates are inaccurate. Please note that "mya" is the abbreviation for "million years ago."

```
Earliest indication of life (4,100 mya)
Prokaryota (3,900 mya)
Eukaryota (2,100 to 1000 mya)
Podiata
Unikonta (also known as Amorphea)
Obazoa
Opisthokonta
Holozoa (1300 mya)
Filozoa
Apoikozoa (950 mya)
Metazoa (760 mya) [14]
Eumetazoa (Dipoblasts, Histozoa, Epitheliozoa) (635 mya)
Parahoxozoa
Bilateria (also known as Tripoblasts) (555 mya)
Nephrozoa (555 mya)
Deuterostomia (also known as Enterocoelomata) (540 mya)
Chordata (530 mya)
Craniata (480 mya)
Vertebrata (500 mya)
Gnathostomata (419 mya)
Euteleostomi (400 mya)
Sarcopterygii
Dipnotetrapodomorpha
Tetrapodomorpha (390 mya)
Tetrapoda (367 mya)
Synapsida (308 mya)
Mammalia (220 mya)
Theria (160 mya)
Eutheria (160-125 mya)
Boreoeutheria (124-101 mya)
Euarchontoglires (100 mya)
Euarchonta (99-80 mya)
Primates (75 mya)
Haplorrhini (63 mya)
Simiiformes (40 mya)
Catarrhini (30 mya)
Hominoidea (28 mya)
Hominidae (15 mya)
```

```
Homininae (8 mya)
Hominini (5.8 mya)
Homo (2.5 mya)
Homo sapiens (0.3 mya)
```

Section 5.3 Eukaryotes to Obazoans

After the game, the King and the pawn go into the same box.

Italian proverb

In this section, we will discuss the first four classes of eukaryotic (i.e., nucleated) organisms.

```
Eukaryota (2,100 to 1000 mya)
Podiata
Unikonta (also known as Amorphea)
Obazoa
```

Eukaryota

On the simplest level, all cellular life can be divided into two forms: the eukaryotes, which have a membrane-bound organelle, known as the nucleus, that contains the cell's genetic material; and the prokaryotic life forms (Eubacteria and the Archaeans, also known as the Archaebacteria) that have no nucleus.

Nobody knows when the first eukaryotes appeared on earth. One school of thought estimates that eukaryotes may have been here as early as 2.7 billion years ago, about one billion years following the first appearance of prokaryotic life forms. This theory is based on finding sterane molecules in shale rocks, dating back nearly 3 billion years. Eukaryotic cells are the only known source of naturally occurring sterane molecules. Their presence in shale is held as evidence that the first eukaryotic organisms must have existed no later than 2.7 billion years ago [15]. Other biologists tie their estimate of the beginning or eukaryotic life to the epoch in which the first eukaryotic fossil remains are found, about 1.7 billion years ago. This leaves a one billion year gap between estimates for the origin of the eukaryotes (i.e., 1.7–2.7 billion years ago).

Although there is an enormous range between the smallest and the largest eukaryotic cells, it is reassuring to note that most eukaryotic cells look very similar to one another and are of about the same size (i.e., 25–50 μm in diameter). The largest single-celled eukaryotic organism is 20 cm in length. This is *Syringammina fragilissima*, a member of Class Foraminifera, found off the coast of Scotland. The smallest eukaryotes are picoplankton that have a diameter as small as 0.2–2 μm in diameter [16]. There are many eukaryotic species that have never been adequately studied, and these include pico- and nano-sized stramenopiles [17]. The heaviest eukaryotic cell is the egg of the ostrich (*Struthio camelus*), which typically weighs between 3.5 and 5 pounds (Fig. 5.2).

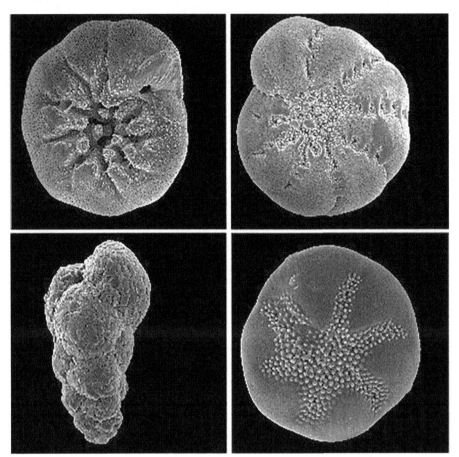

FIG. 5.2 Single-celled eukaryotes of Class Foraminifera. Clockwise starting at the top left: *Ammonia beccarii, Elphidium excavatum clavatum, Buccella frigida,* and *Eggerella advena.* Foraminifera are among the largest known single-celled organisms. *Source: Wikipedia, a public domain image produced by the US Geological Survey.*

The eukaryotes have a number of chemical and metabolic features in common. For example, several closely related filamentous molecules are components of the cytoskeleton of every eukaryote, and are only found in eukaryotes [18]. In addition, the earliest eukaryotes had distinctive membrane-delimited internal organelles (i.e., mitochondria and nuclei), and at least one protruding undulipodium. Let's review these characteristic eukaryotic structures.

1. Mitochondria

Mitochondria are membrane-delimited organelles, with their own genome, and they proliferate within eukaryotic cells. Current theory holds that mitochondria developed as an obligate intracellular endosymbiont from an ancestor of Class Rickettsia. All existing eukaryotic organisms descended from early eukaryotes that contained mitochondria. Furthermore, all existing eukaryotic organisms, even the so-called amitochondriate classes (i.e., organisms lacking mitochondria), contain vestigial forms of mitochondria (i.e., hydrogenosomes and mitosomes) [19–22].

The mitochondria provide eukaryotes with a source of internal energy, via oxidative phosphorylation. If fuel is abundant, then the amount of energy produced by a cell is proportionate to the number of mitochondria, and the cells that utilize the most energy contain the greatest number of mitochondria. For example, nearly half the cytoplasmic volume of heart muscle cells is composed of mitochondria, with several thousand mitochondria per cell.

The mitochondria open up lots of evolutionary options. Eukaryotic cells can grow very large, employing their mitochondria to provide the energy needed to maintain a large organism. In addition, cells can expand the size of their genome, accommodating lots of junk DNA (i.e., nonfunctioning DNA), and the high energy costs of replicating the enlarged genome will be paid by the mitochondria. Cells that can create their own energy can afford to specialize, producing muscle cells that aid in movement, and sensory cells that aid in finding food. The mitochondria enabled single-celled animals to evolve into the large, multicellular, energy-inefficient organisms that occupy our planet.

Of course, with the mitochondria came the mitochondriopathies; diseases whose underlying cause is mitochondrial pathology (i.e., dysfunctional mitochondria, or an abnormal number of mitochondria). Mitochondriopathies can be genetic or acquired. Most of the root causes of genetic mitochondriopathies are due to gene mutations occurring in the nucleus of the cell. Though mitochondria live outside the nucleus and have their own genomes, mitochondrial DNA codes for only 13 proteins of the respiratory chain. All the other proteins and structural components of mitochondria are coded in the nucleus. [Glossary Underlying cause]

As we might expect, mitochondriopathies affect the cells that are most dependent on their mitochondria for their functionality. It is not surprising that many of the mitochondriopathies are characterized by multisystem disorders that yield muscle weakness, cardiomyopathy, and ataxia. Additional types of mitochondriopathies may include: pigmentary retinopathy, ocular atrophy, deafness, gut motility disorder, and sideroblastic anemia, among many others. A mitochondriopathy should be in the differential workup for any unexplained multisystem disorder, especially those arising in childhood [23].

Isolated deafness (i.e., deafness as the only symptom) is observed in some forms of inherited mitochondriopathy in humans [24]. Sometimes, isolated deafness is present in acquired conditions, such as seen with deafness following antibiotic usage (e.g., aminoglycosides). Why would an antibiotic produce a mitochondriopathy? We must remember that the mitochondria evolved as a captured bacteria that adapted to an intracellular existence within eukaryotes. Our mitochondria, true to their bacterial origins, are susceptible to toxicity induced by some antibacterial agents [25]. Why do such agents produce an isolated form of deafness? Presumably, the mitochondriopathic effect in these cases is systemic, affecting every cell of the body, to some small extent. The cells involved in hearing happen to be the most sensitive. A specific example of isolated deafness due to the A1555G mtDNA mutation was discussed in Section 3.2, "The Epigenome and the Evolution of Cell Types" [26]. It has been observed that about a quarter of individuals receiving aminoglycoside therapy have some loss in hearing, as measured by audiometry testing.

2. Undulipodia

Prokaryotes and eukaryotes have rods that protrude from the organism; their undulating motion propels cells forward through water. Such rods are commonly known as flagella (from the singular flagellum, the Latin word meaning whip). Aside from a superficial resemblance, the flagella of eukaryotes have no relationship to the flagella of prokaryotes. Eukaryotic flagella are orders of magnitude larger than the prokaryotic flagella, contain hundreds of species of proteins not present in the flagella of prokaryotes, have a completely different internal structure, anchor to a different cellular location, and do not descend phylogenetically from prokaryotic flagella [27]. Specifically, prokaryotic flagella are composed primarily of flagellin. Undulipodia, the eukaryotic rods, are composed primarily of tubulin and contain more than 100 other identified proteins, including dynein. Flagella have a diameter of 0.01–0.025 µm. Undulipodia have a much larger diameter (0.25 µm).

Biologists provided the eukaryotic flagellum with a more accurate name: undulipodium. Perhaps they chose a term with a few too many syllables. Most biologists continue to stamp eukaryotes with the misleading short term "flagellum" (plural, "flagella"). Terminology aside, it is important to recognize that every existing eukaryote descended from an organism with a undulipodium. The undulipodium is a highly conserved feature of eukaryotes, and all descendant eukaryotic classes contain undulipodia or structures that evolved as modified forms of undulipodia. For example, we humans have cilia on the surface of mucosal lining cells, and these cilia are abbreviated forms of undulipodia. As previously mentioned, our spermatocytes have a single, long, undulipodial tail, that propels the mature sperm cell along its undulating voyage toward the oocyte.

A variety of structures in eukaryotic organisms have evolved from the undulipodium and their homologous derivatives, all of which are composed of tubulins [28]. These include pericentriolar bodies, centrioles, kinetids, specialized receptors, haptonemes of coccolithophorids, and undulating membranes of trypanosomes. In Section 6.2, "Vertebrates to Synapsids," we'll be looking at the primary cilium, a derivative of the undulipodium found exclusively in vertebrates, and the root cause of a newly characterized family of human diseases known as the ciliopathies.

3. Nucleus

As far as anyone knows, the very first eukaryote came fully equipped with a nucleus. Theories abound, but nobody really knows from whence the nucleus came. There are many commonalities between the eukaryotic nucleus and archaean cells, in terms of the structure and organization of DNA, RNA, and ribosomes. Here are few examples:

- Only eukaryotes and archaeans have a TATA box (a sequence of thymidine-adenine-thymidine-adenine that specifies where RNA transcription can begin). Bacteria have the so-called Pribnow box, consisting of a TATAAT sequence.

- Eukaryotes and archaeans have histone proteins attached to their DNA [29]. [Glossary Histone]

- The RNA polymerase and ribosomes of eukaryotes and archaeans are very similar.

Based on striking similarities between the archaean and eukaryotic genome, it has been hypothesized that the eukaryotic nucleus was derived from an archaean organism [15].

Recently, following the discovery of giant viruses that rival eukaryotes in terms of genome size and complexity, and encouraged by the observation that genes of giant virus origin have been found in eukaryotes, a viral origin for the eukaryotic nucleus has been proposed [30–32].

Along with the nucleus came eukaryote-specific methods of transcribing DNA into RNA. In eukaryotes, DNA sequences are not transcribed directly into full-length RNA molecules, ready for translation into a final protein. There is a pre-translational process wherein transcribed sections of DNA, so-called exons, are spliced together, and a single gene can be assembled into alternative spliced products. As mentioned earlier, errors in normal splicing can produce inherited disease, and it has been estimated that 15% of disease-causing mutations involve splicing [33, 34].

Subclasses. The very top division of Class Eukaryota has been the subject of intense interest over the past few decades, and there is as yet no general consensus as to how the split from the top eukaryote to its subclasses should be drawn. Formerly, it was thought that all eukaryotes had either one undulipodia or two, and the two major subdivisions of eukaryotes were Class Unikonta and Class Bikonta [35]. The wisdom of this simple morphologic division was bolstered by genetic findings that three fused genes (carbamoyl phosphate synthase, dihydroorotase, and aspartate carbamoyltransferase) characterize Class Unikonta. Two fused genes (thymidylate synthase and dihydrofolate reductase) characterize Class Bikonta. Hence, the morphologic property dividing Class Eukaryota into unikonts and bikonts was shadowed by a genetic property that draws the equivalent taxonomic division.

Further studies indicated that this simple split did not achieve monophyletic subclasses (i.e., could not ensure that all members of either division had the features that defined its assigned division and lacked the features that defined its sister division). Of late, the eukaryotes are divided into Class Bikonta, Class Podiata, and Class Excavata. The bikonts contain Class Archaeplastida (plants) plus Class Hacrobia (single-cell eukaryotes that includes the cryptomonads) plus the SAR supergroup (unicellular eukaryotes including the stramenopiles, also known as heterokonts plus the alveolates, and Class Rhizaria. Class Excavata is another major group of unicellular eukaryotes. Other than Class Archaeplastida (plants), the bikonts and the excavates dwell in relative obscurity. Nonetheless, they must not be ignored, as they represent many of the species of organisms populating planet earth, and they account for a fair portion of the infectious diseases affecting humans and other animals.

Class Podiata, containing the unikonts, is the major eukaryotic division from which all animals, including *Homo sapiens*, will eventually evolve.

Podiata (also known as Sulcozoa and as Sarcomastigota and roughly equivalent to Class Unikonta)

As you might surmise from its several aliases and plesionyms, Class Podiata is a somewhat controversial class, and its taxonomy is unstable. At least for this moment, the podiates accommodate Class Unikonta plus a few smaller groups such as Diphyllatea and Rigifilda, Breviata, Ancyromonas (Planomonas), Mantamonadida, and Apusomonadida) [14]. [Glossary Problematica, Unclassifiable objects, Unstable taxonomy]

Subclasses. Podiata contains two subclasses: Class Varisulca and Class Unikonta (contains humans). Class Varisulca contains several lineages of single-celled eukaryotes.

Unikonta (same as Amorphea)

Just a decade ago, it was relatively easy to characterize the unikonts; they all had a single undulipodium. This feature, it seems, does not characterize all members of its class, and the proposed new name for Class Unikonta is now Class Amorphea, indicating the lack of any distinguishing morphologic feature. A better clue is the aforementioned fusion of three genes involved in pyrimidine nucleotides. All of the descendants of the early unikonts seem to have preserved this fusion event in their genomes.

Subclasses. Class Unikonta has two subclasses: Class Amoebozoa and Class Obazoa (includes humans). Class Amoebozoa contains about 2400 known species of amoeboid single-celled eukaryotes.

Obazoa

Class Obazoa is a class created for the purposes of separating Class Amoebozoa from the lineage of the subclasses of Class Obazoa. Put bluntly, if it were not for the taxonomist's insertion of Class Obazoa, humans would be descendants of Class Amoebozoa. Artifactual classes are created when we don't know enough to fully understand the defining features that determine membership within a class, but we do know enough to exclude organisms from the class. This is the best taxonomists can do, under the present circumstances.

Subclasses. Class Obazoa contains three subclasses: Class Breviatea and the two sister classes, Class Apusomonadida and Class Opisthokonta (from which Class Fungi and Class Metzoans will emerge). The breviates are single-celled eukaryotic organisms that were formerly included in Class Amoebozoa, until phylogenetic evidence placed them as a sister class to the apusomonadids and the opisthokonts. The apusomonadids contain a number of unicellular eukaryotes.

Section 5.4 Opisthokonts to Parahoxozoa

The evolutionary relationships between the earliest branches of the animal kingdom—bilaterians, cnidarians, ctenophores, sponges and placozoans—are contentious.

Maximilian J Telford [36]

The stretch of evolution extending from the opisthokonts to the parahoxozoans account for two of the three major classes of multicellular organisms (Class Fungi and Class Metazoa), the third being Class Archiplastidae (plants), which split from the lineage leading to humans back at the bikont division of eukaryotes. In addition, the evolutionary innovations that define the animal kingdom were achieved prior to the appearance of the parahoxozoans, and prior to the Cambrian explosion, in a some-what vague time frame that may have extended from about 1.5 billion years ago to about 600 million years ago.

```
Opisthokonta
Holozoa (1300 mya)
Filozoa
Apoikozoa (950 mya)
Metazoa (760 mya) [14]
Eumetazoa (Dipoblasts, Histozoa, Epitheliozoa) (635 mya)
Parahoxozoa
```

Opisthokonta

Members of Class Opisthokonta, like all unikonts, have a single undulipodium (com-monly but erroneously referred to as a flagellum). The undulipodium of the opisthokonts protrudes from the posterior pole, and serves to propel the organism through water. The undulipodia (one or two in number) of other eukaryotic classes are most often anterior. Although the posterior undulipodium characterizes Class Opisthokonta, the cells of many of the descendant opisthokonts, that have evolved from formerly aquatic organisms (i.e., fungi and animals) lack the undulipodium. However, among the fungi, the aquatic chy-trids produce gametes (i.e., fungal spores) that have an undulipodium [17]. As previously indicated, among the animals, the undulipodium is retained in spermatocytes. Because each spermatocyte has a undulipodium, we can infer that the genes for building a undu-lipodium are retained in animals.

In addition to their characteristic posterior undulipodium, members of Class Opistho-konta have the ability to synthesize chitin, a long-chain polysacccharide synthesized throughout Class Fungi and in many but not all members of Class Metazoa (e.g., chitin is not synthesized in mammals). All extant organisms that produce chitin are opistho-konts. Chitin is the opisthokont equivalent of cellulose, another long-chain polysaccha-ride, which is found in members of Class Plantae. Chytrids, unlike all other known opisthokonts, can synthesize both cellulose and chitin.

Subclasses. Class Opisthokonta has two subclasses: Class Holomycota and Class Holozoa (contains humans). Class Holomycota includes the fungi plus a class of single-celled organisms known as Class Rozellida (also known as Class Cryptomycota and as Class Rozellomycota). Unlike fungi, the rozellids lack chitinous cell walls. At this moment, the precise classification of the rozellids is somewhat tentative.

Class Fungi were originally assigned as a subclass of Class Archaeplastida (plants), based on a few shared similarities (e.g., both live in soil and both grow as immobile, sessile multicolored structures). The early taxonomists who placed the fungi among the plants should have known better. Fungi synthesize chitin, like all other opisthokonts and unlike all plants; and fungi lack chloroplasts, which are found in nearly all plant species. The important structural role of chitin in fungi and animals should have been a clue to the close relationship between these two classes. Furthermore, fungi and animals are both heterotrophic, acquiring energy by metabolizing organic compounds obtained from the environment. Plants, unlike animals and fungi, are phototropic autotrophs, producing organic compounds from light, water, and carbon dioxide. Classifying the fungi as a type of plant was a big mistake, and demonstrates the perils of matching organisms by superficial similarities. We now know that fungi are much more closely related to humans (another example of an opisthokont) than to the plants that fungi superficially resemble.

As previously mentioned, the only fungal subclass known to have a undulipodium is Class Chytrid, considered to be the first fungal class [17]. The chytrids are primarily aquatic, unlike most other fungi that have evolved to live in soil. Chytrids are currently ravaging the amphibian population. The chytrid Batrachochytrium dendrobatidis is capable of infecting thousands of different amphibian species and threatens many of these species with extinction.

Holozoa

Class Holozoa consists of all animals plus all single-celled opisthokonts that are more closely related to animals than to fungi.

Subclasses. Class Holozoa has three subclasses: Class Mesomycetozoea (formerly known as Class Ichthyosporea), Class Pluriformea, and Class Filozoa (which contains humans). Class Mesomycetozoea consists of single-celled organisms, most of which are parasites of fish and other animals. Class Pluriformea consists of several species of single-celled organisms.

Class Filozoa

Class Filoza is a contrived class intended as a grouping for its subclasses.

Subclasses. Class Filozoa has two subclasses: Class Filasterea and Class Apoikozoa (contains humans). The filasteria are small amoeboid organisms covered with spiky protrusions called filipodia.

Apoikozoa (also known as Choanozoa)

Class Apoikozoa is a contrived class intended as a grouping for its two closely related subclasses.

Subclasses. Class Apoikozoa has two subclasses: Class Metazoa (contains humans) and Class Choanoflagellatea. The choanoflagellates are single-celled eukaryotes that live in

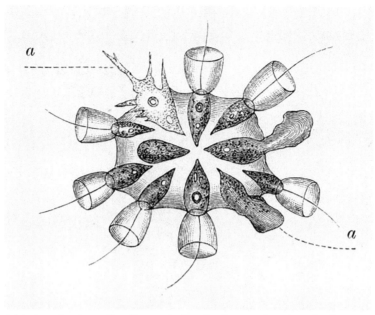

FIG. 5.3 Historical drawing of a choanoflagellate, a single-cell eukaryote thought to be the sister class to the metazoans. *Source: Wikipedia, an illustration by Ilia Mechnikov, from "Embryologische Studien an Medusen. Ein Beitrag zur Genealogie der Primitiv-organe," 1886.*

colonies that simulate organismal entities. Choanoflagellate cells have striking morphologic similarity to cells in sponges (a metazoan species), and are sometimes credited as the parent organism of the sponges (Fig. 5.3).

Metazoa (Animalia)

Metazoans, also known as animals, are multicellular organisms that develop from a blastulating embryo.

Animals are thought to have evolved from simple, spherical organisms floating in the sea, called gallertoids. These hypothesized proto-animals were lined by a single layer of cells enclosing a soft center in which fibrous cells floated in extracellular matrix. As the gallertoids evolved to extract food from the seabed floor, they flattened out. It is presumed that gallertoids evolved a novel way of locking cells together: the desmosome attachment (Fig. 5.4).

Animals, unlike plants and fungi, have nonrigid cell walls (i.e., not hardened by chitin or cellulose). Animals create water-tight membranes, ducts, glands, and epidermis using only baggy, fluid-filled, spherical cells; an unlikely building materia. How is this even possible?

We think of epithelial cells, such as those that line ducts and glands, as being rigid, polygonal cells that fit together, like Lego pieces. This is absolutely untrue. Each animal cell that appears to be shaped like a polygon is actually a sphere, like the soft, round, single-celled organisms belonging to all the sister and close cousin classes of Class Metazoa (Fig. 5.5).

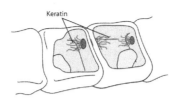

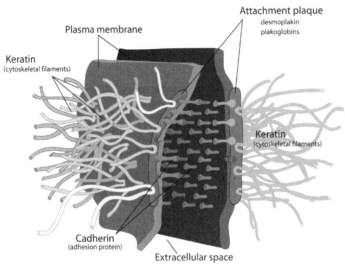

FIG. 5.4 Graphic of a desmosome. Desmosomes act like buttons that hold together the flat-surfaced areas where cells touch one another. The net effect is to produce a water-tight epithelial membrane (i.e., composed of polyhedral cells). Desmosomes are characteristic of cells that develop from blastocysts (i.e., all metazoan organisms). *Source: Wikipedia, created and released into the public domain by Mariana Ruiz, LadyofHats.*

FIG. 5.5 Animal cells are essentially soft bags filled with water. When they are suspended in fluid, they assume their natural spherical shape. The illustration depicts how three epithelial cells might appear, if they simply touched one another. *Source: Jules J. Berman, image created with Pov-ray rendering software.*

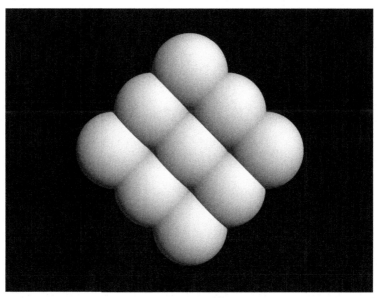

FIG. 5.6 Round animal cells, when pushed together, form straight edges at the surfaces of intersection. Touching spheres represent epithelial cells that are pushed together in a growing tissue. Notice that where spheres touch, flat surfaces are produced. The effect of many spheres pushing together is a polyhedral network, simulating an epithelium. *Source: Jules J. Berman, image created with Pov-ray rendering software.*

When two soft spheres are pushed together, the conjunction of the cells is a flattened surface. As multiple cells of the same size are pushed together, each cell takes the shape of a regular polygon, and the net effect is a honeycomb-shaped network of cells called an epithelium (Fig. 5.6).

An epithelial structure built from a group of crowded, soft spheres will disassemble into individual, round cells, unless they are somehow fastened together. The evolution of the desmosome and the tight junction permitted animals to button-up epithelial tissues, without the aid of external building materials (such as cellulose or chitin). Desmosomes and tight junctions, unique to Class Metazoa, create a leak-proof continuum of epithelial cells. All metazoans contain epithelial cells that line the external surface of the animal (i.e., the skin), the gastrointestinal tract, and most of the internal organs.

There are three classes of eukaryotes that produce complex living organisms: Class Plantae, Class Fungi, and Class Metazoa. These three classes account for virtually every organism that we can see with the naked eye. Consequently, prior to the invention of the microscope (about AD 1590) and the advent of scientific observations of the microscopic world (about AD 1676), these three classes accounted for the totality of the observed living world. Fungi, plants and animals can be distinguished by the method by which they develop. Fungi develop from spores. Plants, like animals, develop as embryos, but they do not have a blastula phase. Animals, unlike plants and fungi, develop from an embryonic blastula (Figs. 5.7 and 5.8).

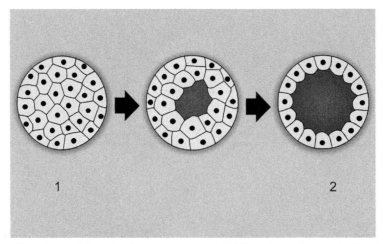

FIG. 5.7 Graphic of blastulation. The early, solid, embryo secretes fluid into a central viscus, the developing blastocyst. Blastulation is accomplished with specialized membrane channels that transport ions and water, with desmosomes that provide a water-tight boundary between adjacent cells. Blastocysts are characteristic of Class Metazoa. *Source: Wikipedia, created and released into the public domain by Pidalka44.*

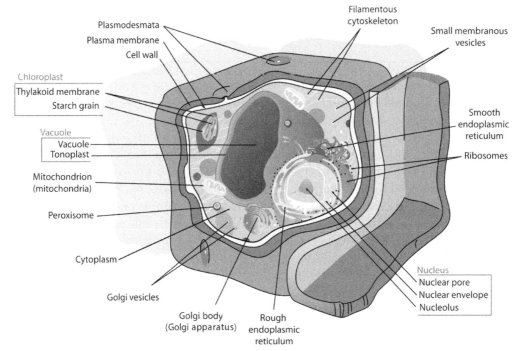

FIG. 5.8 Plant cells lack the desmosomes that characterized animal cells and that are primarily responsible for fastening soft and round epithelial cells into epithelial sheets. Instead, plant cells attain their shapes by their encasement in a rigid cellulose frame (*green* walls in diagram—*gray in print version*). *Source: Wikimedia, entered into the public domain by its author, LadyofHats.*

The blastula is the fundamental cellular feature that defines animals, and distinguishes animals from all other organisms. The evolution of the blastula can be traced to specialized junctions. Presumably, early gallertoids a center core of jelly-like fluid that was surrounded by a lining of soft epithelial cells made water-tight by desmosomes [37]. Upon the evolution of fluid-filled spaces lined by epithelium, it became possible to evolve a variety of fluid-filled compartments in animals, beginning with the blastocoel (i.e., the cavity of the blastula), and moving onwards to the internal coelom and the stomach. We can infer that the stomach was one of the earliest metazoan organs, insofar as all metazoans, other than sponges and placozoans, are equipped with a stomach.

As with most evolutionary advancements, the arrival of the desmosome brought with it a variety of diseases. For example, many types of cardiac arrhythmias can be traced to defects in the desmosome. How so? The heart is a muscle with rhythm. It is the synchronized rhythm of connected myocytes that pumps out about 70 mL of blood with each completed beat. Desmosomes and gap junctions are concentrated at the longitudinal end of each myocyte, at the point of contact with the next myocyte. The desmosome provides a tight continuum between cells, and the gap junctions mediate the passage of electrolytes between cells. Knowing this, it comes as no surprise that mutations in desmosomal proteins can produce conditions conducive to cardiac arrhythmias; the so-called arrhythmogenic cardiomyopathies [38]. Specifically, arrhythmogenic right ventricular dysplasia-8 is caused by a mutation in DSP, the gene encoding desmoplakin. Arrhythmogenic right ventricular cardiomyopathy 12 is caused by a mutation in the gene encoding junction plakoglobin (i.e., JUP gene). A form of arrhythmogenic right ventricular cardiomyopathy/dysplasia is caused by a heterozygous mutations in the PKP2 gene, which encodes plakophilin-2, a protein of the cardiac desmosome. [Glossary Dysplasia]

In addition to their importance to cardiac myocytes, desmosomes play an essential role in the epidermis, where they appear in high concentrations on interlocking squamous cells and at the interface between the epidermis and the dermis. Desmosomes keep the epidermis from rubbing off when friction is applied to skin. Desmosomal skin disorders produce diseases characterized by blistering, acantholysis (e.g., epidermal cells falling apart from one another), and keratoses characterized by a thickened and hardened epidermis.

Because desmosomal mutations have been associated with cardiac disease and with skin disease, it should not be surprising that mutations of desmosomal proteins produce rare syndromes characterized by a combination of cardiac and skin pathology. Naxos syndrome, also known as diffuse non-epidermolytic palmoplantar keratoderma with woolly hair and cardiomyopathy is caused by a mutation in the plakoglobin gene. Dilated cardiomyopathy with woolly hair and keratoderma is caused by a mutation in the DSP gene, coding for desmoplakin, as is a similar syndrome in which the dermatologic features include a pemphigus-like skin disorder [39].

Subclasses. Class Metazoa has two subclasses: Class Parazoa and Class Eumetazoa. Class Parazoa contains two relatively small subclasses: Class Porifera (the sponges) and Class

Placozoa. Class Placozoa currently contains one named species, *Trichoplax adhaerens*, to which every member of the class is tentatively assigned. Sponges and *Trichoplax adhaerens* are exceedingly simple animals, consisting of a layer of jelly-like mesoderm sandwiched between simple epithelia. Like all metazoans, the parazoans develop from blastulated embryos [40]. Neither sponges nor placozoans have stomachs, nor any other specialized organs.

As the gallertoids evolved to extract food from the detritis-lined seabed floor, they flattened out. The modern animals most like the gallertoids are the placozoans, discovered in 1833, first identified plastered against the wall of a seawater aquarium. These organisms are just under a millimeter in length and are composed of about 1000 epithelial cells.

Among the subclasses of Class Porifera is Class Demospongiae. Species of Class Demospongiae are the only living organisms known to methylate sterols at the 26-position, a fact used to place the earliest demosponges to a date preceding the age of the first observed fossils [41, 42]. Some sponges may live for thousands of years, suggesting that evolution of sponges predates the evolved trait known as aging, that arose later in the eumetazoan lineage. This concept of aging as an evolved trait will be discussed further in Section 7.4, "The Evolution of Aging, and the Diseases Thereof" (Fig. 5.9).

As a precautionary notice, the phylogenetic position of the parazoans, in particular the placozoans, is a hotly contested subject. It is possible that Class Ctenophora and Class Cnidaria (vida infra) may need to be repositioned next to Class Metazoa, with the locations of the poriferans and the placozoans very much in play. To avoid internecine squabbles,

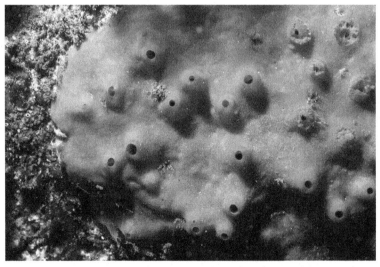

FIG. 5.9 *Acarnus erithacus*, the red volcano sponge, an example of one of the 8800 known species of Class Demospongiae. *Source: A public domain work of the US National Oceanic and Atmospheric Administration, photographed by SIMoN/MBNMS.*

we fall back on the traditional groupings of the basal animals, cognizant that future adjustments are inevitable.

Eumetazoa (also known as Dipoblasts, Histozoa, Epitheliozoa)

The eumetazoans are characterized by tissues organized into germ layers. The eumetazoan embryo achieves a gastrula phase, and neurons are formed. All eumetazoans have a gastrointestinal tract that includes a stomach.

Subclasses. Class Eumetazoa has two subclasses: Class Ctenophora (comb jellies) and Class ParaHoxozoa (includes humans) (Fig. 5.10).

Parahoxozoa

Class Parahoxozoa have a set of genes that is common to all its members. Among these genes are the Hox/ParaHox transcription factor genes.

Subclasses. Class Parahoxozoa has two subclasses: Class Bilateria, and Class Cnidaria. The cnidarians were formerly grouped with the ctenophorans (comb jellies) but have now been assigned their own class. The cnidarians contain over 21,000 extant species of aquatic animals, including sea anemones and corals, jellyfish, and hydra-like animals. The cnidarians are now known to include the obligate parasitic myxozoans, a modified and simplified jellyfish that infects fish [43]. The myxozoans contains species having extremely small genomes, as small as 22.5 megabases in the case of Kudoa iwatai. It is possible that the myxozoans are the first obligate animal parasites (Fig. 5.11).

Like the long-lived sponges, the cnidarians also enjoy long lives, owing to their ability to regenerate. The adult medusan form of the organism can revert back to the polyp, and fragments of the polyp can regenerate to create new organisms, with no apparent limit to the cycles of reversion and regeneration (Fig. 5.12).

Because the cnidarians have the ability to regenerate as needed, they are effectively ageless and immortal. The bilaterians, with possibly one exception, seem doomed to grow old and to die. Hence, if we wanted to determine the phylogenetic point marking the evolution of aging as a newly acquired phylogenetic trait, then we should probably choose the cnidarian/bilaterian split, about 600 million years ago.

Section 5.5 Bilaterians to Chordates

Vita in motu. (Life is in motion.)

Latin motto

```
Bilateria (also known as Triploblasts) (555 mya)
Nephrozoa (555 mya)
Deuterostomia (also known as Enterocoelomata) (540 mya)
Chordata (530 mya)
```

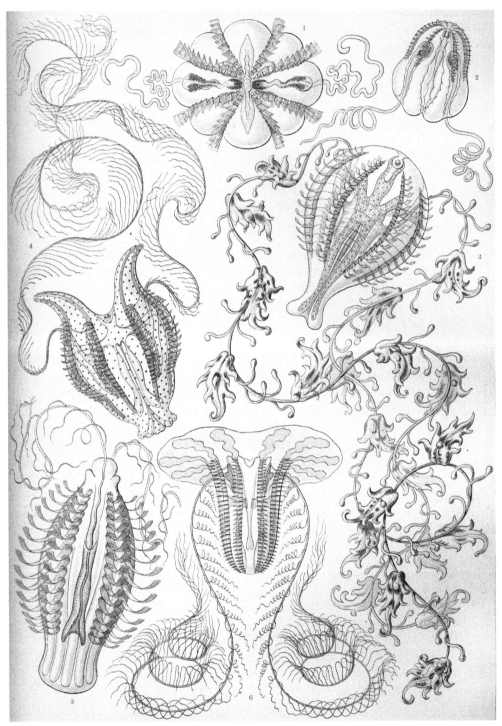

FIG. 5.10 Drawing of ctenophorans (comb jellies). All comb jellies have prominent cilia, derived from undulipodia, to help propel themselves through water. *Source: Wikipedia, from a drawing by Ernst Haeckel, first published in 1853 and now part of the public domain.*

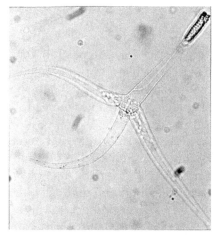

FIG. 5.11 *Myxobolus cerebralis*, a parasitic myxozoan and member of Class Cnidaria. *Source: Wikipedia, from a public domain US Government work.*

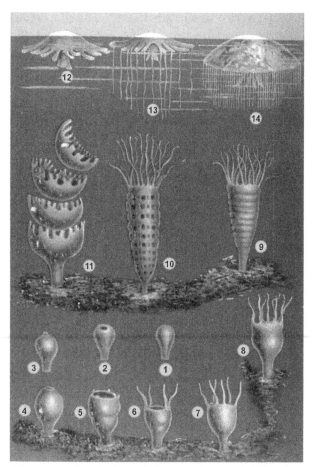

FIG. 5.12 Drawing of the life cycle observed in some cnidarian organisms. Starting at the bottom of the figure, we see polyps transforming into the adult medusa (upper right). *Source: Wikipedia, from a figure in Schleiden M. J. "Die Entwicklung der Meduse," in: "Das Meer," 1869.*

Bilateria (Tripoblasts)

Bilateria, also known as tripoblasts, have three embryonic layers (endoderm, meso-derm, and ectoderm). Animals of Class Bilateria display bilateral symmetry (i.e., their bod-ies can be divided into two symmetrical halves by a plane that runs along a central axis). Hence, the bilaterians have laterality, with an anterior head, a posterior tail, a dorsal back, and a ventral belly.

As discussed in Section 3.1, "Tight Relationship Between Evolution and Embryology," the acquisition of bilateral symmetry is achieved in the embryo, as a phylogenetic trait. Bilateral symmetry does not necessarily extend to the fully developed bilaterian organ-isms. As an example, the starfish is an animal that has bilateral symmetry in embryos, but which loses its symmetry as a five-armed adult. There are numerous examples in which embryonic symmetry is lost by twists and turns of embryonic tissues, during the fetal period: an axial twist of the early brain, producing the optic chiasm [44]; a twist of the intestines, shifting the liver to the right side of the chest; a fusion and twist of the embryonic aortic arches, producing the heart.

Subclasses. Class Bilateria has two subclasses: Class Nephrozoa and Class Xenacoelomor-pha. Class Xenacoelomorpha contains very simple, small animals all of which lack a true gut (alimentary system).

Class Nephrozoa

The coelom, the excretory organs, and nerve cords are found in animals of Class Nephrozoa. The coelom is the body cavity that contains the digestive tract and the organs deriving from endoderm and mesoderm. The coelom is lined by either a specialized mesodermal cell (the mesothelial cell in humans) or by apparently undifferentiated meso-dermal cells (as observed in molluscs). Nerve cords are the evolutionary precursors of the central nervous system and the spinal cord, as found in vertebrates. A primitive nerve cord would be a tract of nervous tissue, mostly bundled axons, that touch upon or emanate from ganglia (clusters of neuronal bodies) located at either extremity of the organism.

One of the few nephrozoan subclasses lacking a coelom is Class Platyhelminthes, the flatworms (from Greek "platys," flat, "helmins," worm). The inclusion of a coelom in most other subclasses of Class Nephrozoa suggests that the coelom is important. In humans, the lungs are suspended in a chamber of the original embryonic coelom (i.e., the pleural cavity), and the pleural coelomic space permits the lungs to expand and contract. Having no coelom, members of Class Platyhelminthes lack lungs. Oxygen is absorbed by simple diffusion. We can speculate that platyhelminthes are flat to maximize oxygen absorption and transfer.

There is some evidence to suggest that smooth and striated muscle cells originated in Class Nephrozoa, as these cell types were present in an ancestral class of both protostomes and deuterostomes [45]. Hence Class Nephrozoa, being the direct parent class of both pro-tostomes and deuterostomes, is a candidate for the first class of organisms to be endowed with both smooth and striated muscles. The protostome/deuterostome split from Class Nephrozoa is thought to have occurred about 558 years ago, just before the Cambrian explosion. As we discussed previously, the Cambrian explosion marks the period when predatory/prey activities moved into full swing; when animals needed muscles for chasing, seizing, and eating their prey. It makes sense that muscle cells would be largely unnecessary much earlier than the appearance of Class Nephrozoa. Lastly, it would not seem that muscle cells have been reported in animals other than those that have descended from Class Nephrozoa.

Subclasses. Class Nephrozoa has two subclasses: Class Deuterostomia and Class Protostomia. In most protostomes, the mouth forms first, and the anus follows. In deuterostomes, the anus comes first, then the mouth.

One additional early developmental feature contributes to the dichotomy between the protostomes and the deuterostomes. All protostome blastulas derive from an 8-cell morula wherein the cells are spirally cleaved from one another and in which each of the 8 cells are committed to a specific differentiation lineage. All deuterostome blastulas derive from an 8-cell morula wherein the cells are radially cleaved, and commitment of early embryonic cells to any particular differentiated lineage is delayed until the blastocyst is fully formed and the first two layers of the embryo appear.

The dichotomy of protostomes and deuterostomes, the two largest classes of animals, serves as another indicator of the tight biological relationship between embryologic development and evolutionary advances in phylogenetic classes.

Class Protostomia marks the departure, from the human lineage, and includes the arthropods (ancestors of insects, chelicerata, and crustaceans), as well as mollusks (including octopi and squid), annelids, nematodes, and platyhelminthes.

Deuterostomia (also known as Enterocoelomata)

In deuterostomes, a dent forms, at an early stage of development, which eventually forms the anus, from which endoderm tunnels to produce a mouth, at the opposite end of the embryo. Humans and all vertebrates are deuterostomes.

Class Deuterostomia is also known as Class Enterocoelomata, the reason being that every deuterostome develops its central coelom (body cavity) through a process known as enterocoely. In enterocoely, the coelom is formed by pinched-off pouches of endoderm. In classes other than the deuterostomes, the central coelom is formed by cavitations occurring within the mesoderm, a process called schizocoely.

Class Deuterostomia has the distinction of being the first major class of animals to contain no members that are infectious or parasitic in humans. The descendants of Class Deuterostomia may be eager to attack and eat humans (lions and tigers and bears, oh my!), but none of these descendant species infect humans (as do bacteria and unicellular eukaryotes, and fungi) or parasitize humans, as do protostomal nematodes (roundworms) and platyhelminthes (flatworms).

Subclasses. Class Deuterostomia has two subclasses: Class Chordata (in the human lineage) and Class Ambulacraria. Class Ambulacraria contains Class Echinodermata (includes starfish) and Class Hemichordata (includes so-called acorn worms, most of whom reside on ocean beds).

Chordata (530 mya)

Class Chordata consists of the deuterostomes that have evolved a notochord, a flexible rod of cells in a cartilaginous matrix that runs the length of the back, and which serves as the primary support of the endoskeleton. All members of Class Chordata develop an embryonic notochord, but we will soon see that in vertebrates, the adult notochord is almost entirely replaced by a segmented bony rod known as the vertebra. In adult vertebrates, it is widely believed that remnants of the embryonic notochord are retained in the nucleus pulposus, the elastic material that in the center of the intervertebral disks. The notochordal derivation of intervertebral disks of vertebrates has been recently contested, another indication that embryology and evolution, like all vital disciplines of science, are perpetually under scrutiny [46].

All chordates are capable of developing a tumor known as the chordoma. Obviously, non-chordates, lacking the specialized cells that form the notochord, cannot develop chordomas. Chordomas occur somewhat commonly in ferrets. In humans, chordomas are rare cancers that can occur in any age group. The chordoma is an exception to the general rule that types of tumors that occur in children tend not to occur in adults; and vice versa. The zebrafish, a species of Class Chordata, is an animal model for chordoma formation. As in human chordomas, zebrafish chordomas show positive immunohistochemical staining for brachyury, considered a specific diagnostic marker for chordomas and for normal notochord [46–48]. As zebrafish are easy to study in the laboratory, it has been proposed that the zebrafish model may serve to screen potential therapeutic agents effective against human chordomas [47].

Subclasses. Class Chordata contains three extant subclasses: Class Craniata (includes the human lineage), Class Tunicata (includes sea squirts), and Class Cephalochordata (includes lancelets, also known as amphioxus).

Glossary

Apomorphy A new trait, appearing in a species, that can be passed to all the descendant species of the class. The apomorph is the new trait that is called the synapomorphy in the descendant species. There really isn't much difference between the terms apomorph and synapomorphy, other than positional. As a menominic, you may find it convenient to remember that the apomorph sits at the apex.

Child class The direct or first generation subclass of a class. Sometimes referred to as the daughter class, or as the immediate subclass.

Cladistics The technique of creating the schema for a hierarchy of clades, wherein each clade is a monophyletic class (i.e., a class whose subclasses contain every descendant species of the class and excludes species that are not descendants of the class).

Classification versus ontology A classification is a system in which every object in a knowledge domain is assigned to a class within a hierarchy of classes. The properties of superclasses are inherited by the subclasses. Every class has one immediate superclass (i.e., parent class), although a parent class may have more than one immediate subclass (i.e., child class). Objects do not change their class assignment in a classification, unless there was a mistake in the assignment. For example, a rabbit is always a rabbit, and does not change into a tiger.

A classification should be distinguished from an ontology. In an ontology, a class may have more than one parent class and an object may be a member of more than one class. A classification can be considered a restrictive and simplified form of ontology wherein each class is limited to a single parent class and each object has membership in one and only one class [49].

Diagnosis versus classification Diagnosis is the process by which a disease is assigned a name (i.e., a taxonomic term) belonging to a class of diseases. When a pathologist examines a tumor specimen and declare that the tumor is a squamous cell carcinoma of the skin, she is rendering a diagnosis (i.e., finding a specimen's location in a preexisting classification). Classification is different from diagnosis, the former referring to the process of inventing the class structure into which the taxonomy (i.e., the list of names of diagnosed diseases) must fit. You'll occasionally hear sentences such as "The pathologist classified this lesion as a squamous cell carcinoma of skin." This is an inaccurate use of terminology. Strictly speaking, pathologists diagnose lesions; they do not classify them.

Dysplasia The term means abnormal growth, and it is used in different ways in different biomedical specialties. Developmental biologists, and pediatricians use the term "dysplasia" to refer to organs or parts of organs that have not grown properly. Stunted growth of an organ, or morphologically abnormal tissues within an organism would be types of developmental dysplasia. Oncologists (i.e., cancer specialists) use the term "dysplasia" to describe the kinds of cellular changes that characterize neoplasms. Cellular dysplasia is found in precancerous, cancerous, and benign tumors.

Histone The major protein in chromatin (i.e., the material composing chromosomes). A segment of DNA, wound around a histone protein core, is called a nucleosome, the basic unity of DNA packaging in eukaryotes. When histones are modified by deacetylation (via deacetylases), the nucleosome tightens, reducing normal transcription by blocking transcription factors from attaching to target sites.

Metazoa Class Metazoa is equivalent to Class Animalia.

Monophyletic class A class of organisms that includes a parent organism and all its descendants, excluding organisms that did not descend from the parent. When a subclass of a parent class omits any of the descendants of the parent class, then the parent class is said to be paraphyletic. When a subclass of a parent class includes organisms that did not descend from the parent, then the parent class is polyphyletic. The goal of cladistics is to create a hierarchical classification that consists exclusively of monophyletic classes (i.e., no paraphyly, no polyphyly).

Multiclass classification A term indicating that an instance (e.g., species or object instance) has been assigned to more than one class. Classifications, as defined in this book, impose one-class rule (i.e., an instance can be assigned to one and only one class). It is tempting to think that a ball should be included in class "toy" and in class "spheroids," but multiclass assignments yield taxonomies of enormous size, consisting of many replicate items. Ontologies are a type of classification that generally permit multiclass assignments.

Multiclass inheritance In ontologies, multiclass inheritance occurs when a child class has more than one parent class. For example, a member of Class House may have two different parent classes: Class Shelter and Class Property. Multiclass inheritance is generally permitted in ontologies but is forbidden in classifications which restrict inheritance to a single parent class (i.e., each class can have at most one parent class, though it may have multiple child classes).

Medical taxonomists should understand that when multiclass inheritance is permitted, a class may be an ancestor of a child class that is an ancestor of its parent class (e.g., a single class might be a grandfather and a grandson to the same class). An instance of a class might be an instance of two classes, at once. The combinatorics and the recursive options become computationally absurd.

Taxonomists who rely on simple, uniparental classifications do so on epistemological grounds. They hold that an object can have only one nature, and therefore can belong to only one defining class, and can be derived from exactly one parent class. Taxonomists who insist on uniparental class inheritance believe that assigning more than one parental class to an object indicates that you have failed to grasp the essential nature of the object [7, 10, 50].

Ontology An ontology, like a classification, is a class-based organization of objects in a domain of knowledge; the primary difference between an ontology and a classification is that classes in an ontology may have more than one parent class (i.e., multiclass inheritance is permitted) [7, 51]. Hence, all classifications are ontologies, but not all ontologies are classifications.

Paradoxes of classification The rules for constructing classifications seem obvious and simplistic. Surprisingly, the task of building a logical, and self-consistent classification is extremely difficult. Most classifications are rife with logical inconsistencies and paradoxes. Let's look at a few examples. In 1975, NIH's Building 10 (the Clinical Center), located in Bethesda, Maryland, was touted as the largest all-brick building in the world, providing a home to over 7 million bricks. Soon thereafter, an ambitious construction project was undertaken to greatly expand the size of Building 10. When the work was finished, building 10 was no longer the largest all-brick building in the world. What happened? The builders used material other than brick, and Building 10 lost its place in the class of all-brick buildings. This poses something of a paradox; objects in a classification are not permitted to move about from one class to another. An object assigned to a class must stay in its class (i.e., the nontransitive property of classifications).

Apparent paradoxes that plague any formal conceptualization of classifications are not difficult to find. Let's look at a few more examples. Consider the geometric class of ellipses; planar objects in which the sum of the distances to two focal points is constant. Class Circle is a child of Class Ellipse, for which the two focal points of instance members occupy the same position, in the center, producing a radius of constant size. Imagine that Class Ellipse is provided with a class method called "stretch," in which the foci are moved further apart, thus producing flatter objects. When the parent class "stretch" method is applied to members of the Class Circle, the circle stops being a circle and becomes an ordinary ellipse. Hence the inherited "stretch" method forces members of Class Circle to transition out of their assigned class, violating the intransitive property of classes (i.e., that objects cannot switch over from one class to another, over time).

Let's look at the "Bag" class of objects. A "Bag" is a collection of objects, and the Class Bag is included in most object-oriented programming languages. A "Set" is also a collection of objects (i.e., a subclass of Bag), with the special feature that duplicate instances are not permitted. For example, if Kansas is a member of the set of the US States, then you cannot add a second state named "Kansas" to the set. If Class Bag were to have an "increment" method, that added "1" to the total count of objects in the bag, whenever an object is added to Class Bag, then the "increment" method would be inherited by all of the subclasses of Class Bag, including Class Set. But Class Set cannot increase in size when duplicate items are added. Hence, inheritance creates a paradox in the Class Set.

The suggested upper merged ontology (SUMO) in an ontology designed to contain classes for general types of objects that might be included in other, more specific knowledge domains. The SUMO, as one might expect from an ontology, allows multiple class inheritance. For example, in SUMO, the class of humans is assigned to two different parent classes: Class Hominid and Class CognitiveAgent. "HumanCorpse", another SUMO class, is defined in SUMO as "A dead thing which was formerly a Human." Human corpse is a subclass of Class OrganicObject; not of Class Human. This means that a human, once it ceases to live, transits to a class that is not directly related to the class of humans. Basically, a member of Class Human, in the SUMO ontology, will change its class and its ancestral lineage, at different timestamped moments.

One last dalliance. Consider these two classes from the SUMO ontology, both of which happen to be subclasses of Class Substance: Subclass Natural Substance, Subclass Synthetic Substance. It would seem that these two subclasses are mutually exclusive. However, diamonds occur naturally, and diamonds can be synthesized. Hence, diamond belongs to Subclass Natural Substance and to Subclass Synthetic Substance. The ontology creates two mutually exclusive classes that contain some of the same objects, thus creating a paradox [7].

Parent class The immediate ancestor, or the next-higher class (i.e., the direct superclass) of a class. For example, in the classification of living organisms, Class Vertebrata is the parent class of Class Gnathostomata. Class Gnathostomata is the parent class of Class Teleostomi, and so on.

Problematica The term "problematica" is used by taxonomists to indicate a class of organism that defies robust classification [52]. The very existence of this term tells us that taxonomy is a delicate and tentative science. We must always be prepared to examine and test our current classification, and to make corrections wherever warranted.

Pseudoscience A set of beliefs or theories that cannot be tested (i.e., cannot be proven to be right or to be wrong). Pseudoscientific beliefs are not necessarily wrong; we just have no way of knowing. Intelligent design is an example of a pseudoscientific belief. If we assert that humans were designed by an intelligent being, such as a superior but permanently absent alien life form, or an all-powerful but invisible entity, then we are making a pseudoscientific statement. There is no way to determine whether the assertion is false.

Rank Synonymous with Taxonomic order. In hierarchical biological nomenclatures, classes are given ranks. In early versions of the classification of living organisms, it was sufficient to divide the classification into a neat handful of divisions: Kingdom, Phylum, Class, Order, Family, Genus, Species. Today, the list of divisions has nearly quadrupled. For example, Phylum has been split into the following divisions: Superphylum, Phylum, Subphylum, Infraphylum, and Microphylum. The other divisions are likewise split. The subdivisions often have a legitimate scientific purpose. Nonetheless, current taxonomic order is simply too detailed for readers to memorize.

Furthermore, the complex taxonomic ranking system for living organisms does not carry over to the ranking systems that might be used for other scientific domains (e.g., classification of diseases, classification of genes, etc.) and creates an impediment for anyone wanting to bridge classifications held within diverse, but related, fields.

Herein, taxonomic complexity is drastically simplified by dropping named ranks and simply referring to every class as "Class." When every class of organism is linked with the name of its parent class, then it becomes possible to computationally trace the complete ancestral lineage for every class, species or organism [8].

Results The term "results" is often mistaken for the term "conclusions." In the strictest sense, "results" consist of the full set of experimental data collected by measurements. In practice, "results" consist of a subset of data distilled from the raw, original data. In a typical journal article, selected data subsets are packaged as a chart or graph that emphasizes some point of interest. Hence, the term "results" may refer, erroneously, to subsets of the original data, or to visual graphics intended to summarize the original data. Conclusions are the inferences drawn from the results. Results are verified; conclusions are validated.

Science Of course, there are many different definitions of science. For me, science is all about finding general relationships among objects. In the physical sciences, the most important relationships are expressed as mathematical equations (e.g., the relationship between force, mass and acceleration; the relationship between voltage, current, and resistance). In the natural sciences, relationships are often expressed through classifications (e.g., the classification of living organisms).

Subclass A subclass is a descendant class of its direct line of ancestral classes. The members of a subclass inherit the properties and methods of the ancestral classes in its direct lineage. For example, all mammals have mammary glands because mammary glands are a defining property of the mammal class. In addition, all mammals have vertebrae because the class of mammals is a subclass of the class of vertebrates. The subclass that is the immediate descendant of a class is called the child class. In common parlance, when we speak of the subclass of a class, we are referring specifically to its child class.

Superclass A superclass is an ancestral class of its direct line of descendants. For example, in the classification of living organisms, the class of craniates is a superclass of the class of mammals. The immediate superclass of a class is its parent class. In common parlance, when we speak of the super-class of a class, we are referring specifically to its parent class.

Synapomorphy A trait found in all the members of a clade (i.e., shared by the species descending from the ancestral species in which the trait first appeared).

Unclassifiable objects Classifications create an hierarchical collection of classes, and taxonomies assign each and every named object to its correct class. This means that a classification is not permitted to contain unclassified objects; a condition that puts fussy taxonomists in an untenable position. Suppose you have an object, and you simply do not know enough about the object to confidently assign it to a class. Or, suppose you have an object that seems to fit more than one class, and you can't decide which class is the correct class. What do you do? Historically, scientists have resorted to creating a "miscellaneous" or "problematica" or "incertae sedis" (uncertain placement) class into which otherwise unclassifiable objects are given a temporary home, until more suitable accommodations can be provided [52].

The promiscuous application of "miscellaneous" classes have proven to be a huge impediment to the advancement of the biological sciences. In the case of the classification of living organisms, the class of protozoans stands as a case in point. Ernst Haeckel, a leading biological taxonomist in his time, created the Kingdom Protista (i.e., protozoans), in 1866, to accommodate a wide variety of simple organisms with superficial commonalities. Haeckel himself understood that the protists were a blended class that included unrelated organisms, but he believed that further study would resolve the confusion. In a sense, he was right, but the process took much longer than he had anticipated; occupying generations of taxonomists over the following 150 years. Today, its members have been reassigned to their own particular classes. In the meantime, therapeutic opportunities for eradicating so-called protozoal infections, using class-targeted agents, have no doubt been missed [9]. For practical reasons, textbooks still use the term "protozoan," but strictly speaking, Kingdom Protista no longer exists [53].

You may think that the creation of a class of living organisms, with no established scientific relation to the real world, was a rare and ancient event in the annals of biology, having little or no chance of being repeated. Not so. A special pseudoclass of fungi, deuteromycetes (spelled with a lowercase "d," signifying its questionable validity as a true biologic class) has been created to hold fungi of indeterminate speciation. At present, there are several thousand such fungi, sitting in a taxonomic limbo, waiting to be placed into a definitive taxonomic class [9, 54].

Underlying cause The event that initiated the sequence of events leading to some clinical outcome (e.g., disease). Death certificates require physicians to list the underlying cause of death. The World Health Organization, aware of the difficulties in choosing an underlying cause of death, and assigning a sequential list of the ensuing clinical consequences leading to the proximate cause of death, has issued reporting guidelines [55]. Instructions notwithstanding, death certificate data is notoriously inconsistent, giving rise to divergent methods of reporting the diseases that cause death [56–58].

Reluctantly, we must acknowledge that, in any biological system, we can seldom designate an underlying cause with any certitude. One event may lead to many other events, and events which we believe to be initiating may have their own predicate causes.

Unstable taxonomy A taxonomy that is prone to change over time. You might expect that a named species would keep its name forever, and would never change its assigned class. Not so; taxonomists are continually fussing with the classification of living organisms, as new information is received and analyzed. For example, Class Fungi has recently undergone profound changes, with the exclusion of myxomycetes (slime molds, currently assigned to Class Amoebozoa) and oomycetes (water molds, currently assigned to Class Heterokonta), and the acquisition of Class Microsporidia (formerly classed as a protozoan). The instability of fungal taxonomy impacts negatively on the practice of clinical mycology. When the name of a fungus changes, so must the name of the associated disease. Consider "*Allescheria boydii*," Individuals infected with this organisms were said to suffer from the disease known as allescheriasis. When the organism's name was changed to *Petriellidium boydii*, the disease name was changed to petriellidosis. When the fungal name was changed, once more, to *Pseudallescheria boydii*, the disease name was changed to pseudallescheriasis [54]. All three names appear in the literature, past and present, thus hindering attempts to achieve a consistent understanding of the organism [9].

Validation The process whereby a conclusion drawn from results is shown to be repeatable.

Verification The process by which data is checked to determine whether the data was obtained properly (i.e., according to approved protocols), and that the data accurately measured what it was intended to measure, on the correct specimens. Data verification is not easy [59]. In one celebrated case, involving a microarray study, two statisticians devoted 2000 h to the job [60]. Two thousand hours is just about one full man-year of effort. It is important to remember the difference between verification and validation. Verification is a process performed on data. Validation is a process performed on the conclusions drawn from the analysis of the data.

References

[1] Simpson GG. Principles of animal taxonomy. New York: Columbia University Press; 1961.

[2] Simpson GG. The principles of classification and a classification of mammals. Bull Am Mus Nat Hist 1945;85.

[3] Mayr E. The growth of biological thought: diversity, evolution and inheritance. Cambridge: Belknap Press; 1982.

[4] Mayr E. Two empires or three? PNAS 1998;95:9720–3.

[5] Woese CR. Default taxonomy: Ernst Mayr's view of the microbial world. PNAS 1998;95(19):11043–6.

[6] Smith B, Ceusters W, Klagges B, Kohler J, Kumar A, Lomax J, et al. Relations in biomedical ontologies Genome Biol 2005;6:R46. Available at: http://genomebiology.com/2005/6/5/R46 [viewed 09.09.15].

[7] Berman JJ. Data simplification: Taming information with open source tools. Waltham, MA: Morgan Kaufmann; 2016.

[8] Berman JJ. Methods in medical informatics: fundamentals of healthcare programming in Perl, Python, and Ruby. Boca Raton: Chapman and Hall; 2010.

[9] Berman JJ. Taxonomic guide to infectious diseases: understanding the biologic classes of pathogenic organisms. Cambridge, MA: Academic Press; 2012.

[10] Berman JJ. Principles of big data: preparing, sharing, and analyzing complex information. Waltham, MA: Morgan Kaufmann; 2013.

[11] Berman JJ. Biomedical informatics. Sudbury, MA: Jones and Bartlett; 2007.

[12] Berman J. Precision medicine, and the reinvention of human disease. Cambridge, MA: Academic Press; 2018.

[13] Berman JJ. Principles and practice of big data: preparing, sharing, and analyzing complex information. 2nd ed. Waltham, MA: Morgan Kaufmann; 2018.

[14] Cavalier-Smith T. Early evolution of eukaryote feeding modes, cell structural diversity, and classification of the protozoan phyla Loukozoa, Sulcozoa, and Choanozoa. Eur J Protistol 2013;49:115–78.

[15] Brocks JJ, Logan GA, Buick R, Summons RE. Archaean molecular fossils and the early rise of eukaryotes. Science 1999;285:1033–6.

[16] Bazin P, Jouenne F, Friedl T, Deton-Cabanillas A-F, Le Roy B, et al. Phytoplankton diversity and community composition along the estuarine gradient of a temperate macrotidal ecosystem: combined morphological and molecular approaches. PLoS ONE 2014;9:e94110.

[17] Baldauf SL. An overview of the phylogeny and diversity of eukaryotes. J Syst Evol 2008;46:263–73.

[18] Wu D, Hugenholtz P, Mavromatis K, Pukall R, Dalin E, Ivanova NN, et al. A phylogeny-driven genomic encyclopaedia of Bacteria and Archaea. Nature 2009;462:1056–60.

[19] Stechmann A, Hamblin K, Perez-Brocal V, Gaston D, Richmond GS, van der Giezen M, et al. Organelles in Blastocystis that blur the distinction between mitochondria and hydrogenosomes. Curr Biol 2008;18:580–5.

[20] Tovar J, Leon-Avila G, Sanchez LB, Sutak R, Tachezy J, van der Giezen M, et al. Mitochondrial remnant organelles of Giardia function in iron-sulphur protein maturation. Nature 2003;426:172–6.

[21] Tovar J, Fischer A, Clark CG. The mitosome, a novel organelle related to mitochondria in the amito-chondrial parasite *Entamoeba histolytica*. Mol Microbiol 1999;32:1013–21.

[22] Burri L, Williams B, Bursac D, Lithgow T, Keeling P. Microsporidian mitosomes retain elements of the general mitochondrial targeting system. PNAS 2006;103:15916–20.

[23] Finsterer J. Mitochondriopathies. Eur J Neurol 2004;11:163–86.

[24] Kokotas H, Petersen MB, Willems PJ. Mitochondrial deafness. Clin Genet 2007;71:379–91.

[25] Kalghatgi S, Spina CS, Costello JC, et al. Bactericidal antibiotics induce mitochondrial dysfunction and oxidative damage in mammalian cells. Sci Transl Med 2013;5:192ra85.

[26] Raimundo N, Song L, Shutt TE, McKay SE, Cotney J, Guan MX, et al. Mitochondrial stress engages E2F1 apoptotic signaling to cause deafness. Cell 2012;148(4):716–26.

[27] Margulis L. Undulipodia, flagella and cilia. Biosystems 1980;12:105–8.

[28] Margulis L, Sagan D. Order amidst animalcules: the protoctista kingdom and its undulipodiated cells. Biosystems 1985;18(1985):141–7.

[29] Koster MJ, Snel B, Timmers HT. Genesis of chromatin and transcription dynamics in the origin of species. Cell 2015;161:724–36.

[30] Filee J. Multiple occurrences of giant virus core genes acquired by eukaryotic genomes: the visible part of the iceberg? Virology 2014;466-467:53–9.

[31] Forterre P, Gaia M. Giant viruses and the origin of modern eukaryotes. Curr Opin Microbiol 2016;31:44–9.

[32] Forterre P. Defining life: the virus viewpoint. Orig Life Evol Biosph 2010;40:151–60.

[33] Pagani F, Baralle FE. Genomic variants in exons and introns: identifying the splicing spoilers. Nat Rev Genet 2004;5:389–96.

[34] Fraser HB, Xie X. Common polymorphic transcript variation in human disease. Genome Res 2009;19 (4):567–75.

[35] Minge MA, Silberman JD, Orr RJ, Cavalier-Smith T, Shalchian-Tabrizi K, Burki F, et al. Evolutionary position of breviate amoebae and the primary eukaryote divergence. Proc Biol Sci 2009;276:597–604.

[36] Telford MJ. Animal evolution: once upon a time. Curr Biol 2009;19:R339–41.

[37] Tyler S. Epithelium—the primary building block for metazoan complexity. Integr Comp Biol 2003;43:55–63.

[38] Delmar M, McKenna WJ. The cardiac desmosome and arrhythmogenic cardiomyopathies from gene to disease. Circ Res 2010;107:700–14.

[39] Alcalai R, Metzger S, Rosenheck S, Meiner V, Chajek-Shaul T. A recessive mutation in desmoplakin causes arrhythmogenic right ventricular dysplasia, skin disorder, and woolly hair. J Am Coll Cardiol 2003;42:319–27.

[40] Ereskovsky AV. Sponge embryology: the past, the present and the future. In: Porifera research: biodiversity, innovation and sustainability—2007. Rio de Janeiro: Universidade Federal do Rio de Janeiro; 2007. p. 41–52.

[41] Brocks JJ, Jarrett AJM, Sirantoine E, Kenig F, Moczydowska M, Porter S, et al. Early sponges and toxic protists: possible sources of cryostane, an age diagnostic biomarker antedating Sturtian snowball Earth. Geobiology 2016;14:129–49.

[42] Love GD, Grosjean E, Stalvies C, Fike DA, Grotzinger JP, Bradley AS, et al. Fossil steroids record the appearance of Demospongiae during the Cryogenian period. Nature 2009;457:718–21.

[43] Chang ES, Neuhof M, Rubinstein ND, Diamant A, Philippe H, Huchon D, et al. Genomic insights into the evolutionary origin of Myxozoa within Cnidaria. PNAS 2015;48:14912–7.

[44] deLussaneta MH, Osseb JWM. An ancestral axial twist explains the contralateral forebrain and the optic chiasm in vertebrates. Anim Biol 2012;62:193–216.

[45] Brunet T, Fischer AHL, Steinmetz PRH, Lauri A, Bertucci P, Arendt D. The evolutionary origin of bilaterian smooth and striated myocytes. elife 2016;5:e19607.

[46] Vujovic S, Henderson S, Presneau N, Odell E, Jacques TS, Tirabosco R, et al. Brachyury, a crucial regulator of notochordal development, is a novel biomarker for chordomas. J Pathol 2006;209:157–65.

[47] Burger A, Vasilyev A, Tomar R, Selig MK, Nielsen P, Peterson RT, et al. A zebrafish model of chordoma initiated by notochord-driven expression of HRASV12. Dis Model Mech 2014;7:907–13.

[48] Presneau N, Shalaby A, Ye H, Pillay N, Halai D, Idowu B, et al. Role of the transcription factor T (brachyury) in the pathogenesis of sporadic chordoma: a genetic and functional-based study. J Pathol 2011;223:327–35.

[49] Patil N, Berno AJ, Hinds DA, Barrett WA, Doshi JM, Hacker CR, et al. Blocks of limited haplotype diversity revealed by high-resolution scanning of human chromosome 21. Science 2001;294:1719–23.

[50] Berman JJ. Repurposing legacy data: innovative case studies. Waltham, MA: Morgan Kaufmann; 2015.

[51] Berman JJ. Ruby programming for medicine and biology. Sudbury, MA: Jones and Bartlett; 2008.

[52] Jenner RA, Littlewood TJ. Problematica old and new. Philos Trans R Soc B 2008;363:1503–12.

[53] Schlegel M, Hulsmann N. Protists: a textbook example for a paraphyletic taxon. Org Divers Evol 2007;7:166–72.

[54] Guarro J, Gene J, Stchigel AM. Developments in fungal taxonomy. Clin Microbiol Rev 1999;12:454–500.

[55] National Center for Health Statistics. U.S. Vital Statistics System: major activities and developments, 1950–95. Centers for Disease Control and Prevention, National Center for Health Statistics; 1997.

[56] Ashworth TG. Inadequacy of death certification: proposal for change. J Clin Pathol 1991;44:265.

[57] Kircher T, Anderson RE. Cause of death: proper completion of the death certificate. JAMA 1987;258:349–52.

[58] Berman JJ. Rare diseases and orphan drugs: keys to understanding and treating common diseases. Cambridge, MD: Academic Press; 2014.

[59] Committee on Mathematical Foundations of Verification, Validation, and Uncertainty Quantification. Board on mathematical sciences and their applications, division on engineering and physical sciences, National Research Council. Assessing the reliability of complex models: Mathematical and statistical foundations of verification, validation, and uncertainty quantification. National Academy Press; 2012. Available from: http://www.nap.edu/catalog.php?record_id=13395 [viewed 01.01.15].

[60] Misconduct in science: an array of errors. The Economist; 2011. September 10.

6 ▪▪▪ / ▪▪▪ / ▪▪▪

Phylogeny: Craniates to Humans

OUTLINE

Section 6.1 Class Craniata and the Ascent of the Neural Crest

The only interesting thing about vertebrates is the neural crest.

Thorogood

```
Craniata (480 mya)
Vertebrata (500 mya)
Gnathostomata (419 mya)
Euteleostomi (400 mya)
Sarcopterygii
Dipnotetrapodomorpha
Tetrapodomorpha (390 mya)
Tetrapoda (367 mya)
Synapsida (308 mya)
Mammalia (220 mya)
Theria (160 mya)
Eutheria (160-125 mya)
Boreoeutheria (124-101 mya)
Euarchontoglires (100 mya)
Euarchonta (99-80 mya)
Primates (75 mya)
Haplorrhini (63 mya)
Simiiformes (40 mya)
Catarrhini (30 mya)
Hominoidea (28 mya)
Hominidae (15 mya)
Homininae (8 mya)
Hominini (5.8 mya)
Homo (2.5 mya)
Homo sapiens (0.3 mya)
Homo sapiens (modern) 0.07
```

Evolution's Clinical Guidebook. https://doi.org/10.1016/B978-0-12-817126-4.00006-0

Class Craniata is distinguished embryologically by the emergence of a fully functional neural crest, the most versatile and complex of all the embryologic anlagen. With the advent of the neural crest, craniates, and their descendants were imbued with a braincase and a face [1], a peripheral nervous system, cranial nerves, skin pigmentation, and several types of endocrine glands. How can one embryologic innovation account for all of these seemingly unrelated anatomic attributes?

It's impossible for me to think of the neural crest without thinking of the ancient fairy tale of the genie who grants the hero three wishes. As the story goes, after the first two wishes are squandered (they always are), we can only wonder why the hero of the story does not ask, as his third wish, for three more wishes. Apparently, recursive wishes are not permitted in fairy tales. Much to our benefit, nature provided craniates with a recursive opportunity that exceeds the plausibility of fairy tales. Neural crest cells have the uncanny ability to recapitulate cell types that would otherwise be restricted to one of the other embryonic layers (i.e., endoderm, mesoderm, ectoderm, and the ectodermal derivative, neuroectoderm) [2]. In addition, the neural crest offers some fascinating cell types not present in any of the preceding germ layers. We will pretend that an ancient craniate hero, about 480 million years ago (mya), must have asked, as a third wish, for the ability to recursively employ all of the cell types that had previously evolved from the other embryonic layers. [Glossary Neurectoderm]

In embryological development, the neural crest derives from a specialized compartment of cells lying between the ectoderm and the primitive neural tube. The neural crest gives rise to the peripheral nervous system, to the connective tissue of the cranium, to several endocrine glands, and to the connective tissue component of the teeth, and much more. Indeed, the cell type versatility of the neural crest is so vast that it is difficult for us to confidently fathom its biological limitations. Hence, the list that follows, of neural crest derivatives, is somewhat tentative.

- Cranial neural crest

The cranial compartment of the neural crest accounts for the mesenchyme of the face and skull and part of the shoulder [3], including the mesenchyme of several pharyngeal arches (homologous to the gill arches or branchial arches in craniate fish), odontoblasts (of teeth), and possibly the C cells (calcitonin-producing cells) of the thyroid. We can thank the neural crest for teeth, jaws, the bones of the face, and a cranial vault lined by meninges [1, 2, 4]. The cranial trunk of the neural crest also provides the sensory ganglia of the 5th, 7th, 9th, and 10th cranial nerves. Even in those facial organs that predate Class Craniata, such as the eyes, the neural crest makes a profound contribution, providing several accessory parts, including the choroid, sclera, iris, and ciliary body. As a general rule, if an animal has something that we might recognize as a face with two eyes, then it is probably a member of Class Craniata.

The earliest known undisputed craniates are jawless fishes (agnathans) which lived 480 mya. In some fossil jawless vertebrates, the neural crest contributed to the formation of scales and bones, of dermal origin.

- Trunk neural crest

Melanocytes and dorsal root ganglia (of spinal cord)

– Vagal and sacral neural crest

Parasympathetic ganglia

– Cardiac neural crest

Consists of arteries arising from the heart (including the aortopulmonary septum), connective tissue of the pharyngeal arches not produced by the cranial compartment of the neural crest, and melanocytes other than those produced by the trunk neural crest.

If neural crest cells produce bones of the face, cranial vault, and some of the bones of the shoulder, then how can we distinguish the osteocytes (bone cells) of neural crest origin from those of mesodermal origin? The answer to this question is somewhat disappointing insofar as a neural crest osteocyte is morphologically no different from a mesodermal osteocyte. Not surprisingly, the diseases of bones that occur in fully developed adults are essentially equivalent, regardless of the embryonic origin of the bone. An osteosarcoma of a bone of neural crest origin is much like an osteosarcoma arising in a bone of mesodermal origin. Paget's disease of bone can effect bones of the skull (neural crest origin) or bones of the leg (mesodermal origin). However, the neural crest osteocytes have a different developmental origin, and this developmental origin is fully displayed by the manner in which the bones grow. Bones of neural crest origin grow as the so-called dermal bones (also known as intramembranous bones) in which a mesenchymal matrix shaped like a whole bone, becomes gradually ossified, in place. This process works well for flat bones of the skull that fill an anatomic space in the newborn that is not greatly different from its fully mature size in the adult. Bones of mesodermal origin, the so-called endochondral bones, grow outwards from a cartilaginous nidus (the epiphyseal plate) ossifying centrifugally away from the cartilage-based growth zone. This produces a gradually elongating process that works well for long bones, such as the femur.

As the development of neural crest bones is different from the development of mesodermal bones, we can expect to find some developmental disorders of bone appearing in very young children that express themselves in either neural crest bones or mesodermal bones, but not both. In point of fact, there are literally hundreds of developmental anomalies that are confined to the bones and soft tissues of neural crest origin, including cleft palate and cleft lip, craniosynostosis (premature fusion of cranial sutures), macrocephaly, microcephaly, and jaw defects that include agnathia.

Indeed, there is a well-studied groups of inherited genetic disorders of neural crest development, and these are known under the general term neurocristopathies. The neurocristopathies are clinically diverse diseases that include, among others:

– Hirschspring disease (absence of neural crest-derived neurons from varying segments of the gut with consequent lack of peristalsis in the involved segments).

– Treacher Collins syndrome (mandibulofacial dysostosis, characterized by variable involvement by cleft palate, hypoplasia of the mandible and zygomatic arches, notches in the lower eyelids, and malformation of the external ears and middle ear ossicles).

- Piebaldism (patchy interruptions of neural-crest derived melanocytes in skin).

- Waardenburg syndrome (patchy depigmentation, as in piebaldism, along with blue or heterochromatic irides, sensorineural hearing loss, and spina bifida in some cases). [Glossary Irides]

All such diseases involve neural crest development, and all such diseases can only occur in animals of Class Craniata. We now know the underlying genetic cause of nearly 20 different neurocristopathies. **The striking generalization that we draw from these genetic studies is that the genetic mutations responsible for the neurocristopathies have no easily discernible relationship to neural crest development.** For example, the root genetic cause of Diamond Blackfan anemia involves a mutation in any of about a half dozen genes known to play a role in the formation of ribosomes (the structures that translate mRNA into protein). Individuals with Diamond Blackfan anemia have a variety of hematologic disorders as well as malformations of tissues derived from the neural crest. There really is no way by which a medical scientist can take what they know about ribosomes and predict that a generalized ribosomal deficiency would result in the maldevelopment of neural crest-derived tissues.

Much the same observation is true for the other neurocristopathies: identifying the disease gene seldom tells us much about the pathogenesis of the disease. In these cases, medical scientists work their way backwards, beginning with the clinical phenotype that tells us that a disease involves neural crest tissues, proceeding to our knowledge of how the neural crest controls development, moving to the genetic pathways employed by the neural crest, and eventually leading to the role of individual genes in the early pathogenesis of neural crest diseases. [Glossary Neurocristopathy]

- The developmental biology of neural crest tumors

Because all animals of Class Craniata have a neural crest, all craniates are capable of developing tumors of the neural crest. Man and fish are cousins and both belong to Class Craniata. Because man and fish both have neural crests, we both develop neural crest tumors [5–7]. It is easy to find schwannomas and melanomas in fish. It would be impossible to find such tumors in organisms that do not descend from Class Craniata. Do not expect to find schwannomas in a tarantula or a crab or a beetle or in any of the crowd of faceless species (i.e., noncraniates) that dwell among us. [Glossary Schwannoma]

Consider the carcinoid, a tumor arising primarily from gut that was formerly believed to be of neural crest origin. Carcinoid tumors are composed of cells containing small round cyoplasmic organelles known as either neurosecretory granules or as dense core granules. These granules contain hormone-like peptides that are similar to the hormones produced by endocrine cells of neural crest origin. It was presumed that neural crest cells had migrated to the gut during embryonic and fetal development, and that these endocrine-like cells of neural crest origin were the cells of origin of carcinoid tumors (Fig. 6.1).

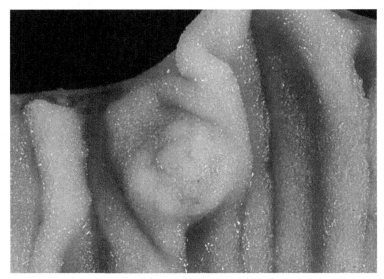

FIG. 6.1 Gross photograph of a small bowel with carcinoid tumor. *Source: Wikipedia, entered into the public domain by Ed Uthman.*

The long-held belief that carcinoid tumors have a neural crest origin has been largely discredited, and it is now believed that carcinoid tumor cells derive from specialized cells of endodermal origin (not neural crest origin).

For comparative embryologists, the strongest arguments against a neural crest origin for carcinoid tumors is based on a simple observation made on crabs. Crabs, like humans, have a gut, and the crab gut is lined by endocrine cells that can secrete neuropeptides, just like carcinoid cells [8]. Crabs unlike humans, are not members of Class Craniata and therefore lack a neural crest. If the crab has no neural crest, it cannot have cells that derive from the neural crest. Therefore, neuroendocrine cells of the gut (such as carcinoid cells) are not a phylogenetic trait of the neural crest, and carcinoid tumor of the gut is not a tumor of neural crest origin.

Here is one more example wherein our knowledge of neural crest evolution helps us better understand cancer. Pathologists occasionally encounter tumors composed of cells that closely resemble muscle cells, mixed with neural cells. We know that neuroectoderm can produce neural cells, but not muscle cells. We also know that mesoderm can produce muscle cells but not neural cells. Of all the embryonic anlagen, only neural crest can produce muscle cells and neural cells. Hence, a tumor, of presumed clonal origin, that differentiates as a mixture of neoplastic muscle cells and neural cells, must have originated from a neural crest cell. As it happens, mixed neural/muscle tumors are rare, but one such case has been shown to express the EWS/FLI1 fusion gene, a marker that seems to be specific for tumors of neural crest origin [9].

Subclasses. Class Craniata has two subclasses: Class Vertebrata (includes the human lineage) and Class Cyclostomata. Class Cyclostomata is very small, and seems to include only

FIG. 6.2 The enigmatic pacific hagfish, a craniate that lacks a vertebra. *Source: Wikipedia, from a public domain US National Oceanic and Atmospheric Administration work.*

the hagfish which has an incomplete braincase and lacks a vertebra. Hence, with the exceptions of maybe one extant species, Class Craniata is equivalent to Class Vertebrata (Fig. 6.2).

Section 6.2 Vertebrates to Synapsids

He is simply a shiver looking for a spine to run up.

Paul Keating

```
Vertebrata (500 mya)
Gnathostomata (419 mya)
Euteleostomi (400 mya)
Sarcopterygii
Dipnotetrapodomorpha
Tetrapodomorpha (390 mya)
Tetrapoda (367 mya)
Synapsida (308 mya)
```

Vertebrata

All members of Class Vertebrata have replaced the notochord (a stiff, unsegmented rod) with a segmented cord composed of vertebral bones separated by soft intervertebral disks. It is generally believed that the intervertebral disks are derived from cellular residua of the embryologic notochord. In addition, a forebrain, with a demarcated diencephalon and telencephalon, is present in all members of Class Vertebrata [10].

At about the time that members of Class Vertebrata appeared, a new evolutionary advance changed the cellular structure of just about every cell of every living vertebrate. This new innovation is found in some cells of nonvertebrates, but the vertebrates seem to be the only class of animals that have evolved the feature as a "must-have" constituent in every cell. It is called the primary cilium, and until a few decades ago, its existence was virtually unknown by the scientific community. The tale of its discovery, and of its role in the development of vertebrates, is one of the most fascinating stories of modern medicine.

Every school child is taught about the structure, cytology, and function of the motile cilia. These organelles protrude from the apical cell wall of epithelial cells lining the respiratory tract, intestines, and of ducts throughout the body, pushing intraluminal materials along their way up a bronchus or down the intestines or through a duct. Motile cilia have a nine-paired microtubule axoneme, with a central pair of microtubules and outer dynein arms. One cell may have many motile cilia.

The primary cilium has the appearance of a deformed motile cilia, lacking, as it does, the central pair of microtubules and outer dynein arms that are essential for normal motility. The primary cilium was first observed by microscopists in the 1950s, but because there was only one primary cilium per cell, and because it didn't seem capable of serving any useful function, it was long-dismissed as an evolutionary relic, much like the coccygeal bone or the vermiform appendix [11].

In the 1990s, it was recognized that the primary cilium, which grows from its tip, has no synthetic machinery within itself to transport growth substrates up through the cilium to the growth zone. Hence, it was presumed that the primary cilium is sustained by intraflagellar transport [12]. About this same time, the genes involved in intraflagellar transport were discovered, and these genes, when knocked out in mice, produced heterotaxy (i.e., left-right organ asymmetry), suggesting that primary cilia play an important role in embryonic and fetal development [11]. [Glossary Genetically engineered mouse, Knockout mice]

Today, we recognize a new class of diseases, the ciliopathies, all of which are associated with dysfunctions of the primary cilium [13, 14]. The ciliopathies constitute one of the most phenotypically diverse classes of diseases and illustrate that the disorders that have related pathogenesis may have highly dissimilar clinical features. For example, primary ciliary dyskinesia is characterized by bronchiectasis, sinusitis, otitis media, infertility, and situs defects. Alstrom syndrome, another ciliopathy, features dilated cardiomyopathy, obesity, sensorineural hearing loss, retinitis pigmentosa, endocrine abnormalities, renal and hepatic disease. It is hard to imagine two diseases less similar to one another than primary ciliary dyskinesia and Alstrom syndrome. Nonetheless, both are caused by aberrations affecting the primary cilia. Many of the ciliopathies produce cystic changes in kidneys and liver. The most anatomically surprising feature of some ciliopathies is situs inversus, wherein the position of organs is reversed (e.g., left lung switched with right lung, liver switched to right side of abdomen) [15]. Despite their diverse clinical phenotypes, there is hope that treatments developed for any member of the ciliopathies might be of benefit for every type of ciliopathic disease [13, 16]. [Glossary Ciliopathies, Pleiotropia]

Subclasses. Class Vertebrata has two subclasses: the jawed vertebrates (Class Gnathostomata) and the jawless vertebrates (Class Agnatha). There are only a few extant members of Class Agnatha, and these are the lampreys (with a very rudimentary vertebra) and some animals related to the lampreys. The mouths of agnathans have horny epidermal structures that function somewhat like the true teeth of Gnathostomes.

Gnathostomata

The first gnathostomes appeared about 419 mya, about 15 million years after the first plants and fungi colonized land, at a time when the first species of land plants were evolving from algae, and before the advent of land animals. Hence, the early gnathostomes were purely aquatic. The gnathostomes, or jawed vertebrates, include about 60,000 living species and account for virtually all living vertebrates.

The jaw of the gnathostomes is a neural crest derivative, and seems to have developed by dermal bone growth within a gill support arch. The early gnathostome jaw may have evolved to serve two purposes: (1) to pump water over the gills, hence aiding oxygenation, and (2) to widen the mouth, thus facilitating the capture of large prey.

In addition to their class-defining jaws, the gnathostomes are all equipped with three additional phylogenetic traits, all of neural crest derivation:

1. All have teeth.
2. All have a horizontal semicircular canal of the inner ear.
3. All have myelinated neurons.

These anatomic features have greatly facilitated the success of Class Gnathostomata, compared with its sister class, Class Agnatha.

Class Gnathostomata is the first class to have acquired a fully functional adaptive immunity. This evolutionary advancement was so successful that all gnathostome descendants have retained their adaptive immunity. The adaptive immune system responds to the specific chemical properties of foreign antigens, such as those that appear on viruses and other infectious agents. Adaptive immunity is a system wherein somatic T cells and B cells are produced, each with a unique and characteristic immunoglobulin (in the case of B cells) or T-cell receptor (in the case of T cells). Through a complex presentation and selection system, a foreign antigen elicits the replication of a B cell whose unique immunoglobulin molecule (i.e., so-called antibodies) matches the antigen. Secretion of matching antibodies leads to the production of antigen-antibody complexes that may deactivate and clear circulating antibodies, or may lead to the destruction of the organism that carries the antigen (e.g., virus or bacteria).

To produce functional B and T cells, each with a uniquely rearranged segment of DNA that encodes specific immunoglobulins or T-cell receptors, recombination and hypermutation take place within a specific gene region. This process yields on the order of a billion unique somatic genes, starting with one germinal genome. This amazing show of genetic heterogeneity requires the participation of recombination activating genes (i.e., RAGs). The acquisition of an immunologically active recombination activating gene is presumed to be the key evolutionary event that led to the development of the adaptive immune system that is present in all gnathostomes.

Before the appearance of the jawed vertebrates, this sort of genetic recombination was unavailable to animals. Our genes simply were not equal to the task. Retroviruses, however, are specialists at cutting, moving, and mutating DNA. Is it any wonder that the startling evolutionary leap to adaptive immunity, made possible by the RAG gene, was acquired from retrotransposons? Thus, the early jawed vertebrates (i.e., gnathostomes) and their descendant classes, owe their most important defense against infections to genetic material retrieved from the vast trove of retrovirally derived DNA carried in our genome [17]. As one might expect, inherited mutations in RAG genes are the root cause of several immune disorders [18, 19]. [Glossary Autoantibody disease versus autoimmune disease]

Subclasses. Class Gnathostomata has two subclasses: Class Placodermi, a class of armored fish having no extant species; and Class Eugnathostomata, which holds all living descendants of Glass Gnathostomata. Class Eugnathostomata has two subclasses: the Acanthodians, a largely extinct class in which the cartilaginous fishes (Class Chondrichthyes) are sometimes included; and Class Osteichthyes, the bony fish, roughly corresponding to Class Euteleostomi. Because the term "Euteleostomi" enjoys common usage in the Bioinformatics community, we will use Class Euteleostomi to indicate the subclass of Class Gnathostomata from which humans will descend.

Euteleostomi (also known as Osteichthyes)

Class Euteleostomi, the bony fish, contains 90% of all living species of vertebrates. When the first species of Class Euteleostomi inhabited the planet, about 400 mya, all of the euteleosts lived in the sea, and all were fish, of one type or another.

Subclasses. Class Euteleostomi has two subclasses: Class Actinopterygii, the ray-finned fish; and Class Sarcopterygii, the lobe-finned fish. Class Actinopterygii account for virtually all of the species of modern fishes. The only extant fish that are not direct descendants of Class Actinopterygii are the few fish species that directly descended from Class Sarcopterygii (i.e., coelacanths and lungfish).

At this point we can digress a bit to discuss the meaning of "survival" and "success," as applied to the two ancestral "fish" classes: Actinopterygii and Sarcopterygii. Class Actinopterygii was much more successful than Class Sarcopterygii, in terms of producing new fish species. Today's oceans, rivers and streams are populated nearly entirely by descendants of Class Actinopterygii. Class Sarcopterygii is a disappointment, fish-wise, but is the ancestral class for all living amphibians, reptiles, birds, and mammals. Basically, Class Actinopterygii staked out the seas, while Class Sarcopterygii captured the land. Both are successful in their own right, but only Class Actinopterygii stayed true to its original form, maintaining the traditional fishy look and habitat. In a sense, all of the sarcopterygians (with the exception of a few extant fish species) simply stropped trying to be fish.

Sarcopterygii

Class Sarcopterygii are the lobe-finned fish. The only living organisms that resemble the early sarcopterygians, are two species of coelacanths and six species of lungfish.

Subclasses. Class Sarcopterygii has two subclasses: Class Dipnotetrapodomorpha (also known as Class Rhipidistia) and Class Actinistia (the coelocanths).

Dipnotetrapodomorpha (also known as Rhipidistia)

Class Dipnotetrapodomorpha contains all the Sarcopterygii except for the extinct Onychodontidae and the Actinistia (coelacanths).

Subclasses. Class Dipnotetrapodomorpha contains two subclasses: Class Dipnomorpha and Class Tetrapodomorpha. Class Dipnomorpha consists of the lungfish and closely related species. We noted, back in Section 2.4, "Genomic Architecture: An Evolutionary Free-For-All," that lungfish have the largest genomes (133 billion base pairs) of any vertebrate (Fig. 6.3).

Tetrapodomorpha

The tetrapodomorpha first appeared about 390 mya, in the Devonian Period, when plants and arthropods colonized land. With the tetrapodomorpha came two evolutionary innovations. We find a modification to their fins, to include a humerus with a convex head fitting into a fossa, constituting a type of shoulder joint. We also find a nostril within the mouth, called a choana, the presumptive evolutionary precursor of the nasopharyngeal opening of humans.

Subclasses. Class Tetrapodomorpha contains Class Tetrapoda plus a collection of extinct animals that are loosely referred to as "half-tetrapod/half-sarcopterygian fish."

Tetrapoda

It is in Class Tetrapoda, the four-limbed animals, that we first glimpse an ancestral class that looks a bit like the land animals that we see today (Fig. 6.4).

Subclasses. Class Tetrapoda has three subclasses: Class Amphibia, Class Sauropsida, and Class Synapsida.

Class Amphibia are anamniotes (i.e., they lay their eggs in water). Class Sauropsida and Class Synapsida are amniotes, laying their eggs on the land. The amphibians, being anamniotes, have that in common with their nontetrapod cousins, the fish of Class Actinopterygii. Occasionally, you will find reference to a class of animals known as Anamniotes (egg-laying animals that lack an amnion). Class Anamniotia is an example of false class, alternately

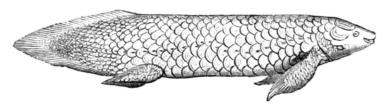

FIG. 6.3 Drawing of a lungfish, one of the few extant members of Class Sarcopterygii that bears any physical resemblance to the earliest members of the class. *Source: Wikipedia from an 1898 drawing by WH Flower.*

FIG. 6.4 Four examples of living members of class Tetrapoda: Rana (amphibian), Opisthocomus (bird), Eumeces (reptile), and Mus (mammal) *Source: Wikipedia, entered into the public domain by the author, Petter Bockman.*

known as an informal class that groups animals based on some common trait, overlooking the fact that the class includes animals with different phylogenies (i.e., different ancestral lineages). Specifically, egg-laying amphibia are tetrapods, and egg-laying ray-finned fish are not. Such informal classes are helpful for zoologists who may be studying aquatic eggs, but vexing for geneticists looking for homologies and phylogenetically acquired genes. So beware!

The three extant classes of amphibia are Anura (the frogs and toads), Urodela (the salamanders, including newts), and Apoda (the caecilians).

Class amphibia evolved a remarkable feature, lacking in its sister classes (i.e., Class Sauropsida and Class Synapsida). It seems as though newts injected with carcinogens will occasionally regenerate new limbs at the sites of injection. For several decades, it was accepted that newts rarely, if ever, develop cancer after the administration of carcinogens [20, 21]. Further studies demonstrated that this same phenomenon could be repeated in urodeles (newts and salamanders) and anurans (frogs and toads).

In the 1950s, cancer researchers developed an interesting theory of cancer based largely on amphibian studies. Because carcinogens caused amphibians to produce local malformations or to regenerate limbs at the site of carcinogen injection, and because these growths occurred instead of cancers, it was hypothesized that all human cancer is a failure of regeneration. According to theory, humans are incapable of regenerating limbs, so they respond to carcinogens with an uncontrolled growth of stem cells. Urodeles and anurans are capable of fully regenerating injured limbs, so they respond to carcinogens with malformations or new limbs; not cancerous growths. Basically, some of the early cancer researchers came to believe that all human cancer resulted from an evolutionary oversight; our inability to regenerate limbs and organs [22].

Actually, urodele and anuran amphibians develop true neoplasms, both spontaneous and cancer-induced [23]. In a large study conducted in 1977, tumors were produced in grass frogs at high frequency (up to 50% of exposed frogs) with several different carcinogens. The induced tumors developed in several months and included tumors of the liver (hepatocellular cancer and hepatic adenomas) and of the hematopoietic system (hemocytoblastosis) [24]. This study demonstrated that carcinogenesis can occur in amphibians much like it occurs in humans. Amphibians develop cancers and benign tumors and malformations and accessory limbs in response to carcinogens. Humans develop cancers and benign tumors and malformations (hamartomas and choristomas) in response to carcinogens, but we just can't regenerate whole limbs. Still, these studies indicate that carcinogenesis, to some extent, recapitulates normal developmental pathways; an observation that has been confirmed in several other model systems [25, 26]. [Glossary Benign tumor]

The amniotes (sauropsids and synapsids) lay their eggs on land or carry the fertilized eggs within the mother. Amniote embryos, whether in eggs laid on land or in developing embryos within the mother, are protected by membranes (the amnion). In humans, the amnion develops into the amniotic sac. It is the amnion plus the absence of a larval phase of development, that distinguishes amniotes (i.e., synapsids and sauropsids and their descendants) from anamniotes such as ray-finned fish and amphibia.

In the amniotes we first see a modern cerebral cortex, wherein a layer of neurons migrates outwards from the periventicular zones and extends their axons behind them to form a hemisphere of white matter (sheathed extensions of neurons) crowned by a layer of neuronal cell bodies (grey matter) [10]. All amniotes have a homologous and anatomically similar cerebral cortex, that develops from the same embryologic recipe.

Class Sauropsida contains all the living reptiles and birds, and the extinct dinosaurs. It is fascinating to find that birds (Class Aves) occupy a place in the direct descent from Class Dinosauria (as shown below). This is just another way of saying that all birds are true dinosaurs; and not just distant cousins to the dinosaurs.

```
Sauropsida -> Eureptilia -> Diapsida ->
Sauria -> Archelosauria -> Archosauria ->
Dinosauria -> Saurischia -> Theropoda ->
Coelurosauria -> Aves
```

As it happens, the closest living cousin class to birds is Class Crocodilia. The intimate relationship between birds and crocodiles reminds us once again that classifications are built on relationships, not on similarities.

Synapsida

Class Synapsida, alternately known as theropsids, are sometimes called the mammal-like reptiles or the stem mammals. They are distinguished from other amniotes by a temporal fenestra, an opening in the skull roof behind each eye, producing bilateral bony arches (the cheekbones in humans). These fenestra permitted the attachment of strong jaw muscles (Fig. 6.5).

Subclasses. Although Class Synapsida produced numerous subclasses of animals well known to paleontologists, nearly every descendant of Class Synapsida is extinct, with the major exception of Class Mammalia. At the end of the Permian, about 245 mya, there was a major extinction, at which time most of the theretofore flourishing subclasses of synapsids died off. From the surviving classes of amniotes came the first dinosaurs (descendants of Class Sauropsida) and the first mammals (descendants of Class Synapsida).

We must acknowledge that an awful lot of evolution intervened between the basal class of synapsids and the emergence of the first mammals, and many of the evolutionary breakthroughs that we will be attributing to the mammals almost certainly evolved in preceding subclasses of synapsids, particularly the therapsids and the therapsid subclass known as the cynodonts, which were soon followed by the mammals, about 220 mya, in the late Triassic period.

FIG. 6.5 An artist's rendition of a synapsid (in the now-extinct subclass Sphenacodontidae). *Source: Wikipedia, and released into the public domain by its author.*

Section 6.3 Mammals to Therians

Most of our ancestors were not perfect ladies and gentlemen. The majority of them weren't even mammals.

Robert Anton Wilson.

Mammalia

Now that we have reached the mammals, we must take stock of how far we've come, in a relatively short time. The first mammals lived in the Late Triassic Period, 220 mya, at about the same time as the very first dinosaurs made their appearance (225 mya). This preceded the flowering plants by 90 million years, all of which waited until 130 mya for their first blooms. We mammals are older than the earliest salamanders and newts (170 mya) and we were scurrying in the underbrush for 140 million years prior to the appearance of the first ants and termites (80 mya).

As previously mentioned, the end of the Permian period, about 245 mya, marked a major extinction, following which many varieties of synapsid disappeared, and evolution proceeded at a remarkable pace. When the mammals emerged, about 220 years ago, they came equipped with a set of innovative evolutionary adaptations, including endothermia (i.e., the internal control of body temperature), a new region of the brain, occupying two frontal lobes (i.e., the neocortex), hair, three middle ear bones, and mammary glands.

– Mammals process pseudogenes

Mammals seem to be the only animals that process their pseudogenes. Pseudogenes are DNA sequences that have degenerated over time, and do not code for functioning proteins. Some pseudogenes may have arisen from genes that failed to serve a useful purpose and whose sequences were not conserved. Others represent replicates of genes that coded for functional proteins that were permitted to degenerate, over time, to the point that they no longer code for a functional protein. The processed pseudogenes are transcribed, in whole or in part, into mRNA. It would seem that some pseudogenes arose through the insertion of reverse-transcribed mRNA (i.e., cDNA) into chromosomes of germ cells [27].

The purpose or destiny of pseudogenes is not fully understood, but there is good reason to suspect that pseudogenes serve as an additional layer of genomic regulation. As one of many species of mRNA floating in cells, they act as molecular sponges for miRNA (micro-RNA). In so doing, they derepress the target genes of miRNAs, thus enhancing gene expression. [Glossary MicroRNA]

Perhaps they serve as the raw ingredient from which new genes may arise. In any case, processed pseudogenes seem to be a genetic innovation peculiar to members of Class Mammalia. Of course, with the innovation of processed pseudogenes comes a host of negative consequences, found exclusively in mammals. Processed pseudogenes have been implicated in neurodegenerative diseases [28], and there has been speculation that they play a role in carcinogenesis [29].

– Among the vertebrates, only mammals employ imprinting

Back in Section 3.2, "The Epigenome and the Evolution of Cell Types," we discussed erasure, the process by which the epigenome is wiped clean in preparation for the early embryo's gradual construction of a new epigenome during cell-type differentiation.

Erasure is not always a totally thorough process. In mammals, there are about 100 known genes that retain their parental epigenetic patterns (i.e., remain highly methylated after erasure). We can roughly estimate that there are about 50 paternally-derived genes that remain methylated after erasure, and another set of about 50 maternally-derived highly methylated genes. These highly methylated genes (i.e., imprinted genes) are under-expressed in a mono-allelic fashion (e.g., the imprinted paternal gene does not affect the expression of the unimprinted maternal gene, and vice versa). This can lead to nearly a 50% reduction of expression in the involved genes.

Several lines of reasoning have established that in mammals, successful embryonic development requires contributions from both the paternal genes and the maternal genes [30–32].

-1.   Mammals never have instances of birth by parthenogenesis (birth following the fertilization of an egg with two of the mother's haploid genomes) or androgenesis (birth following the fertilization of an egg with two male haploid genomes). Both parthenogenesis and androgenesis may occur in metazoans, but never in metazoans that employ imprinting.

-2.   Studies of mouse have demonstrated that both maternal and paternal chromosomes are necessary for normal embryonic development. When there is a deficiency of maternal chromosomes and a redundancy of paternal chromosomes in a mouse conceptus, growth of extraembryonic tissues (trophectoderm) is exaggerated, and growth of embryonic tissue is stunted [30, 31]. [Glossary Extraembryonic cells and tissues, Trophectoderm]

-3.   In humans, we occasionally observe instances wherein an anucleate ovum is fertilized by two sperms. In these instances, a so-called complete hydatidiform mole may grow in the uterus (instead of a normal embryo). The complete hydatidiform mole is characterized by edematous placental villi and little or no embryonic tissue, again indicating the need for both maternal and paternal chromosomes. [Glossary Gestational trophoblastic disease, Hydatidiform mole, Invasive mole]

These observations prompted the hypothesis that imprinting evolved as a defense against parthenogenesis [32]. To test this hypothesis, a group of researchers achieved parthenogenesis in mutant mice that expressed a set of genes that would otherwise have been suppressed by imprinting [33–35]. They did so by finding mutant mice that had a deletion of two imprinting control regions. The successful birth of viable parthenogenetic (i.e., bi-maternal) mice broke a tradition that extended back to the earliest mammalian species, about 220 mya!

Along with imprinting, mammals have also acquired a set of rare diseases of imprinting. When imprinted genes contain disease-causing mutations, the disease that develops will express a phenotype that is influenced by parental lineage. For example, Prader-Willi syndrome is a genetic disease characterized by growth disorders (e.g., low muscle tone, short stature, extreme obesity, and cognitive disabilities). Angelman syndrome is a genetic disease characterized by neurologic disturbances (e.g., seizures, sleep disturbances, and hand-flapping), and a typifying happy demeanor. Both diseases can occur in either gender and both diseases are caused by the same microdeletion at 15q11-13. When the microdeletion occurs on the paternally-derived chromosome, the disease that results is Prader-Willi syndrome. When the microdeletion occurs on the maternally-derived chromosome, the disease that results is Angelman syndrome.

Just as mammals may develop diseases due to imprinting, mammals may also develop deleterious effects due to loss of imprinting. When there is an acquired loss of this normal imprinting, the affected gene may be over-expresssed. The first discovered example of loss of imprinting was the over-expression of insulin-like growth factor-2 in Wilms tumor. In this case, a mutation disrupts the imprinting control region that would normally silence the maternally inherited Igf2 allele.

—   Mammals have a corpus callosum

It seems as though only mammals have evolved a corpus callosum, that curved sheet of white matter swinging like a hammock between the right and left cerebri, connecting neurons of the right hemisphere with neurons of the left [36]. All other vertebrates rely on the relatively small anterior commissure for their inter-hemispheric connections. Mammals have a corpus callosum and an anterior commissure.

Rarely, mammals are born without a corpus callosum, a condition known as ACC (Agenesis of the Corpus Callosum), associated with a wide range of functional deficits and abnormalities. These may include visual impairment, poor motor coordination, swallowing difficulties, and numerous cognitive impairments that can be misdiagnosed as autism. A variety of inherited syndromes may include ACC as part of their clinical phenotype (e.g., Aicardi syndrome, Andermann syndrome, Shapiro syndrome). Currently, it is thought that ACC arises as an inherited ciliopathy (discussed in Section 6.2, "Vertebrates to Synapsids").

—   Only mammals have hair

All mammals have hair. Other classes of animals may have stringy skin appendages, but only mammals are equipped with the complex anatomic structure known as "hair." Consequently, only mammals develop cancers that arise from the growing hair root. This simple inference helps us to understand much of what we observe in basal cell carcinomas, a skin tumor that was once thought to arise from epidermal cells located at the bottommost layer of the epidermis (i.e., the basal cells of the skin).

The two most common cancers of humans are basal cell carcinoma and squamous cell carcinoma of skin. Each accounts for about 600,000 new cases each year in the United States. Together, these two tumors account for 1.2 million new cases each year, which happens to be just about the total of every other type of human cancer, combined! Fortunately, neither of these tumors are likely to metastasize, particularly basal cell carcinomas of skin. A metastasis from a basal cell carcinoma of skin is so unlikely that only a handful have been reported in the scientific literature. Hence, deaths due to either squamous cell carcinoma of skin, or from basal cell carcinoma, are rare. These tumors occur almost exclusively on sun-exposed skin and are believed to result from DNA mutations induced by ultraviolet light (Fig. 6.6). [Glossary Metastasis]

Despite the absence of distant metastases, all basal cell carcinomas are invasive. In fact, there is no recognized noninvasive or precancerous stage of basal cell carcinoma. The growth of basal cell carcinoma is always characterized by the downward growth of nests or strands of tumor cells into the dermis (the connective tissue underlying the epithelium of the epidermis). [Glossary Precancer]

Carcinogenesis (the process by which cancers develop) progresses through a series of sequential events. A noninvasive growth phase precedes an invasive phase. Once invasion occurs, the likelihood of metastasis increases over time. If that is the case, then how is it possible that basal cell carcinomas lack a noninvasive or precancerous phase? And how is it possible that basal cell carcinomas, a tumor that grows by invading the dermis, cannot metastasize?

The peculiar behavior of basal cell carcinomas is explained by its histogenetic origin. Unlike squamous carcinomas, which arise from the basal layer of the epidermis, basal cell

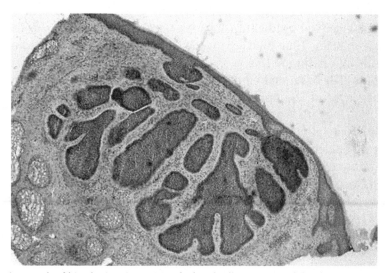

FIG. 6.6 Photomicrograph of histologic preparation of a basal cell carcinoma of the skin. Nests of tumor cells (*blue*—dark gray in print version), below the epidermal layer of skin, are typical of this tumor. *Source: Wikipedia, released by John Hendrix.*

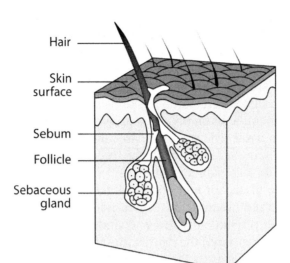

FIG. 6.7 Diagram of a hair in its follicle. Hair follicles are characteristic of Class Mammalia. *Source: Wikipedia, released into the public domain by its copyright holder.*

carcinomas arise from the hair follicle epithelium. Hairs grow by a cyclic process whereby the follicle of the hair invades into the dermis, providing a growing hair shaft that eventually sheds, leading to another round of invasion. The cycle consists of telogen (follicle resting phase), catogen (follicle involution phase), and anogen (follicle growth and invasion phase). Invasion is a fundamental property of normal hair epithelial cells. A tumor composed of hair-derived epithelium invades because invasion is the mechanism of growth for hair follicles. Hair follicles invade, but they do not metastasize. The same applies to basal cell carcinoma (Fig. 6.7).

How do know that basal cell carcinomas arise from the hair follicle, and not from the lower-most layer of the epidermis, as was previously believed? There are three reasons.

–1.   Basal cell carcinomas grow in the dermis (like hair roots) and are not found in the overlying epidermis.
–2.   Squamous cell carcinomas can arise from many locations in the body, wherever there is a squamous, pavemented epithelium. Basal cell carcinomas only arise from locations where hair grows.
–3.   Basal cell carcinomas are found exclusively in mammals, the only hair-bearing class of animals. Nonmammalian species do not develop basal cell carcinomas, but they do develop squamous cell carcinomas of the skin.

The third item in the list illustrates a general point. The types of tumors that we encounter in man and other animals corresponds to the limited set of biological phenotypes available to the species. Furthermore, the cell types of any species are derived from the cell types that were present in their phylogenetic lineage.

Subclasses. Class Mammalia has two extant subclasses: Class Theria and Class Proto-theria. Class Prototheria, also known as Class Yinotheria, contains a few fossil organisms dating back to the Mesozoic era and includes five extant species of monotremes. Mono-treme fossils have been found in England, China, Madagascar, and Argentina. The surviv-ing species of monotremes, all of which have retreated to Australia and New Guinea, are the duck-billed platypus and four species of echidna (spiny anteaters). The Prototherians lay eggs, which must be tended until they hatch.

Back in Section 1.2, "Bootstrapping Paradoxes," we observed that every animal and every plant is actually two organisms, each serving its own biological role. There is the diploid somatic organism and the haploid gametic organism. In the case of the bryophyte plants, the diploid and the haploid organism have their own free-living structures. In most animals and plants, the haploid organism lives within the diploid organism most of the time. This dichotomy between haploid and diploid life is mentioned once again here, because the prototherian haploid gamete is organized quite differently from the haploid therian gamete, as though the haploid forms were following their own evolutionary des-tinies. Without going into the details, meiosis, and the subsequent haploid phase of the platypus, and presumably the few other monotremes in existence today, uses a variety of regulatory mechanisms that are different from those of the therians, particularly in those steps that protect the male and female sex chromosomes from transcription and from recombination events [37].

Variations in meiosis among classes of animals may seem a trivial point, but the mei-otic mechanism is an evolutionary construct that shapes the identity of a class of organ-isms, and of all its descending classes. In this case, we can say that the haploid form of the monotremes is fundamentally different from the haploid form of the therians.

Theria

Therian mammals unlike their egg-laying sister class, the Prototherians, give birth to live young. Egg-laying can take its toll on the mother if the eggs are in danger of being eaten by predators, or if the eggs must be kept at the temperature of the mother's body. Carrying the fetus in utero frees the mother to move about, to protect herself from pred-ators, and to eat. Dispensing with egg-laying, and freeing the mother to search for food, knowing that the developing fetus is safe within its womb, was a major evolutionary achievement, requiring two innovations:

–1. The suppression of the mother's immune system

As you recall, the adaptive immune system arose in Class Gnathostomata, an ancestor of Class Theria. An immune system designed to detect and destroy all "non-self" invaders of the human body would, as a matter of course, destroy fertilized ova, and blastocysts, and developing embryos, all of which are half-composed of foreign gene products derived from the father.

–2. The production of an extraembryonic placenta

All species of Class Theria are distinguished by the development of the placenta, an organ that grows into the uterine cavity (i.e., the endometrium). Traditionally, however, discussions of the placenta are reserved for a subclass of therians, known as the placental mammals (i.e., Class Eutheria). The remaining subclass of Class Theria, the marsupials (i.e., Class Metatheria) have placentas that are rudimentary and which do not endure for the entire period of fetal maturation. We will follow tradition and wait until we come to Class Eutheria, before discussing the evolution of the placenta, and its attendant diseases.

Subclasses. Class Theria has two subclasses: Class Eutheria (the placentals) and Class Metatheria (the marsupials). As previously noted, all therians have a placenta, but the metatherian (marsupial) placenta is rudimentary and not strictly homologous with the eutherian placenta. In metatherians, the yolk sac and chorian fuse to form a chriovitelline placenta. In eutherians (placental mammals) the allantois and chorion fuse to form a cho-rioallantoic placenta [38].

While still fetuses, marsupial babies crawl from the womb and into the mother's pouch, where they finish their gestation months later. The mother marsupial nourishes the baby marsupial, called the joey, in its pouch until it can survive on its own. Today, most of the extant marsupials live in Australasia and its environs, though fossils of metatherian ancestral species are found in other continents. Strikingly, there are very few native eutherians in Australia and Oceania. The exceptions are bats (arriving in Australia about 55 mya) and rodents (arriving about 5 mya). Any other eutherians you may find in Australia and New Zealand were almost certainly introduced by humans [38]. Members of the sister class of Theria (i.e., the prototherians, which include the platypus and the echidnas) have taken up exclusive quarters in Australia, perhaps preferring the company of metatherians (marsupials) to that of eutherians (placentals).

Some of the metatherians of Australia look very much like their eutherian sisters. Skeletons of the recently extinct thylacines look much like dog skeletons. The skeleton of the metatherian (thylacine) is distinguished from the its eutherian doppelganger (dog) by the exclusive presence of two holes in the roof of the thylacine mouth.

Section 6.4 Eutherians to Humans

Our greatest responsibility is to be good ancestors.

Jonas Salk

The age at which the first eutherian appeared is somewhat vague. The oldest eutherian fossil belongs to Juramaia sinensis, and is dated at 160 million years old. Other sources date the eutherians at about 125 mya. Perhaps the source of the confusion can be attributed to the paucity of eutherian fossils, owing to the paucity of eutherian animals, prior to the extinction of the big dinosaurs (about 66 mya). Until that time, eutherians were hairy little creatures that scurried through the underbrush, hunting at night, and trying their best to avoid detection by carnivorous dinosaurs. After the dinosaur extinction event, the surviving eutherians emerged from the undergrowth to become earth's dominant predators and champion climate warmers.

With the appearance of the eutherians, a clear line of ancestry leads us to modern humans, as shown:

```
Eutheria (160-125 mya)
Boreoeutheria (124-101 mya)
Euarchontoglires (100 mya)
Euarchonta (99-80 mya)
Primates (75 mya)
Haplorrhini (63 mya)
Simiiformes (40 mya)
Catarrhini (30 mya)
Hominoidea (28 mya)
Hominidae (15 mya)
Homininae (8 mya)
Hominini (5.8 mya)
Homo (2.5 mya)
Homo sapiens (0.3 mya)
Homo sapiens (modern) 0.07 mya
```

In this section, we will be reviewing the ancestral lineage of `Homo sapiens`, up to our eutherian ancestors.

Eutheria (placental mammals)

"Let That Be Your Last Battlefield" is the title of the fifteenth episode of the third season of the original American science fiction television show Star Trek. It was first broadcast January 10, 1969, and featured two feuding aliens, played by Lou Antonio and Frank Gorshin, who had managed to embroil the crew of the Starship Enterprise in their planet's war. It seems that aliens who were black on one side of the body, white on the other, were at war with aliens whose color scheme was left-right reversed. One was the mirror image of the other, and the difference was sufficient to provoke eternal enmity.

Basically, all the eutherians, past and present, are about the same. We enjoy drawing sharp distinctions between one eutherian species and another (e.g., cows go "moo" while sheep go "baaa"), but to visitors from another planet, the differences among species of eutherians would be trivial; no more significant than the difference between Lou Antonio and the late Frank Gorshin.

The eutherians (placentals) are much like their sister class, the metatherians (marsupials), with the obvious exception of the placenta that nourishes the developing fetus during its long (compared with marsupials) gestation. With the advent of the large placenta comes the necessary elimination of epipubic bones, which are present in all the noneutherian mammals (i.e., marsupials and monotremes). Otherwise, eutherians are distinguished from noneutherians by small variations in feet, jaws and teeth. As mentioned previously, the marsupials and monotremes are more or less confined to Australasia (the American opossum being an exception). The eutherians live in Australasia plus everywhere else.

By the time we've reached the eutherians, we've developed the basic skeleton that every eutherian possesses; the various eutherian species differing from one another in size and shape of the bones, but not much else. All eutherians have anatomically similar skin, hair, blood, organs, and brains. Our genomes are all similar, and that includes our genes, the systems that control the expression of those genes, and the repair systems that keep our genome intact. The celltypes in one eutherian are much the same as the cell types in other eutherians, as determined by morphology and function. We can infer that the metabolic pathways in all eutherians are mutually similar; otherwise, the cell types would need to have evolved differently, for each eutherian species.

At this point, let's stop a moment and ask ourselves a simple question: If all eutherian species are fundamentally alike, then what accounts for all the "nonfundamental" differences that we see when we survey the remarkable array of extant eutherians? For example, what is the difference between men and mice?

We can infer that the difference among the different eutherians are chiefly regulatory in nature (i.e., not determined by differences in genes) for the following reasons:

-1.   Man and mouse must pass through the exact same embryonic and fetal developmental stages, if they are to have the same organs and tissues (which they do). Both man and mouse have been provided with these same embryonic recipes by the ancestral classes (of the eutherians). Hence, we would expect that man and mouse and every other eutherian species will employ nearly the same set of genes to build their embryos and fetuses.

-2.   In fact, we observe that most of the genes of any two metazoans are largely the same, and the gene similarities are even more pronounced among the eutherians (see Section 1.3, "Our Genes, for the Most Part, Come From Ancestral Species").

-3.   The differences between any two eutherian species comes down to traits—quantitative differences in morphologic features. Humans have bigger bones than do mice, and many of the bones have mildly different shapes, but a human osteocyte is much like a mouse osteocyte. The difference is one of degree: number and arrangement of cells. The genomic mechanism that controls small differences in gene expression is the epigenome, along with the other genetic regulators.

-4.   When we try to account for variations between closely related species or between subgroups of one species, we typically find differences in transcription factor binding, one of the many mechanisms controlling the genome [39, 40]. We should note that there are no apparent qualitative differences in the manner by which the genome in regulated, among all the different eutherians. The differences among the eutherians are quantitative, with transcription factors and other regulatory processes having more or less the same activity in one species compared to another.

The characterizing feature of eutherians is the placenta, which is one of the first structures to develop in the early embryo, and which persists throughout gestation, until such time as the young organism can safely breathe, move, and suckle; untethered from its umbilical connection to the mother.

Soon after the birth of the infant, the placenta must detach from the uterus. You can imagine the delicate balancing act between attaching firmly to the wall of the uterus, and detaching cleanly from the wall of the uterus. During placental development, large, flat cells called cytotrophoblasts form the interface between placenta and uterus. To create the thin membrane that forms the delicate and impermanent border between the lining of the uterus and the blood percolating through the spaces of the placental villi, the cytotrophoblasts must somehow fuse into a syncytium (i.e., a multinucleate collection of cells lacking individual cytoplasmic membranes that would otherwise separate one cell from another).

There is one task at which all animals excel; maintaining a clear separation between one cell and another. As previously mentioned, the most distinctive difference between animal cells and all other cells of eukaryotic origin, happens to be the presence of cell junctions, whose purpose is to bind cells to one another without fusing cells (i.e., without dissolving the walls separating one cell from another). This being the case, you can see that the normal direction of animal evolution would preclude the appearance of a gene intended to form a huge syncytium of placental cells. Whereas animal cells are poor at fusion, viruses are champions. One of the most often-deployed methods by which viruses invade cells is through fusion, at the cytoplasmic membrane. It happens that retroviral envelope genes, captured and preserved in the eutherian and metatherian genomes, can do a very good job at fusing membranes [41]. Animals captured a retroviral fusogenic envelope gene and adapted it as syncytin, involved the development of the syncytial layer of trophoblasts lining the placenta. Apparently, this acquisition worked out so well for mammals that later-evolving mammalian classes made additional retrovirus gene acquisitions to obtain additional syncytins, thus refining the placenta for their own mammalian classes [41, 42] (Fig. 6.8). [Glossary Syncytium]

Subclasses. Class Eutheria has two subclasses: Class Boreoeutheria and Class Atlantogenata. Members of Class Atlantogenata originated in and radiated from the South American and African continents, and include African Elephants and giant anteaters.

Boreoeutheria

Boreoeutheria, from the Greek root, roughly meaning "northern true beasts" are a large class of placental mammals. Most male boreoeutherians (including humans) have a scrotom, which serves to cool spermatocytes.

Subclasses. Class Boreoeutheria has two subclasses: Class Laurasiatheria and Class Euarchontoglires (the supraprimates). Species of Class Laurasiatheria originated on the northern super-continent of Laurasia about 99 mya. It includes most hoofed mammals and most pawed carnivores, specifically shrews (other than treeshrews), pangolins, bats, whales, carnivorans (e.g., housecats), odd-toed and even-toed ungulates.

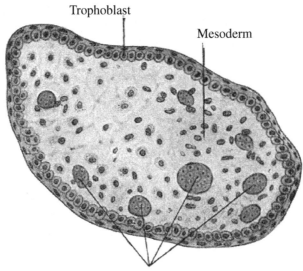

Trophoblast

Mesoderm

Branches of umbilical vessels

FIG. 6.8 The placental villi have an outer trophoblastic lining that has two cell layers: the cytotrophoblast cells (the individual round cells in the figure) and the syncytiotrophoblasts (the outermost cell layer, having no sharp borders separating cells). *Source: Wikipedia, from Henry Gray's "Anatomy of the Human Body," 1918.*

Class Euarchontoglires (85 mya)

The euarchontoglires, also known as the supraprimates, comprise a class that has been defined by its shared DNA sequences, particularly by a collection of retrotransposon markers. There are no apparent morphological features that specifically characterize the euarchontoglires.

Subclasses. Class Euarchontoglires contains two subclasses: Class Glires and Class Euarchonta. Class Glires contains rodents and lagomorphs (e.g., rabbits) and some currently extinct species.

Class Euarchonta (85 mya)

Class Euarchonta, like its parent Class Euarchontoglires, is another class that was created to hold its related but distinctive subclass members. Hence, we note that the approximate dates of origin of the Class Euarchontoglires and Class Euarchonta are the same.

Subclasses. Class Euarchonta contains four subclasses: Class Scandentia (the treeshrews), Class Dermoptera (the colugos), Class Plesiadapiformes (currently extinct), and Class Primates.

Class Primates (75 mya)

The earliest found primate fossils date to about 55 mya. The molecular clock studies would suggest that the primate branch may have appeared as early as 85 mya. For no particularly compelling reason, we'll compromise and round-off our guess for the origin of primates to a time 75 mya.

Primates are the first ancestral class of animals that look something akin to humans. They are characterized by a large brain, and by stereoscopic vision. Compared to other mammals, primates are less sensitive to odors, and it has been suggested that we traded our olfactory acuity for superior stereoscopic vision. The early primates were arboreal, and virtually all living primates spend at least part of their lives in trees. With the exception of apes (including humans), primates have tails. Most primates have opposable thumbs.

Subclasses. Class Primate has two subclasses: Class Haplorrhini and Class Strepsirrhini. Class Strepsirrhini ("wet nosed" primates) includes the lemuriform primates, which consist of the lemurs of Madagascar, the galagos ("bushbabies") and the pottos from Africa, and the lorises from India and southeast Asia.

Haplorrhini (sometimes spelled Haplorhini)

Haplorrhini, meaning "dry nose" are the subclass of primates that lack the ability to synthesize Vitamin C. In addition, the genomes of all haplorrhines contain five characteristic short interspersed nuclear elements (SINEs). Due to having upper lips untethered to the nose, all haplorrhines have the ability to grin (if they wish it so).

Subclasses. Class Haplorrhini has two subclasses: Class Tarsiidae and Class Simiiformes (the anthropoids). Class Tarsiidae consists only of species of tarsiers, all of which inhabit islands in Southeast Asia.

Simiiformes

Class simiiformes, better known as simians, are the monkeys and the apes.

Subclasses. Class Simiiformes contains two subclasses: Class Platyrrhines (New World monkeys) and Class Catarrhini (Old World monkeys plus apes). The playtyrrhines (meaning "flat-nose") monkeys consist of five families of monkeys living in Central and South America.

Catarrhini

The catarrhine (meaning "down nose") monkeys and apes are native to Africa and Asia.

Subclasses. Class Catarrhini contains two subclasses: Class Cercopithecoidea (Old World monkeys) and Class Hominoidea (apes). Class Cercopithecoidea is the sister group to new world monkeys and contains the monkeys from Africa and Asia. There are numerous physical distinctions between New World monkeys (from South America), Old World monkeys (from Asia and Africa), and Apes. Zoologists distinguish monkeys from apes by the shape of their teeth; and they distinguish New World monkeys from Old World monkeys by the shapes of their noses. As a fairly easy rule to remember, monkeys, unlike apes, have long tails. The tails of Old World Monkeys are not prehensile, while the tails of New World monkeys are prehensile.

Hominoidea

Subclasses. Class Hominoidea has two extant subclasses. Class Hylobatidae and Class Hominidae. Class Hylobatidae consists of 18 species of gibbons.

Hominidae (the great apes)

Class Hominidae contains two subclasses: Class Ponginae and Class Homininae. Class Ponginae has three extant species, all of which are critically endangered orangutan (also known as pongos).

Homininae

Class Homininae are the hominids or African apes. They are distinguished from old world monkeys by the absence of a long tail and by shoulder joints that afford a wide range of motion.

Subclasses. Class Homininae contains two subclasses: Class Hominini and Class Gorillini (the two living species of gorillas plus the extinct Chororapithecus).

Hominini

Class Hominini is an informal class contrived to hold the closely related humans, chimpanzees, and bonobos.

Subclasses. Class Hominini contains two subclasses: Class Homo (humans) and Class Pan (chimpanzees and bonobos).

Homo

Class Homo comprises one extant species *Homo sapiens* (modern humans), plus several extinct species classified as ancestral to or closely related to modern humans (e.g., *Homo erectus*). Historically, humans have claimed for themselves a special status among the terrestrial organisms. Many such claims are grandiose and hardly worth serious discussion (e.g., least savage, most empathetic, only species with a soul). A satisfactory summary of the distinction between humans and other primates has been provided by the writer Richard Preston: "Humans appear to be the only primates that I know of that are afraid of heights. All other primates, when they're scared, they run up a tree, where they feel safe."

Subclasses. Class homo has no subclasses (yet) and only one extant species.

Homo sapiens (315,000 years ago)

With the advent of *Homo sapiens*, we have come to the end of the line, so to speak; but just how far have we come? Despite the rich assortment of ancestral classes, there is scarcely any phenotypic difference between *Homo sapiens* from any other eutherian species. For that matter, we are pretty much standard issue eukaryotes. We have no trouble finding single-celled eukaryotes having about the same size as human cells, with the same set of organelles, having similarly sized nuclei, and carrying a similar amount of DNA. When we look at single-celled eukaryotes, microscopically, we see some morphological differences that distinguish them from human cells, but we also see morphological differences among the different human cell types (e.g., hepatocytes compared with melanocytes, or red blood cells, spermatocytes, or oocytes).

We think of ourselves as being special, but we haven't found the gene that accounts for our special cognitive abilities; and it is doubtful that we ever will. If humans are the most intelligent animals on earth, we can be fairly certain that our intelligence is a trait, and not the result of a mutation in a single gene. The reasoning behind this assertion is that all animals have intelligence. Humans seem to have quantitatively more intelligence than other eutherians. Traits that can be expressed as quantitative differences between species, or between individual members of a species, are generally the result of small polygenic variations in existing genes and gene regulators.

For the most part, the genes of humans are scarcely any different from the genes of any other mammalian species, and if we were to search for genes that most clearly distinguish men from mouse, we are almost certainly going to find the greatest differences among the genes that regulate our vastly different responses to microorganisms and other environmental stressors. Contain your skepticism. We will justify this assertion when we read Chapter 8, "Animal Models of Human Disease: Opportunities and Limitations."

Glossary

Autoantibody disease versus autoimmune disease The two terms are often used interchangeably, but they represent pathogenetically distinct conditions. Autoantibody diseases occur when the adaptive immune system synthesizes antibodies against some normal body constituent. This may occur when an antigen in an infectious organism elicits antibodies that happen to cross-react with a normal cellular protein; occasionally producing some adverse clinical consequence. In most cases, as far as anyone can tell, individuals with the common autoantibody diseases have normal functioning immune systems. In autoimmune diseases, there is a primary dysfunction of the immune system, often producing an array of clinical sequelae, including the synthesis of one or more antibodies that react against self-antigens [43].

Benign tumor With a few exceptions, benign tumors are tumors that grow continuously, without invading or metastasizing. As a general rule, benign tumors are often diploid. Some benign tumors are stable-aneuploid (i.e., varying from the normal chromosome number, without becoming increasingly aneuploid over time). There are rare examples of benign tumors that are progressively aneuploid [44].

Ciliopathies The ciliopathies are a genetically and phenotypically diverse set of diseases that involve inherited disorders of the primary cilium (also called the nonmotile cilium). The primary cilium is a single flagellum extending from virtually every cell in vertebrates, that is distinguished from motile cilia by the absence of a central pair of microtubules and outer dynein arms that are essential for normal motility. Included in the ciliopathies are: Joubert syndrome, nephronophthisis, Senior-Loken syndrome, orofaciodigital syndrome, Jeune chondrodysplasia syndrome, autosomal dominant polycystic kidney disease, recessive polycystic kidney disease, Leber congenital amaurosis, Meckel-Gruber syndrome, Bardet-Biedl syndrome, Usher syndrome, Alstrom syndrome, McKusick-Kaufman syndrome, Ellis van Creveld syndrome, cranioectodermal dysplasia (Sensenbrenner syndrome), short rib polydactyly, some forms of retinal dystrophy, and heterotaxy (including visceral situs anomalies, asplenia or polysplenia, congenital heart defects, biliary atresia, and midline defects) [13, 45]. There are only a few organs and functional systems of the human body that escape involvement by this strange collection of related rare diseases.

Extraembryonic cells and tissues Cells that grow from the early embryo but which are not destined to be retained in the embryo as it develops. The extraembryonic tissues include the placenta, the umbilical cord, the yolk sac and the amniotic sac.

Genetically engineered mouse Abbreviation: GEM. Mice whose genomes contain inserted sequences of DNA that provide a new gene function or that produce the loss of function of a gene. New gene functions are achieved by inserting a nonendogenous gene into a fertilized mouse egg, along with a promoter that drives the expression of the gene. When the promoter is organ-specific, the functionality of the new gene can be observed in a single organ.

Loss of function GEMs are usually created by culturing embryonic stem cells from a mouse, introducing a nonfunctional variant of a gene into the nucleus of a cultured embryonic stem cell using a electroporation, and hoping that the foreign DNA will replace its closely related native DNA sequence through homologous recombination. A marker gene is used to determine if the recombination was successful. The genetically modified embryonic stem cell can be inserted into a mouse blastocyst. If the modified embryonic stem cell differentiates into a germ cell, and if the germ cell contributes to a next-generation conceptus, the offspring will have the altered genotype.

Gestational trophoblastic disease Refers to neoplasms and malformations arising from cells of the trophectoderm (i.e., the extraembryonic tissue of the conceptus that gives rise to the placenta and extraembryonic tissues). The most common gestational trophoblastic diseases are: hydatidiform mole (complete and partial types), invasive mole, choriocarcinoma, placental site trophoblastic tumor, and epithelioid trophoblastic tumor.

Hydatidiform mole Also called gestational mole. A hydatidiform mole is a malformed conceptus and is characterized by the presence of edematous placental villi and a variable amount of malformed, nonviable embryonic tissue. There are two types of hydatidiform mole; complete mole, and partial mole. Hydatidiform moles are associated with a risk of developing choriocarcinoma.

Partial moles typically contain more embryonic tissue than is found in complete moles. A partial mole, like a complete mole, can develop into an invasive mole or a choriocarcinoma, but the incidence of conversion is much smaller (about 20% for complete mole and about 3% for partial mole). Nearly all partial moles are diandric (two sets of paternal chromosomes) and triploid [46]. In partial moles, the fertilization of an ovum by two sperms produces two sets of paternal chromosomes and one set of maternal chromosomes.

In a complete mole, there is little or no embryonic tissue attached to the placental villi. Complete moles occur after an anucleate ovum is fertilized by two sperm. The cells of a complete mole are diploid with both sets of chromosomes having a paternal origin. The occurrence of complete moles, and the absence of occurrences of androgenetic humans (i.e., humans having only paternally-derived chromosomes) indicates that humans are incapable of androgenesis.

The use of the word "mole" in this term should not be confused with the use of the word "mole" as a synonym for nevus (a benign melanocytic lesion of skin).

Invasive mole Previously called chorioadenoma destruens. Occurs when the growing placental villi of a hydatidiform mole invades into the myometrium (i.e., wall of the uterus). The presence of invading molar villi, and the occasional co-occurrence with morphologically similar noninvasive molar tissue, helps to distinguish an invasive mole from a choriocarcinoma.

Irides Plural of iris.

Knockout mice Strains of laboratory mice in which a specific gene has been knocked out or replaced. Every mouse in the strain has the identical gene deletion in every cell of its body. The primary purpose of producing knockout mouse strains is to show us how the organism behaves in the absence of the gene, thus providing some insight to the normal regulation and system-wide function of the gene.

About 15% of gene knockouts are developmentally lethal, which means that the genetically altered embryos cannot grow into adult mice (i.e., no knockout mouse strain can be created). As a rule, highly conserved genes play some important role during the development of the organism or the reproductive life of the organism; otherwise the genes would not be conserved. Hence, we would expect that the highly conserved genes would not serve well when trying to produce new strains of knockout mice. As it happens, all of the oncogenes are highly conserved. When you try to create a knockout mouse strain with an oncogene or knock-in a dysfunctional oncogene, we would expect, and we sometimes see, lethality in the embryo [47–49].

Neurectoderm Alternate spelling of neuroectoderm. An embryonic derivative of the ectodermal layer that produces the neural tube and neural plate, from which the central nervous system (primarily brain and spinal cord) derive.

Neurocristopathy A disease of neural crest cells. Examples include MEN2 (multiple endocrine neoplasm syndrome type 2), aganglionic diseases of the GI tract, and neurofibromatosis.

Pleiotropia Refers to an effect wherein one gene influences more than one phenotypic trait. A gene that exhibits pleiotropia is said to be pleiotropic (alternate spelling pleiotrophic).

Precancer Lesions preceding the development of a cancer, and from which the cancer ultimately arises. Precancers, generally, are neither invasive nor metastatic, and can be cured by excision. As all cancers seem to be preceded by an identifiable precancer stage, a successful strategy to eliminate all precancers would prevent the occurrence of all cancers [26, 50].

Schwannoma A tumor composed of neoplastic Schwann cells that are normally found wrapped around the axonal extensions of peripheral nervous system neurons (i.e., of neural crest origin). Schwannomas of the acoustic spinal nerves occur in neurofibromatosis type 2.

Syncytium A collection of fused cells in which the cytoplasm and organelles of the adjacent cells can, to some extent, mix together. Syncytia are found in fungal mycelial growths and in the plasmodial slime molds observed in mycetozoans. They are also found in the syncytial layer of the eutherian placenta, wherein specialized trophoblastic cells (the syncytiotrophoblasts) fuse with one another.

Trophectoderm The trophectoderm is formed in the blastocyst stage and consists of cells derived from the conceptus that are not part of the inner cell mass. The trophectoderm gives rise to the placenta and the amniotic membranes. Because these tissues are not incorporated into the developed animal, they are referred to as extraembryonic tissue. Gestational trophoblastic neoplasia (e.g., tumors that arise in the pregnant mother's uterus, such as gestational choriocarcinoma), all come from the trophectoderm.

References

[1] Kuratani S. Craniofacial development and the evolution of the vertebrates: the old problems on a new background. Zool Sci 2005;22:1–19.

[2] Trainor PA, Melton KR, Manzanares M. Origins and plasticity of neural crest cells and their roles in jaw and craniofacial evolution. Int J Dev Biol 2003;47:541–53.

[3] Matsuoka T, Ahlberg PE, Kessaris N, Iannarelli P, Dennehy U, Richardson W, et al. Neural crest origins of the neck and shoulder. Nature 2005;436:347–55.

[4] Jiang X, Iseki S, Maxson RE, Sucov HM, Morriss-Kay GM. Tissue origins and interactions in the mammalian skull vault. Dev Biol 2002;241:106–16.

[5] Spitsbergen JM, Kent ML. The state of the art of the zebrafish model for toxicology and toxicologic pathology research—advantages and current limitations. Toxicol Pathol 2003;31(Suppl):62–87.

[6] Smolowitz R, Hanley J, Richmond H. A three-year retrospective study of abdominal tumors in zebrafish maintained in an aquatic laboratory animal facility. Biol Bull 2002;203:265–6.

[7] Beckwith LG, Moore JL, Tsao-Wu GS, Harshbarger JC, Cheng KC. Ethylnitrosourea induces neoplasia in zebrafish (*Danio rerio*). Lab Invest 2000;80:379–85.

[8] Christie AE, Kutz-Naber KK, Stemmler EA, Klein A, Messinger DI, Goiney CC, et al. Midgut epithelial endocrine cells are a rich source of the neuropeptides APSGFLGMRamide (*Cancer borealis* tachykinin-related peptide Ia) and GYRKPPFNGSIFamide (Gly1-SIFamide) in the crabs *Cancer borealis, Cancer magister* and *Cancer productus*. J Exp Biol 2007;210:699–714. Pt 4.

[9] Sorensen PH, Shimada H, Liu XF, Lim JF, Thomas G, Triche TJ. Biphenotypic sarcomas with myogenic and neural differentiation express the Ewing's sarcoma EWS/FLI1 fusion gene. Cancer Res 1995;55:1385–92.

[10] Goffinet AM. The evolution of cortical development: the synapsid-diapsid divergence. Development 2017;144:4061–77.

[11] Satir P. CILIA: before and after. Cilia 2017;6:1.

[12] Gerdes JM, Davis EE, Katsanis N. The vertebrate primary cilium in development, homeostasis, and disease. Cell 2009;137:32–45.

[13] Novarino G, Akizu N, Gleeson JG. Modeling human disease in humans: the ciliopathies. Cell 2011;147:70–9.

[14] Jakobsen L, Vanselow K, Skogs M, Toyoda Y, Lundberg E, Poser I, et al. Novel asymmetrically localizing components of human centrosomes identified by complementary proteomics methods. EMBO J 2011;30:1520–35.

[15] Tobin JL, Beales PL. The nonmotile ciliopathies. Genet Med 2009;11:386–402.

[16] Tang Z, Zhu M, Zhong Q. Self-eating to remove cilia roadblock. Autophagy 2014;10:379–81.

[17] Kapitonov VV, Jurka J. RAG1 core and V(D)J recombination signal sequences were derived from Transib transposons. PLoS Biol 2005;3:e181.

[18] Zhang J, Quintal L, Atkinson A, Williams B, Grunebaum E, Roifman CM. Novel RAG1 mutation in a case of severe combined immunodeficiency. Pediatrics 2005;116:445–9.

[19] de Villartay JP, Lim A, Al-Mousa H, Dupont S, D chanet-Merville J, Coumau-Gatbois E, et al. A novel immunodeficiency associated with hypomorphic RAG1 mutations and CMV infection. J Clin Invest 2005;115:3291–9.

[20] Tsonis PA, Eguchi G. Abnormal limb regeneration without tumor formation in adult newts directed by carcinogens 20-methylcholanthrene and benzo(a)pyrene. Develop Growth Differ 1982;24:183–90.

[21] Breedis C. Induction of accessory limbs and of sarcoma in the newt (*Triturus viridescens*) with carcinogenic substances. Cancer Res 1952;12:861–6.

[22] Prehn RT. Regeneration versus neoplastic growth. Carcinogenesis 1997;18:1439–44.

[23] Anver MR. Amphibian tumors: a comparison of anurans and urodeles. In Vivo 1992;6(4):435–7.

[24] Khudoley VV. Tumor induction by carcinogenic agents in anuran amphibian *Rana temporaria*. Arch Geschwulstforsch 1977;47:385–99.

[25] Berman J. Precision medicine, and the reinvention of human disease. Cambridge, MA: Academic Press; 2018.

[26] Berman JJ. Neoplasms: principles of development and diversity. Sudbury: Jones & Bartlett; 2009.

[27] Vanin EF. Processed pseudogenes: characteristics and evolution. Annu Rev Genet 1985;19:253–72.

[28] Costa V, Esposito R, Aprile M, Ciccodicola A. Non-coding RNA and pseudogenes in neurodegenerative diseases: "The (un)Usual Suspects" Front Genet 2012;3:231.

[29] Hu X, Yang L, Mo YY. Role of pseudogenes in tumorigenesis. Cancer 2018;10:e256.

[30] McGrath J, Solter D. Completion of mouse embryogenesis requires both the maternal and paternal genomes. Cell 1984;37:179–83.

[31] Barton SC, Surani MAH, Norris ML. Role of paternal and maternal genomes in mouse development. Nature 1984;311:374–6.

[32] Kaneko-ishino T, Ishino F. Mammalian-specific genomic functions: newly acquired traits generated by genomic imprinting and LTR retrotransposon-derived genes in mammals. Proc Jpn Acad Ser B Phys Biol Sci 2015;91:511–38.

[33] Kono T, Obata Y, Wu Q, Niwa K, Ono Y, Yamamoto Y, et al. Birth of parthenogenetic mice that can develop to adulthood. Nature 2004;428:860–4.

[34] Kawahara M, Kono T. Roles of genes regulated by two paternally methylated imprinted regions on chromosomes 7 and 12 in mouse ontogeny. J Reprod Dev 2012;58:175–9.

[35] Kawahara M, Wu Q, Takahashi N, Morita S, Yamada K, Ito M, et al. High-frequency generation of viable mice from engineered bi-maternal embryos. Nat Biotechnol 2007;25:1045–50.

[36] deLussaneta MH, Osseb JWM. An ancestral axial twist explains the contralateral forebrain and the optic chiasm in vertebrates. Anim Biol 2012;62:193–216.

[37] Daish TJ, Casey AE, Grutzner F. Lack of sex chromosome specific meiotic silencing in platypus reveals origin of MSCI in therian mammals. BMC Biol 2015;13:106.

[38] Archibald JD. Eutheria (placental mammals). encyclopedia of life sciences. London: MacMillan; 20011–4.

[39] Kasowski M, Grubert F, Heffelfinger C, Hariharan M, Asabere A, Waszak SM, et al. Variation in transcription factor binding among humans. Science 2010;328(5975):232–5.

[40] Zheng W, Zhao H, Mancera E, Steinmetz LM, Snyder M. Genetic analysis of variation in transcription factor binding in yeast. Nature 2010;464:1187–91.

[41] Cornelis G, Vernochet C, Carradec Q, Souquere S, Mulot B, Catzeflis F, et al. Retroviral envelope gene captures and syncytin exaptation for placentation in marsupials. Proc Natl Acad Sci U S A 2015;112:e487–96.

[42] Patel MR, Emerman M, Malik HS. Paleovirology: ghosts and gifts of viruses past. Curr Opin Virol 2011;1(4):304–9.

[43] Lleo A, Invernizzi P, Gao B, Podda M, Gershwin ME. Definition of human autoimmunity–autoantibodies versus autoimmune disease. Autoimmun Rev 2010;9:A259–66.

[44] Joensuu H, Klemi PJ. DNA aneuploidy in adenomas of endocrine organs. Am J Pathol 1988;132:145–51.

[45] Ware SM, Aygun MG, Hildebrandt F. Spectrum of clinical diseases caused by disorders of primary cilia. Proc Am Thorac Soc 2011;8:444–850.

[46] Banet N, DeScipio C, Murphy KM, Beierl K, Adams E, Vang R, et al. Characteristics of hydatidiform moles: analysis of a prospective series with p57 immunohistochemistry and molecular genotyping. Mod Pathol 2014;27:238–542.

[47] Mercer K, Giblett S, Green S, Lloyd D, DaRocha Dias S, Plumb M, et al. Expression of endogenous oncogenic V600E B-raf induces proliferation and developmental defects in mice and transformation of primary fibroblasts. Cancer Res 2005;65:11493–500.

[48] Saleque S1, Cameron S, Orkin SH. The zinc-finger proto-oncogene Gfi-1b is essential for development of the erythroid and megakaryocytic lineages. Genes Dev 2002;16:301–6.

[49] Jacks T, Fazeli A, Schmitt EM, Bronson RT, Goodell MA, Weinberg RA. Effects of an Rb mutation on the mouse. Nature 1992;359:259–300.

[50] Berman JJ. Precancer: the beginning and the end of cancer. Sudbury: Jones and Bartlett; 2010.

7

Trapped by Evolution

Section 7.1 Spandrels, Pendentives, Corbels, and Squinches

If you want to make an apple pie from scratch, you must first create the universe.

Carl Sagan

Spandrels, pendentives, corbels, and squinches are arcane terms that apply to medieval architectural solutions that were created to compensate for various design choices. The idea, made popular by the late evolutionary biologist Stephen Gould, in his essay, "The spandrels of San Marco and the panglossian paradigm: a critique of the adaptationist programme" is that our design choices (e.g., a dome, or a column, or a flying buttress, or an oropharynx, or an external ear), will result in secondary adaptations [1]. We see spandrels all around us. A spandrel for a bolt is a nut. A spandrel for a shoe is a sock. A spandrel of a computer is a monitor. A spandrel for a television is a remote control.

Spandrels, in the context of biology, are not strictly necessary items. They are secondary by-products of an evolutionary advance that may, or may not, rise to the level of characteristic traits of the organism and its descendants. Epipubic bones serve as an example of an abandoned spandrel. All nonplacental mammals (marsupials and monotremes) have an epipubic bone that extends forward from the pelvis to strengthen the lower body during locomotion. When the placentals evolved, they needed the pelvis to expand to accommodate the growing amniotic sac. As a consequence, the epipubic bone is absent in eutherians [2].

Just for fun, let's look at another example of a spandrel by asking ourselves whether an evolutionary adaptation may account for the large differences in gestation lengths among the different mammals. For example, a rat has a gestation period of about 22 days, while a human has a gestation period of 280 days, and a horse has a gestation period of about 340 days. The usual explanation is that large eutherians need more time to develop than do small mammals. This facile explanation does not fully satisfy, insofar as large mammals are no more complex than small mammals. They all have about the same cell types, the same organs and the same anatomy. Regarding size, a 1000-fold difference in cell number can be accounted by 10 days of cell division, assuming that cells divide once per day, and that there is no attrition due to dying cells. Even allowing for cell death, which we know occurs in embryonic development, the 258 day delay between the rat gestation and the human's seems overly long. What really accounts for the differences in gestation among the mammals?

Some cellular processes cannot be hurried. As it happens, axonal growth rate is much the same in rats, humans, and horses. Axons can grow, at best, a few millimeters each day. In the case of rats, humans, and horses, the legs of the newborn must be innervated at the time of birth; otherwise, the newborn would be born paralyzed. The most distal point on the mammalian leg is innervated by the longest nerve in the body, stretching from the lowermost position on the spine, down the thigh, to the very end of the leg. In the case of the newborn rat, the length of the longest nerve cell innervating the leg is just a few millimeters. In the case of the foal, born with legs roughly 90% the length of the legs of an adult horse, an axon may exceed 35 in. (889 mm) in length. At an axonal growth rate of 3 mm per day, a foal's leg could have achieved full innervation in not much less than 300 days; in the same ballpark as the observed 340 day gestation of horses.

Animals cannot have a gestation time shorter than the time required to grow an axon that reaches its most distal anatomic site. Hence, we infer the gestation period of a species is limited by the length of the newborn's leg. We can find noneutherians, such as the ostrich, with extraordinarily long legs. The gestation period of the ostrich is short; just 6 weeks. But ostrich chicks are born with tiny legs. Their legs grow to adult lengths postnatally. In the case of Class Eutheria, the species-specific gestation period is an adaptation to the evolution of long legs in newborns. Thus, we can imagine that the lengths of eutherian gestation periods is another form of spandrel (Fig. 7.1).

Section 7.2 Evolving Backwards

In fact there is no real doubt that the direction of evolution has changed very markedly within phyla that single characters do sometimes return to an ancestral condition, and that lost characters can be regained.

George Gaylord Simpson [3]

Every simplifying principle in biology, no matter how basic, represents compelling ideas and is potentially applicable to other areas of thought.

Alain Berthoz and Giselle Weiss [4]

You will all recall that in Episode 19, season 7, of Star Trek The Next Generation, Picard, and Data returned to the Enterprise, after a short away mission, to find the entire crew had devolved into large but primitive creatures: amphibians, insects, and what-not. Worf morphed into something that spat venom. Spot, Data's cat, had transformed into an iguana. If the show were not limited to its 1 hour time slot, Captain Picard would have

FIG. 7.1 The gestation period of a horse is 11–12 months. The length of a foal's leg is about 90% the length of the leg of a full-grown horse. *Source: Wikipedia, entered into the public domain by its author, Vassil.*

joined the lemurs. According to the storyline, a therapeutic misadventure, intended to stimulate Lieutenant Barclay's immune system, had unlocked the crew's introns. Lieutenant Data saved the day, in the nick of time, by administering a retrovirus that reset everyone's genomes back to their normal status.

At the risk of conflating reality with make-believe, let's take a moment to examine the premise of this episode. Do we humans contain in our genomes the genetic material necessary to reconstruct any animal of any evolutionary lineage? Sadly, no. Insects, iguanas, venomous animals, and birds, all followed evolutionary paths that split from the path followed by humans. We humans lack the genes that arose in evolutionary pathways divergent from our own. We cannot inadvertently activate genes that we do not possess. Hence, Star Trek fans will be shocked to learn that Federation science is not always based on reality. Still, when Counselor Troi submerged herself in her bathtub, as some sort of primitive amphibian, she posed an interesting question. Can evolution move backwards?

Up to this point, we have reviewed evolution as though it were a process that has a direction. Basically, we think of the living organisms as being the result of small improvements to ancestral organisms, occurring over billions of years. In this section, we will be considering whether evolution has a direction, and if so, whether the direction of evolution is sometimes reversed.

Let's look at two examples of devolution.

– Class Microsporidia

Class Microsporidia is not your typical fungus. Unlike all other fungal classes, the members of Class Microsporidia are obligate intracellular parasites that have adapted themselves to parasitic lives in a wide range of eukaryotic organisms. Unlike virtually all other members of Class Fungi, the members of Class Microsporidia lack fully functional mitochondria, retaining only remnant forms [5]. In addition, unlike most other fungi, the microsporidia lack a hyphal form and do not produce multicellular tissue structures.

With all these nonfungal properties, it was never apparent, based on morphology or physiology, that the microsporidia are fungi. Not until the advent of rigorous molecular phylogenetic studies have we come to accept microsporidians into Class Fungi [6].

All members of Class Microsporidia form spores, thick-walled cells that can survive in the environment. Infected animals (usually mammals, birds, or insects) pass spores in their urine and feces, and the spores infect humans by direct contact, by water contamination, or by inhalation. The spores pass to the intestines, where they extrude a polar tube into the intestinal lining cells. Through the polar tube, the cytoplasm of the spore (sporoplasm) enters the host cell and organizes into a cell capable of division. Eventually, more spores are formed. When the host cells lyse (break open and die), spores are released into the intestine, and are passed with feces into the environment. Preliminary evidence suggests that microsporidial infections are common [7], with full-blown disease occurring primarily in immune suppressed individuals. About 14 species of Class Microsporidia account for a variety of clinical disorders in humans, ranging from keratoconjunctivitis to diarrhea, to cholecysticis to myositis. (Fig. 7.2).

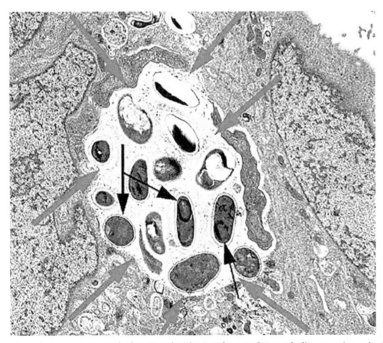

FIG. 7.2 Transmission electron micrograph showing developing forms of *Encephalitozoon intestinalis* (Class Microsporidia), demonstrating spores within a vacuole. *Source: A public domain work provided by the US Center of Disease Control and Prevention.*

Because all fungi evolved from an ancestor belonging to Class Opisthokonta, we can infer that the members of Class Microsporidia devolved from an ancestral species having a single posterior undulipodium and fully functional mitochondria and the capacity for mitosis and meiosis. Lacking such ancestral traits, at least in their fully functional forms, the microsporidia must have evolved to lose them (i.e., devolved).

– Class Myxozoa

The Myxozoans are a class of parasitic cnidarians and were described in Section 5.4, "Opisthokonts to Parahoxozoa." Their membership in Class Cnidaria is based on phylogenetic studies and various class-specific anatomic traits, including the cnidarian body plan [8]. The myxozoans are the only parasitic members of a Class that includes a diversity of free-living aquatic animals. In their evolutionary transition as endoparasites in fish, they stripped a good deal of their genomes. The myxozoans have the smallest known animal genomes; Kudoa iwatai having a genome that is only 22.5 million base pairs in length. Genome surveys indicate that myxozoans do not express many cnidarian genes related to cellular development, cell differentiation, intercellular signaling, and many transcription factors [8].

Like the microsporidians of Class Fungi, the Myxozoan animals have demonstrated the ability to shed themselves of much of their ancestral genomes, as needed.

No substitutions, no returns

The microsporidians and the myxozoans have shown us that organisms can devolve, losing much of their genomes during the process. Nonetheless, microsporidians and myxozoans retain their phylogenetic identities. If we think of the evolution of a species as a voyage on a jet plane, then we have the opportunity to lose our baggage, but we don't have the right to reverse directions, return the jet to its runway, and hop on a different plane having a different destination.

A good example that illustrates the impossibility of going backwards to correct an evolutionary error is found in RuBisCo (ribulose-1,5-bisphosphate carboxylase/oxygenase), the most abundant enzyme on earth. RuBisCo catalyzes the carboxylation and cleavage of ribulose-1,5-bisphosphate into two molecules of 3-phosphoglycerate. RuBisCo is ancient, predating the first synthesis of oxygen by cyanobacteria, and found in every major class of life, including archaea, bacteria, and eukaryotes. Nearly half of the soluble protein in plant leaves is RuBisCo.

There are lots of enzymes involved in oxygenic photosynthesis. Why do we have so much RuBisCo? The reason for the superabundance of RuBisCo has less to do with its function in carbon fixation and oxygenic photosynthesis than with its remarkable inefficiency. It is strongly inhibited by oxygen (a pathway product), saturates poorly, and catalyzes slowly. Its inhibition by oxygen may relate to the fact that it evolved prior to the appearance of any appreciable oxygen, at a time when every organism was an anaerobe. Despite its many shortcomings, RuBisCo had, at a very early epoch, secured its role in terrestrial organisms [9]. To compensate for its inefficiency, ancillary enzymes evolved to join RuBisCo and enhance its catalytic output, forming a complex known as the RuBisCosome. Plants went into high gear to synthesize RuBisCo and its associated proteins in great abundance.

Will earth organisms evolve an efficient enzyme that can replace RuBisCo? If more than a billion years of history is any indicator, the outlook for the future looks dim, but not absolutely dismal. We have seen examples wherein alternate enzymes have evolved, to substitute for ancient enzymes, when absolutely necessary. The proof for this assertion is somewhat convoluted, but well worth the effort to discuss here. It seems that the minimum number of genes required to support the least complex forms of bacterial life is about 250. This is approximately the number of genes in *Bacillus subtilis*, *Mycoplasma genitalium*, *Mycoplasma pneumoniae*, and *Haemophilus influenzae*. About 80 of the genes (from the 250 gene set) seem to be present, as homologs, in all nonviral life on earth. To see just how minimal the minimal gene set may go, researchers went about knocking out individual genes to see which specific genes were absolutely necessary for viability. Much to their surprise, they could knock out individual genes that belonged to the universal gene set (i.e., were present as homologs in every nonviral organism) without destroying viability. This means that there are conserved genes that are present in every organism but not absolutely necessary for cell survival. How can this be? The conclusion drawn is that alternative genes must have evolved that can substitute for the universal genes, in a pinch (e.g., when the universal gene is experimentally knocked out) [10]. This conclusion can be challenged, but it serves to remind us that evolution produces diversity, and it is diversity that provides biological options that may come in handy, in unexpected situations.

The path of evolution

When we compare present-day organisms, such as members of Class Eutheria, to ancient single-celled eukaryotes, it is hard to avoid thinking that evolution always marches forward, toward increasingly complex forms of life. Perhaps we should reconsider. Most of the species of eukaryotes are single-celled organisms, much like their most ancient ancestors. These extant single-celled organisms have had just as much evolution as any other organism, but they haven't changed much over the past billion years or so. Evolution did not make these organisms more complex.

We can also surmise that the genes in extant single-celled organisms are, more or less, the same genes that were present in the ancient ancestors. The reason we can draw this conclusion is that all of the animal organisms that have been studied are known to contain a great number of core genes that are shared among all eukaryotes. It would be absurd to think that all these different organisms evolved the same genes independently. Hence, we can infer that all living eukaryotes (metazoans, plants, fungi, or single-celled eukaryotes) inherited the same set of genes from the same ancestral species. This being the case, we can infer that evolution toward more complex forms of organism is exceptional, insofar as most eukaryotic species are simple one-celled organisms, and all organisms have settled for about the same core set of genes.

Finally, we have seen ample evidence that some classes of organisms devolve by losing complexity and by losing many of the ancestral traits that related class members have retained (e.g., Class Microsporidia and Class Myxozoa). If evolution does not always produce species of constantly increasing complexity, and if devolution is observed in complex organisms, then can we really say anything about the direction of evolution?

In hindsight, we should never have assigned a purpose, destiny, or direction to a process such as natural selection that is little more than a mathematical truism. Evolution does not produce better and better species. Evolution is a mindless bookmaker that favors species that are most likely to survive.

As for the emergence of highly complex species, we can thank diversification. As discussed back in Section 4.3 "The Diversity of Living Organisms," all successful species speciate, producing more and more diverse species. Some of these increasingly numerous and diverse species will be complex, and some of these complex species may be endowed with particularly versatile central nervous systems. Basically, this general observation acknowledges that every possibility must occur in an infinitely diverse universe. Whereas evolution may or may not lead to greater and greater intelligence, we can be certain that diversification will lead to every possible biological outcome, including intelligence.

Evolution teaches us that we cannot escape our past. Because we are evolved organisms, not designed organisms, each of us develops from an embryo that repeats a sequence of steps that had evolved in our ancestors, and this sequence of steps endows us with certain inalienable properties. Insofar as we are all eukaryotes, we have a nucleus, and mitochondria, and some of our cells replicate via meiosis while other cells replicate via mitosis. Insofar as humans are deuterostomes, we all have coeloms and a gut that viscerates from anus to mouth. Because our human development employs ancestral genes, in a process that evolved through our ancestral lineage, the process of backwards evolution would require us to change the history of our species.

Once a complex organism has evolved, everything is interconnected and cannot be safely reversed. When you evolve forward, you're not just modifying one subsystem (i.e., the one with the mutation), you're creating a system change in the way that every part of the system interacts with every other part. There is no do-over because you've always got to start with the whole genome that you've created. [Glossary Dollo's law]

Not convinced? Let's approach the problem of backwards evolution from a different viewpoint. Suppose, just for fun, we wanted to produce a new breed of humans that lack a right and left clavicle (the bone that struts between the sternum and the shoulder). Would we start our project by identifying the gene (or genes) that codes for the clavicle? We would almost certainly fail in our effort because there is no "clavicle" gene. For that matter, there is no set of genes that are devoted to forming the clavicles. The clavicles develop as part of a strict sequence of embryologic processes, with each process preparing the developing embryo for subsequent processes, that continue until the organism if finished.

How can we be so sure that there is no clavicle gene? We can infer that every gene mutation capable of producing a disease, when inherited, must occur (because mutations can occur in any gene). In fact, there are at least 7000 rare genetic diseases, accounting for a sizable portion of the total number of genes present in the human genome. There are many genetic syndromes in which there is maldevelopment of one or more tissues or organs, but there is no inherited disease that results in an isolated missing bone, and nothing else.

The disorders that come the closest to producing an isolated anatomic aplasia are the sporadic or toxic maldevelopments. For example, in anencephaly, there is a specific type of neural tube defect that occurs at a specific moment in embryonic life, that results in the absence of the telencephalon (i.e., both cerebral hemispheres). Like most neural tube maldevelopmental disorders, a portion of the cases of anencephaly are associated with folic acid deficiency [11]. Some cases are sporadic (i.e., we are clueless to cause). When the telencephalon fails to develop, so do those tissues that depend on the developing telencephalon for their development (i.e., meninges, skull, and scalp). Short of toxic destruction of a developing anlage (e.g., thalidomide and limb buds [12]), there seems to be no mechanism to produce isolated aplasia of a single bone. [Glossary Sporadic]

When we look at a list of anatomic variations in humans that are considered "normal," the list is dominated by add-ons (i.e., accessory and supernumerary variants). The list of anatomic absences is much shorter, and we presume that most or all of these missing organs were present, as embryonic anlagen, before being resorbed. In other words, they are absent from the developed individual but were present in a primitive stage in the embryo. For all of us metazoans, it is impossible to reverse the direction of evolution, and this explains why all mammals have essentially the same body plan. The difference between a mouse and a man are primarily differences of size and shape of structures. The functionality is about the same in both species (e.g., a mouse kidney filters blood, much like a human kidney does). Even when one of the species of mammal decides to change its lifestyle drastically (e.g., whales and land turtles returning to the seas), they keep their distinctive mammalian body plans, just tweaking things a bit to suit their accommodations (e.g., moving the snout back to the top of the head, and calling it a blowhole).

What we observe, when we study human embryology and anatomy is that our anatomic restraints have less to do with structure and function than with our evolutionary past. We may never do away with our appendix, and we may never shorten our recurrent laryngeal nerve, and we may never improve the circulation system afforded by our barely sufficient coronary arteries, and we may never discard all of our junk DNA. Nonetheless, there are many instances wherein complex organisms successfully slim down, reducing the size of burdensome anatomic structures, or modifying them to the point that they serve some new and useful purpose. For example, eukaryotes characteristically contain mitochondria; nonetheless, there are single-celled eukaryotic organisms that are amitochondriate. Even in such cases, there is evidence that the classic mitochondria have been modified, not eliminated, and appear in so-called amitochondriate eukaryotes in the form of hydrogenosomes or mitosomes, and other forms of rudimentary mitochondria [5, 13–15]. Likewise, we see that some eukaryotic organisms seem to rely entirely on mitosis, seemingly abandoning the practice of meiosis altogether. Nonetheless, it looks as though the evolved genetic machinery for meiosis remains, in every eukaryote [16].

In summary, it would seem that evolution has no direction, and natural selection is the directionless process that pushes evolution.

Section 7.3 Eugenics: Proceed With Caution

In most cases, the molecular consequences of disease- or trait-associated variants for human physiology are not understood.

Teri A. Manolio and coauthors [17]

Eugenics, which derives from the Greek roots "eu," good and "genos," race, generally refers to any effort aimed at improving the human gene pool. As it is broadly used today, it would also include efforts to eliminate the genetic causes of disease, in individuals or in populations.

We study the past so that we can better understand how to live today, and in the future. Consequently, it is impossible to delve deeply into the history of human evolution without speculating a bit about how we might use our knowledge to influence human destiny. The purpose of this section is to explain that eugenics, as a field is science, has limited value because we are pretty much clueless as to what constitutes "an improved gene pool." Advances in the field of precision medicine do not seem to be bringing us any closer to a solution. Furthermore, it seems as though we won't be eliminating the genetic causes of disease anytime soon, because the genes associated with the common diseases are normal variants, and are constantly reentering the gene pool. Furthermore, many of the genes causing rare, fatal disease are de novo mutations, and we have no way to prevent these mutations from occurring. These assertions are somewhat counter-intuitive, and we ought to review the supporting evidence. [Glossary Precision medicine]

Here are some of the reasons why eugenics will not play much of a role in the near-future of precision medicine [18]:

— No such thing as a superior being

We marvel at the success of animal breeders who seem capable of producing made-to-order animals. The results are deceptive. Breeders select for a single trait (e.g., size, speed, strength, muscle mass, milk production). All traits other than the trait under selection are unpredicted consequences of the process. When we look at any breed of animal, we are not looking at a superior race of the species; we are only looking at a kindred that have certain family traits, one of which was purposely selected by a breeder.

Suppose we wanted to breed a superior human being. How might we go about doing that? You would need to start with some idea of the traits that must be present in the superior person (e.g., strength, endurance, height, longevity, intelligence, and attractiveness). The problem here is that we do not really know how to gauge the quality of any of these traits. A very tall person might be useful for fetching the cereal box from the top shelf of the cabinet, but a short person might be a better choice to stock the bottom shelf in the grocery store. A math wizard may have the kind of intelligence needed to calculate the best trajectory for a space ship traveling to Pluto, but a social worker might have the kind of intelligence needed to keep the space ship's crew from murdering one another on the way over.

Basically, we have no idea what, if anything, makes one person superior to another. Even if we could agree on what we wanted, we would have no way of preparing ourselves for the consequences. If we bred for creativity, then we might produce a population of self-absorbed individuals who had no interest in nurturing children. A race bred for fertility might overpopulate the planet. A race bred for size and strength might have dietary requirements so high that they outstripped food production.

Not convinced? Humans routinely employ a fairly brutal eugenics policy, without knowing it. It's called marriage. The males are on the lookout for the most superior woman, and the women are trying to find the most superior male. If you want to know whether eugenics works, then take a hard look at your own extended family. The unintended consequences of our best efforts at breeding will nearly always outweigh our intended goals.

- Principle of evolutionary frustration

It's a truism that organisms optimally suited to a particular task will be unsuited for some other task [19]. Essentially, when we increase fitness in one instance, we decrease fitness in another. For example, a dog bred for size and strength that is optimally suitable for pulling a sled over arctic ice will be unsuitable for chasing small rodents through narrow passageways in a Florida warehouse.

Whenever we select for a desirable feature, we can be almost certain that there will be an evolutionary price to pay. For example, the Shar-Pei dog was bred for its thick, folded skin. In doing so, the Shar-Pei was endowed with a mutation in the HAS2 gene, which encodes the rate-limiting enzyme for hyaluronan synthesis. The accumulation of hyaluronan in skin accounts for Shar-Pei phenotype. As it happens, hyaluronan, when fragmented, is a strong trigger for the innate immune system. Consequently, Shar-Pei, as a breed, develop a particular disorder of the innate immune system known as periodic fever syndrome [20]. Furthermore, when Shar-Pei breed with other dogs, they can introduce the HAS2 gene mutation to the general population, thus depositing a deleterious mutation into the canine gene pool. [Glossary Innate immunity]

- Breeds are susceptible to being nearly wiped out by a single disease

Ironically, a race of superior beings may be a race that is unfit to survive. The purebreds, particularly those selected for size and strength, have the shortest life spans and carry the greatest burden of disease. While golden retrievers and Bernese mountain dogs have their charms, they also carry a set of genes that predispose them to very high rates of cancer, an unintended consequence of the breeding process [21]. Any veterinarian will tell you that the longest-lived, healthiest dogs are the small mutts (Fig. 7.3).

Imagine, for a moment, that we have just bred a stronger human. The new breed is so popular that within a few generations, everyone on the planet is effortlessly bench-pressing 500 pounds of iron. Genetic variations may have pleiotropic effects. In this case, let's imagine that the strong breed of humans came with an unintended feature: heightened susceptibility to an uncommonly virulent form of the common flu virus. Just when everything was looking swell, a flu epidemic came, wiping out 90% of the earth's human population. Does this seem farfetched? There is an inherited immunodeficiency of cattle caused by a reduction in leukocyte adhesion factor. Affected cattle are homozygous for a gene allele that codes for a substitution in a single amino acid in the protein product. Heterozygotes (i.e., cattle with an unpaired mutant allele) are common in the United States, with a carrier rate of about 10%. Every cattle with a mutant allele is a descendant from one prize bull, whose sperm was used to artificially inseminate cows in the 1950s and 1960s [22]. A disease that was essentially nonexistent in 1950 became a common scourge of the dairy industry within a half-century, all due to the founder effect amplified by modern animal husbandry. Eugenics is a tricky business. [Glossary Susceptibility]

– Polygenic traits cannot, as yet, be eliminated from the human gene pool

A trait is a feature that we all have, but to a different extent, varying from individual to individual. Size, strength, and intelligence are examples of human traits. It has been accepted for nearly a century that traits are seldom determined by a single gene; they are always polygenic [23–25]. You would guess that something like intelligence that has so many different forms and that can scarcely be sensibly defined, could not be the product of a single gene. In point of fact, genome wide association analyses have already identified dozens of genes that correlate with intelligence [26]. Likewise, at least 180 gene variants are known to be associated with variations of normal height. These 180 variants may represent only a fraction of the total number of gene variants that influence the height

FIG. 7.3 The Irish wolfhound, a dog bred for size and strength. Irish wolfhounds live only 7 years, on average, frequently dying from bone cancer, an uncommon cause of death in most other breeds. *Source: Wikipedia, from a National Library of Ireland on The Commons photograph, February 21, 1917.*

of individuals, as they account for only about 10% of the spread [27]. [Glossary Genome Wide Association Study]

The polygenic diseases have a non-Mendelian pattern of inheritance (i.e., neither autosomal dominant, autosomal recessive, nor sex-linked). The reason for this is that each of the variant genes that contribute to the expression of a polygenic disease is inherited independently, from either parent. Hence, the set of genes that together account for the polygenic disease is not present, as a complete set, in either parent. Hence, inheritance of polygenic traits and polygenic diseases cannot be assigned to either parent. Hence, the inheritance pattern in non-Mendelian. [Glossary Autosomal dominance]

Medical researchers could have saved themselves a great deal of effort, over the past few decades, searching for a single causal gene for hypertension, had they simply recognized that hypertension is a quantitative trait that does not have a Mendelian pattern of inheritance. Hence, hypertension is polygenic, and we can presume that some portion of the many different genes that contribute to hypertension are normal variants of genes. Other common diseases, such as type 2 diabetes, are also polygenic [28].

The common diseases of humans can be thought of as traits that are expressed strongly in some individuals (the individuals who get sick), and weakly in other individuals (the ones that are not affected). For example, we all have glucose circulating in our blood; it's a trait of the human species. Some of us have glucose levels high enough to be categorized as diabetic. The lucky ones maintain glucose levels in the normal range. We might be able to produce a breed of humans that have a low incidence of diabetes, if we tried very hard. It wouldn't do us much good, though. Diabetes is a polygenic condition, and natural selection does not provide a way to eliminate a large collection of genes through breeding. This means that the set of genes that can produce type 2 diabetes will remain more or less intact in the gene pool of the diabetes-free, pure-bred population. Over time, conditions may favor the emergence of new gene variants that work in concert with the retained set of diabetes genes heralding the return of diabetes, or the emergence of a new clinical form of diabetes, or that may produce a new disease entirely that nobody anticipated. [Glossary Incidence]

- Screening for disease gene carriers does not lead to improvements in the gene pool

Screening for heterozygotes who carry a recessive disease gene has been proven to be a very effective way of reducing the incidence of disease, with 90%–95% reduction of disease in the cases of Tay-Sachs disease and beta-thalassemia [29–31], and equally promising results for cystic fibrosis and Gaucher disease [32]. [Glossary Thalassemia]

It should be kept in mind that screening for heterozygote carriers does not eliminate the disease gene from the gene pool; it simply reduces the number of individuals born with two affected alleles. In addition, many genetic diseases are caused by de novo mutations for which screening of the parent serves no purpose. Furthermore, most birth defects that are associated with genetic aberrations are characterized by chromosomal anomalies (e.g., trisomies, deletions) and cannot be attributed to any particular gene. Hence, screening has not been an effective method for reducing the number of disease genes from the human gene pool. [Glossary Chromosomal disorder, Congenital anomaly, Contiguous gene deletion syndrome]

- Gene editing techniques, as currently employed, do not eliminate disease-causing mutations from the human gene pool

New methodologies, such as Clustered Regularly Interspaced Short Palindromic Repeats (CRISPR) gene editing have been touted as opportunities to cure and eventually eliminate genetic diseases [33–35]. At present, editing of the human germline DNA (i.e., the DNA present in the zygote that is passed to every cell in the developed organism) will not be funded by NIH and would most likely not meet approval by the FDA [36].

In the United States, the direct clinical uses of CRISPR will be restricted to repairing somatic mutations. This type of procedure might involve taking unhealthy cells from a patient, editing the DNA to repair or eliminate the disease gene, growing the repaired cell in culture, and infusing the cells back into the patient. Such efforts, targeted against somatic cells, though nothing short of medical miracles, will have no effect on the human gene pool. Patients who are cured of their diseases will continue to carry the defective gene in their germ cells and may pass the gene to the next generation.

As discussed in Section 2.1, "Mutation Burden," de novo disease mutations occur in the zygote or in the early embryonic cells of children, and cannot be identified in the parents. Gene editing techniques applied to persons with de novo mutational diseases, may result in cures for affected patients, but will not eliminate new disease-causing gene from the gametes of these patients. This means that every patient cured of de novo mutation disease becomes a carrier of the disease gene, with a 50% chance of passing the gene to his or her children. Hence, gene editing, applied successfully, will result in an increase in the number of disease-causing genes entering the human gene pool.

Even if we were permitted to edit and repair inherited mutations in embryonic cells, a host of unintended consequences may emerge that could add additional disease genes to the human gene pool (e.g., introducing new mutations in the gene-editing process, and producing new somatic mosaics). [Glossary Somatic mosaicism]

If all the technical, ethical, and economic problems raised by CRISPR technology were solved, we would still face the fact that healthy disease-gene carriers, who would not receive the benefits of gene editing, will continue to procreate, thus passing their disease genes to the human gene pool. Furthermore, gene editing on monogenic disorders will have no effect on disease genes involved in polygenic disorders [37].

Historically, important technological advances in molecular biology have had their greatest impact on scientific discovery, and only indirectly benefit treatment. Before we can treat diseases, we must know the biological mechanisms that account for the development of the disease, and the clinical phenotype of the disease. If the past serves as an indicator of the future, the greatest value of CRISPR will be its ability to modify genes, under controlled laboratory conditions, allowing us to carefully observe the consequences [38].

As for the future of eugenics, as a social movement, it suffices to say that the eugenics movement in the 20th century was a horrible travesty of science. In this era, in which genes can be written, revised, and erased, much like words on a piece of paper, we can insert DNA from other organisms into the human germline, and vice versa. Before very long, we'll need to reconsider what it means to be human. If we want to maintain our humanity, we should try our best not to repeat the mistakes of our recent ancestors.

Section 7.4 The Evolution of Aging, and the Diseases Thereof

Life can only be understood backwards; but it must be lived forwards.

Soren Kierkegaard

I didn't attend the funeral, but I sent a nice letter saying I approved of it.

Mark Twain

It is tempting to assume that if all living organisms are subject to death, then all living organisms must age. This is simply not true. Numerous organisms have life expectancies in the thousands of years, perhaps hundreds of thousands of years. Such organisms, though subject to death from physical causes, do not show the physical signs of aging, over long stretches of time.

Here are a few examples of ageless organisms:

- Fungal colonies have no upper bound in terms of size or age. The Malheur National Forest in Oregon is host to a fungus (*Armillarea ostoyae*, the honey mushroom) that covers 2200 acres of land, to an average depth of about 3 ft. It is estimated, based on growth rates, to be about 2400 years old.

- Many species of trees can live hundreds or even thousands of years. Methuselah, a Great Basin bristlecone pine residing in Inyo County, California, is reputed to be about 5000 years old, which seems to be about the observed limit for the lifespan of any individual standing tree.

- Some trees self-clone within a copse, producing a group of trees all having the same genetic identity, with new clonal growths replacing dying trees. It is not unreasonable to consider the copse itself as a single biological organism, characterized by one genome and by a stable collection of growing and dying cells. Such clonal organisms are virtually immortal. A copse of Quaking Aspen (*Populus tremuloides*), living in Fishlake National Forest, Utah, is estimated to be hundreds of thousands of years old [39]. Considered as a single organism, the Quaking Aspen colony occupies 106 acres and has an aggregate weight of about 6 million kilograms. Farmers in ancient and modern times have benefited from the self-cloning nature of plants by developing the agricultural technique known as coppicing. Young trees are cut to near-ground level, and new, clonal trees reshoot from the stump. By repeated cuttings, the trees are maintained as juveniles. Regularly coppiced trees never seem to age or die; they just spread out from the center. Individual coppiced trees have been maintained for centuries.

- Some jellyfish (Class Medusozoa, a cnidarian subclass) may be immortal. The medusae (fully mature form) of the jellyfish *Turritopsis nutricula* has the rare ability to revert back to its immature polyp stage, basically reversing its life cycle. The immature polyp can reproduce by budding. There seems to be no limit to the number of times in which a medusa may form from a budding polyp and revert back to a polyp stage. Hence, the jellyfish *Turritopsis nutricula*, and several other species, including *Turritopsis dohrnii* and *Hydractinia carnea* may be immortal [40–43].

- Estimates based on the growth rates of Antarctic sponges suggest a very long lifespan in these animals. Extrapolating from growth rates, a 2 m high specimen, discovered in the Ross Sea, was estimated to be 23,000 years old. Due to fluctuation in the levels of the Ross Sea, skeptical scientists doubt that the specimen could have lived for much more than 15,000 years. Still, a low-estimate 15,000 year lifespan is impressive enough.

- Planarian flatworms can be cut into pieces, and an entire organism can be regenerated from each of the pieces. Hence, researchers quip that planarians are "immortal under the edge of a knife." Observations of blastemal whole-body regeneration indicate that planarian cells are totipotent and able to form blastemas. We do not know how planaria manage to maintain a large population of totipotent stem cells, but it has been suggested that a key step involves the maintenance of normal telomere length in planaria [44]. Telomeres are sequences of noncoding DNA attached to the termini of chromosomes. Without telomeres, cells cannot divide, and telomeres are typically lost from nondividing somatic cells (the so-called postmitotic cells). Planarian stem cells maintain their telomere length after many replications. [Glossary Postmitotic, Telomere]

Back in Section 5.4, "Opisthokonts to Parahoxozoa," we noted that Class Parahoxozoa has two subclasses: Class Bilateria and Class Cnidaria. Only a small subset of cnidarians could be considered ageless, but all cnidarians have some ability to regenerate from somatic cells, permitting them to recover from injury. If we move backwards up the phylogenetic pathway leading away from the cnidarian/bilaterian split, we find additional examples of ageless animals, such as the sponges (Class Porifera) that live many thousands of years. Backing up to the opisthokonts, we see that some fungi are ageless. Left to their own devices, a colony of fungi may grow continuously, with no apparent limit in terms of size or age. Preceding the opisthokonts are the single-celled eukaryotes, which are also ageless in the sense that their life cycles accommodate endless rounds of replication. Each replication can be thought of as a renewal of the life of the replicating organisms; a form of immortality, if you will.

From the single-celled eukaryotes, let's move sideways, to a phylogenetic lineage that excludes the animals. Doing so, we come to Class Archaeplastida, the root class of all living plants. Plants are multicellular organisms that grow from embryos. Plants, unlike animals, can regenerate whole organisms from somatic cells. Aside from the observation that we find individual trees that have lived 5000 years or more, we also find that trees can be cloned, and re-cloned, with no definite limit. Hence, the plant kingdom is also ageless.

Let's summarize. Single-celled eukaryotes are ageless. Although we haven't discussed them, the prokaryotes and the viruses are ageless in the same sense as the single-celled eukaryotes are ageless: they replicate to produce exact copies of themselves. Of the three major classes of multicellular organisms (i.e., plants, fungi, and animals), the plants and the fungi contain ageless species. Of the remaining class of multicellular organisms (i.e., Class Metazoa), the first several phylogenetic classes are ageless, up until we reach Class Bilateria. **All of these ageless organisms are characterized by continuous growth and or continuous renewal, both of which imply the existence of a nondeclining population of stem cells. In no instance is long life based on the ability of cells to persist as a collection of long-lived nondividing cells.**

From Class Bilateria onwards, we see organisms that age and die, with very few exceptions. One prominent exception to the "get old and die" routine popular among bilaterians is the planarian flatworm, a species in a subclass of the protostomes. As mentioned, planarians can regenerate whole organisms from tiny slices of the original worm. Nonetheless, aging is a phenomenon that is closely associated with bilaterians, and virtually absent from all other organisms on earth. **Hence, we should have a very strong case that aging is an evolved phylogenetic trait that first appeared in Class Bilateria.**

The idea that aging represents an evolutionary advancement seems a bit absurd to those of us who stare in the mirror to see our thinning hair and our turkey necks, as we contemplate our inevitable deaths. It's best to shirk these depressing thoughts and to think of aging as a trait that benefits the species (at the great expense of individual members of the species). The sad fact is that as we age, our gametes accumulate mutations and other forms of cellular degeneration, and the progeny of a mating between two middle-aged adults will, on average, be less fit than a mating between two young adults. From the point of view of a species, the purpose of procreation is to produce new, unique gametes, and to enlarge the gene pool with potentially beneficial genes. Remember, a species is simply an evolving gene pool, and species achieve immortality by speciating, not by simply persisting. For the species, there really is no need to preserve individual organisms that carry damaged gametes. From the perspective of natural selection, aging is an evolved trait whose purpose is to eliminate diploid organisms (e.g., humans) who have outlived their enclosed haploid organisms (e.g., sperm and oocytes). Supposing that this is the case, then we must ask ourselves "**What are cellular processes that account for aging?**"

There are two types of cells in the body: cells that are capable of dividing, and cells that are not capable of dividing. An understanding of the relationship between dividing and nondividing cells tells us a great deal about the physiologic process of aging, including which tissues age, which tissues do not age, and how tissues and organs are likely to be affected by the rare diseases of premature aging.

Humans grow rapidly in utero. After birth, growth continues through adolescence, tapering off as we enter early adulthood. Ideally, humans maintain about the same height and weight in late adulthood as they had in early adulthood. Though our bodies plateau and attain something like a steady state during adulthood, our tissues are undergoing constant renewal. Vigorous, continual cell renewal is most evident in three tissues: the epidermis of the skin, the mucosal lining cells of the gut, and the blood forming cells of the bone

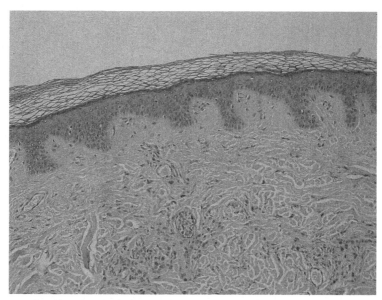

FIG. 7.4 Biopsy specimen of largely unremarkable skin, showing epidermis overlying the fibrous tissue of the dermis. The epidermis is thinner than the dermis, and contains multiple layers of cells, each layer having characteristic morphologic features. The epidermis has an undulating lower border, with papillae known as rete pegs, jutting down into the dermis. Incidentally, some nevus cells are scattered in the dermis, suggesting that the biopsy was taken to confirm the presence of a benign intradermal nevus. *Source: Wikipedia, and entered into the public domain by its author Kilbad.*

marrow. In each of these three tissues, cellular renewal proceeds according to a simple principle: a stem cell divides to produce another stem cell plus a postmitotic fully differentiated cell that lives for a while, doing whatever it was intended to do, and then dies. Because the stem cell replaces itself with a new stem cell when it divides, the total number of stem cells stays more or less constant throughout adult life (Fig. 7.4).

The skin is covered by a thin epidermis that lies atop a continuous connective tissue sheath known as the dermis. The bottom layer of the epidermis, directly adjacent to the underlying dermis, is called the basal layer and contains regenerating cells that divide to produce another regenerating cell and a nondividing epidermal cell that is incapable of further division. The nondividing cells are referred to a postmitotic cells. With the exception of the bottom layer of regenerating cells, the full thickness of the epidermis is postmitotic. These postmitotic epidermal cells gradually fill their cytoplasm with keratin and flatten out to produce a protective barrier covering our bodies. Flattened epidermal cells are squamous, from the Latin root, meaning scale. Aside from serving as bricks in a wall, the postmitotic squamous cells are "dead men walking." Their fate is to rise to the top layer of the epidermis, where they slough off into the environment. The dancing house dust that we see in a beam of light is composed of sloughed postmitotic squamous cells (Fig. 7.5).

Similarly, the entire gastrointestinal tract is lined by a mucosal surface consisting primarily of nondividing enterocytes. Under normal conditions, cell division is confined to the cells

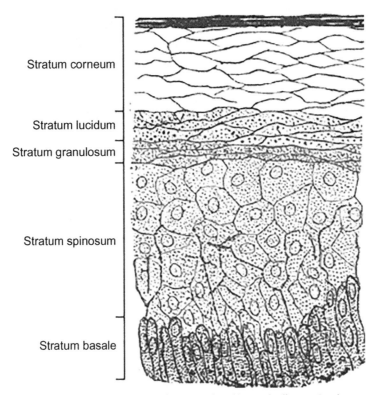

Stratum corneum

Stratum lucidum

Stratum granulosum

Stratum spinosum

Stratum basale

FIG. 7.5 A drawing of the layers of the epidermis. The lowest, or basal layer of cells contains the generative cells of the epidermis, the only cells of the epidermis capable of cell division. Basal layer cells, when they divide, produce one basal cell (able to divide again), and one postmitotic cell (unable to divide). The postmitotic cell is pushed up into the next higher cell layer, and it continues to rise through the epidermis as it is pushed by successive postmitotic progeny of the basal layer. As it rises, it flattens out, fills with keratin, and eventually loses its nucleus. At this point, it is little more than a squamous flake, sitting atop the epidermis. The cells at the very top of the epidermis eventually slide off into the air. A good portion of house dust consists of sloughed epidermal squamous cells. Dandruff is little more than clumps of squamous cells, sloughed from the scalp. *Source: Wikipedia, a modified version of a plate from Gray's Anatomy of the Human Body, 1918, entered into the public domain.*

at the very bottom of the crypts and glands that line the alimentary tract. The postmitotic enterocytes eventually slough into the gut lumen, and add to the bulk matter of stool.

In the bone marrow, a cascade of stem cells produces the fully differentiated red cells, white cells, and platelets that circulate in our blood. The circulating blood cells are postmitotic. The red cells have shucked their nuclei and their mitochondria, reducing themselves to little more than bags of hemoglobin. The circulating postmitotic red blood cells persist in the blood for a few months, after which they are phagocytized by the spleen and by other constituents of the reticuloendothelial system, the physiological equivalent of municipal garbage collectors. Phagocytized red blood cells are replaced by new red blood cells, so that the total number of circulating red blood cells stays fairly constant.

Why is it important to know how the epidermis, the gut, and the bone marrow produce postmitotic cells from a subpopulation of continuously renewing cells?

Rule: The tissues of the body that are constantly renewing do not show signs of aging.

The epidermis, the gut, and the bone marrow have evolved to regenerate continuously throughout life. It is not unusual to find elderly individuals with no histopathological signs of degeneration or atrophy in these tissues. If we look at the gut, epidermis and bone marrow of elderly persons, we find that the basal layer of the epidermis, the regenerating cells of the gut crypts and gland, and the progenitor hematopoietic cells are all dividing normally; just as they had in youth.

We can think of the adult human body as consisting of two types of cells: aging cells (i.e., the nondividing cell populations); and ageless cells (i.e., the stem cells that continuously renew throughout life). The degenerative processes that we call "aging" occurs exclusively in cells that have lost the ability to divide. While the skin, gut, and bone marrow are self-renewing systems, there are a variety of tissues in the body that become postmitotic early in life, and remain so. These would include cartilaginous cells, and oocytes. Other cells such as neurons and muscle cells, and connective tissue cells have a limited ability to divide in adulthood. The long-lived postmitotic cells are all slowly degenerating throughout life [45]. When we think of the infirmities of aging, we are almost always referring to pathology in nondividing cells populations (e.g., osteoarthritis secondary to cartilage degeneration, dementia secondary to neuronal degeneration, and sarcopenia secondary to muscle cell degeneration).

Oocytes represent an extreme example of nonregeneration. All of the oocytes that a woman will produce in her lifetime are produced in utero, before she is born. The number of oocytes reaches a peak of about 7 million cells at 5 months gestation. After the peak is reached, about 3 months before birth, the oocytes begin to die; they are not replaced. The number of live oocytes declines until the number falls below a threshold of 1000, triggering menopause [46]. In this instance, as in every other example of human tissues undergoing aging, the process involves cells that cannot regenerate.

Contrariwise, we have seen that organisms that grow continuously (e.g., sponges, redwood trees, and giant fields of mushrooming fungi) never really age. They simply keep growing until some calamitous event, such as an ice age or a supervolcano eruption, ruins their day. The general rule that regenerating tissues do not age seems to apply to all the classes of living organisms.

There are times when the self-renewal process malfunctions, always to the detriment of the organism. Disease processes that target the renewing cells of skin, gut or bone marrow always have adverse consequences. For example, Fanconi anemia (vide infra) produces bone marrow failure, apparently by interfering with the normal renewal of hematopoietic stem cells [47]. Experiences with rare disorders of self-renewing tissues, such as Fanconi anemia and dyskeratosis congenita, remind us that any cellular system can be disrupted. We are not alone; whether in immortal nematodes or in short-lived mice, conditions that deplete stem cells will produce premature aging [45, 48–50].

Diseases of Aging Versus Diseases of Old People

If aging is an evolutionary advancement, presumably achieved through a genetic acquisition, then is it possible that we can find, isolate, and eliminate the "aging gene," and thus achieve immortality, like the planarium or the jellyfish, or the sponge? The answer to this question is "No!," there is no single gene that accounts for the aging process, and we can prove this logically. **If aging were caused by a single gene, you would expect rare occurrences of loss-of-function mutations of the gene, leading to instances of human immortality. Outside of science fiction, immortal humans do not exist. Therefore there is no "aging gene."**

Still, we can infer that genes play some role in aging because long-lived parents tend to have long-lived children [51], and monozygotic twins tend to have closer life span concordances than unrelated individuals [52]. What, then, is the relation between genes and aging?

Aging is a trait, meaning that it is a quantitative feature. Like all quantitative traits, we assume that many different genes influence aging. Hence, if you have the good fortune to live a long life, it will not be because you have inherited any particular gene from a parent; but because you have inherited lots of genes that contribute to producing a long-lived individual.

The ancient Romans had a pessimistic but elegant way of summarizing life: "Omnes vulnerant, ultima necat. (All hours wound; the last one kills.)" Their brutal motto emphasizes the notion that all the passing moments in our lives have their negative effects and bring us closer to our deaths. The Romans may have been correct, but not every negative effect can be sensibly attributed to the aging process. Most of us gauge aging by looking for visible features that always seem to be present in older individuals and that are absent in youth. For the most part, these signs have very little to do with the biological aging process. The most familiar example is wrinkling, and sagging. Wrinkling is a condition produced by chronic exposure to ultraviolet (UV) light. Over time, the UV light denatures the connective tissue in the dermis, producing a condition called senile elastosis or, more accurately, solar elastosis. The skin changes that we erroneously associate with aging, such as cracking, leathery texture, and poor elasticity (i.e., the ability of skin to regain its flat, tight surface after being stretched or pinched) are the result of chronic UV toxicity.

The other obvious change observed in older individuals is skin sagging. In many individuals, this is most pronounced in the folds of skin that grow under the chin and down the neck. Sagging flesh on older individuals is due entirely to two phenomenon. The first is skin growth; humans grow their skin throughout life. This skin accumulates to different degrees in different individuals, depending on their genetically determined propensities for skin growth. The other phenomenon is gravity. Without the effects of gravity, our skin would grow evenly over our body contours. We develop pendulous skin at sites with the least skeletal support (e.g., under chin, under breasts, under our arms).

The changes we see in the skin of older individuals are due to the chronic effects of UV light, and gravity. They are not fundamental features of biological aging, because they do not occur in the absence of toxic conditions. Is there any evidence to support this claim? One piece of evidence lies in differences in skin damage among races. The heavily pigmented races have much less wrinkling than the less pigmented races, because they are better shielded from UV light. Yet there is no corresponding extension of life expectancy among the less-wrinkled races; suggesting that damaged skin is unrelated to the aging process. Aside from that, any elderly person can do a simple experiment that will doubtless settle the issue for themselves. Strip off your clothes and inspect the parts of your body that are not exposed to light and that are not hanging from an anatomic prominence. For some, this would be the lower back. In almost every case, you will be gratified to learn that this region of skin is unwrinkled, youthfully elastotic (i.e., will snap back in place when pinched), and relatively flat. Aside from a bit of softness due to skin growth, there really is not much difference between these protected regions of skin in elderly individuals and in young individuals.

If wrinkling and sagging are not part of the aging process, then what physiological processes characterize aging? In the animal kingdom, the Pacific salmon provides a stark example of senescence. These organisms enjoy a maritime existence that can extend for many years, followed by a brief and tumultuous counter-current swim to its river-of-birth. Here, the exhausted salmon spawns and dies. Rapid senescence is characterized by multiorgan deterioration and immunosuppression. In humans, aside from frailty (always accompanied by a decrease in muscle mass) and cachexia (a general lack of vigor and resistance to stress) [53, 54], humans don't have any obvious physiologic milestones to demarcate the stages of aging. It would seem that our most earnest observations of elderly individuals have failed to explain any of the underlying cellular mechanisms that account for the human aging process. In our confusion, we reflexively jump to the conclusion that any disease that occurs predominantly in an elderly population must arise through the aging process. One such example is cancer, a disease that is rare in the young, and common in the aged.

Cancer is a disease of the elderly, but a disease process of the young

In a news release, issued by the US National Institutes of Health, and dated February 3, 2014, the following statement appears: "Scientists have known for years that age is a leading risk factor for the development of many types of cancer, but why aging increases cancer risk remains unclear [55]." The mysterious relationship between cancer and aging was reemphasized 3 years later, in another official NIH announcement: "The incidence of most human cancers increases dramatically with advanced age. Cancer incidence is ten times higher in adults 65 or older compared to younger populations and this age group also has the highest mortality rate from cancer. Aging is characterized by the impairment of multiple cellular (including genome stability, epigenetic regulation, protein quality control, nutrient sensing, mitochondrial function, etc.) and systemic (response to paracrine or endocrine cues) functions. However, how aging-associated cellular impairments affect cancer etiology is not clear. Despite progress toward understanding the mechanisms that lead to cancer initiation and progression, how aging-associated processes contribute to cancer is largely unknown and a better understanding of the aging impact on the trajectory of cancer is needed to address the looming public health concern associated with populations that are increasingly living longer [56]."

The officials at NIH do not seem to understand that cancer is a disease that takes years, sometimes decades, to develop. All diseases, without exception, that require many years to develop will occur preferentially in an older age group. To illustrate with an example, a disease that develops over a period of 20 years cannot occur in individuals who are under 20 years of age. To illustrate using a specific case, if an individual is exposed to a carcinogenic agent at age 50, and the process of carcinogenesis in this individual unfolds over the next 20 years, at the minimum, then the cancer will not clinically emerge until the patient is 70 years of age, or older. Cancer researchers have been known to summarize the relationship between cancer and age with the following pithy remark: "**Cancer is a disease primarily of the old, but carcinogenesis is a disease process of the young**."

How can we be certain that cancer is not related to aging, and that the high occurrence rate of cancers in the elderly population is simply a statistical consequence of the lengthy carcinogenesis process?

−1.   We find precancers in a younger population than we find cancers

If every invasive cancer of humans is preceded by a well-characterized precancerous condition, then we can infer that every precancer must occur in individuals at a younger age than the age at which the cancer occurs. What applies to an individual must also apply to populations of individuals. The average population age at which a cancer develops must be older than the average population age at which the precancer lesions developed. This inference may seem obvious, but it frees us to follow the progression of disease in population data. Using publicly available cancer data, we can determine whether precancers fit into a timeline of events that lead to cancer.

For example ductal breast cancer, the most common cause of breast cancer in women, is thought to develop through a series of morphologically distinct precursors, specifically: intraductal hyperplasia precedes atypical intraductal hyperplasia, which precedes ductal carcinoma in situ, which precedes invasive breast carcinoma. The published average ages of occurrence for these different lesions bear out the hypothesized chronological precedence for precancers [57–59]. [Glossary Hyperplasia, In situ]

```
Median age, women with intraductal hyperplasia is 45 years

Median age, women with atypical intraductal hyperplasia is 50 years

Median age, women with ductal carcinoma in situ is 60 years

Median age, women with invasive breast cancer is 61 years
```

Similar observations have been noted for the age of occurrence of uterine cervical cancers [60]. The observed timelines document a sequence of lesions that precedes the development of invasive cancer, and that begins in relatively young individuals.

−2.   Inherited cancer syndromes occur disproportionately in the young

When genetic mutations that predispose to the development of cancer occur in germ cells, the cancers that result nearly always occur in children and young adults [61]. This observation informs us that tumor development does not wait until the individual reaches late adulthood before age-related susceptibility factors kick in. The biological steps leading from mutation to cancer proceed just as rapidly, if not, all the more so, in infants than they do in the elderly.

-3.   It is much easier to induce cancer in young animals than in old animals

In animal models, rainbow trout embryos are exquisitely sensitive to carcinogen-induced cancers [62]. Likewise, young rats are much more susceptible to carcinogen-induced neoplasms than are older rats [63]. Furthermore, we can infer that it would be impossible to induce cancer in elderly rats. How so? If an animal were close to the end of its natural life expectancy, and if an experimentally induced cancer requires months to years to develop, then there really is no chance of the cancer developing prior to the animal's demise.

In summary, there simply is no mystery as to why cancer must occur disproportionately in the elderly; it is because carcinogenesis is a biological process that develops over many years [18, 64]. If we want to block the development of cancers, we must look for treatments and preventions in the earliest steps of disease development. **We won't learn very much about how cancer begins by studying how it ends**.

As we might expect, cancer is not the only disease that occurs disproportionately in the elderly population. Any disease wherein the early causal events can occur at any age (e.g., smoking, exposure to a toxin, and trauma), and which has a pathogenesis that plays out over many years, will occur disproportionately in the elderly. Heart disease, osteoarthritis, autoantibody diseases, and chronic obstructive pulmonary disease (COPD) are common examples. [Glossary COPD]

The pathogenesis of every inherited disease always begins at the time of conception, when the mutated gene enters the germline. The appearance of clinical symptoms is determined by the length of pathogenesis. For example, most cases of Charcot-Marie-Tooth disease have their clinical onset in adolescence or early adulthood. In one extraordinary case, the first symptoms of Charcot-Marie-Tooth disease (an inherited neuropathy) appeared in a 65 year old individual. We can presume that the length of pathogenesis, in this case, was 65 years plus 9 months (length of gestation). **We must not assume that the patient's susceptibility to the disease was heightened by "old age."**

Aging, like every inherited condition, begins at conception

Imagine, if you will, that some clever geneticist identified all the genes responsible for aging and somehow managed to correct these genes, in germline cells, without adversely affecting cellular functions (a very tall order). In this case, we would expect to see babies who do not age, because the aging process begins at birth. If every newborn was ageless, then the human race would die out in a single generation. We'd be a world of helpless babies, with no aging parents to care for and love them.

The notion of ageless humans raises an important medical point. Aging is a lifelong process that begins in the embryo. It's silly to look in older people if you intend to find the roots of the aging process. Superannuated individuals give us some idea of the end-stage of the aging process, not its origin or its pathogenesis. If we want to examine diseases of the aging process, then we must expect to find these diseases where and when the aging process commences; in the embryo. [Glossary End-stage condition]

Let's look at the diseases of aging that begin at conception, and that emerge clinically in infants.

– Bloom syndrome

Bloom syndrome is caused by a defect in a gene encoding a member of the RecQ helicase family of genes that play an important role in DNA replication, repair, recombination, and transcription [65]. Signs of premature aging include reduced immune competence, predisposition to developing diabetes, and early menopause in women. The most striking clinical feature of Bloom syndrome is a heightened risk of developing a wide range of cancers, and this cancer predisposition seems to be directly related to helicase-related DNA instability.

– Cockayne syndrome

Cockayne syndrome is characterized by a failure to grow, impaired development of the brain, physical signs of premature aging, photosensitivity, and leukodystrophy (degeneration of the white matter of the brain). The underlying genetic cause of the disease is a mutation in either the ERCC6 gene or the ERCC8 gene, that code for a DNA-binding protein involved in DNA excision repair.

– Dyskeratosis congenita

Dyskeratosis congenita is characterized by three striking morphologic features: abnormal skin pigmentation, nail dystrophy, and leukoplakia (i.e., white patches) in the oral mucosa. Clinically, the most deleterious feature of dyskeratosis congenita is progressive bone marrow failure, which occurs in about 90% of cases. Bone marrow failure seems to result from a defect in cellular telomerase, leading to shortened telomeres, and to limitations on the replicative potential of bone marrow stem cells [66].

An understanding of the relationship between telomere length and continuous replication of stem cells is key to understanding the biology of dyskeratosis congenita. Chromosomes are built with a long padding sequence of repetitive DNA, at the chromosome tips, and this sequence is called the telomere. Animal cells lose a fragment of DNA from the tip of the chromosome, with each cell division. This is because one strand of DNA is replicated as sequential fragments, with each fragment requiring a template sequence beyond its end to initiate replication. The last fragment in the DNA strand has no template for itself and is not replicated. By providing a DNA padding at the tips of chromosomes, the telomere sequence sacrifices fragments of itself for the sake of preserving the coding sequences of the chromosome. As all good things come to an end, the telomere exhausts itself, after about 50 rounds of mitosis. At this time, the cell ceases further replication and will eventually die.

Cells that continually renew throughout life, such as bone marrow stem cells, epidermal cells, and hair follicle cells can restore their telomeres with an enzyme, telomerase. When such cells contain a mutation in genes encoding for components of the telomerase complex, their ability to divide throughout the lifetime of the organism is shortened. Mutations in the telomerase-associated genes are the underlying cause of many cases of dyskeratosis congenita, and account for the progressive bone marrow failure associated with this syndrome [67]. About half of the cases of dyskeratosis congenita are molecularly undefined [68]. Telomerase gene mutations have also been found in some cases of acquired bone marrow failure [69]. This would suggest that telomerase deficiency, attained through the clonal expansion of a somatic mutation in a bone marrow stem cell, can lead to premature bone marrow aging.

– Fanconi anemia bone marrow failure

Mutations in any of at least eight genes, all involved in one way or another with recognizing and repairing DNA damage, account for Fanconi anemia. Individuals with Fanconi anemia have a high likelihood of developing bone marrow failure with consequent aplastic anemia. Bone marrow failure can precede the development of acute myelogenous leukemia and myelodysplastic syndrome, a type of preleukemia. Fanconi anemia does not produce the general features of premature aging, such as frailty and cachexia. It is included here because bone marrow failure in Fanconi anemia seems to develop as the result of the organ-specific aging of bone marrow stem cells [47]. As in the more typical diseases of premature aging, the inability to continually renew tissue cells leads to a reduction in organ functionality. The reduction of normal bone marrow stem cells provides an opportunity for the clonal expansion of preexisting abnormal stem cells, leading to hematologic disorders such as leukemia and myelodysplastic syndrome. [Glossary Aplastic anemia, Myelodysplastic syndrome]

– Hutchinson-Gilford progeria syndrome

Hutchinson-Gilford progeria syndrome is a prototypical disease of premature aging characterized by wrinkled skin, atherosclerosis, renal failure, reduction in visual acuity, alopecia, scleroderma (i.e., skin tightening), and a high risk of heart attacks and strokes occurring at a young age. The underlying cause of Hutchinson-Gilford progeria syndrome is the production of progerin, a mutant form of lamin A. Lamin A is a nuclear protein that has important roles in maintaining the shape of the nucleus and in organizing DNA and RNA synthesis. Progerin, the mutant form of lamin A, produces striking abnormalities in the shape of the nucleus, featuring blebs, folds, and herniations of the nuclear envelope [70]. Also found are abnormalities in chromatin structure and increased DNA damage [71, 72]. Individuals with Hutchinson-Gilford progeria seem to have a dysfunction of stem cells, limiting their ability to renew differentiated cells [71, 73] [Glossary Progeria]

– Werner syndrome

Werner syndrome is a form of progeria with fewer severe symptoms than those associated with Hutchinson-Gilford progeria syndrome. It is characterized by scleroderma-like skin changes (i.e., skin tightening) with calcifications, cataracts, premature atherosclerosis, diabetes, and facial aging. Werner syndrome is caused by a defect in the WRN gene encoding RecQ helicases [65]. DNA helicases play a role in DNA replication, repair, recombination, and transcription. With multiple deficiencies in the DNA processing activities, it is not surprising that cells from individuals with Werner syndrome demonstrate chromosomal instability and a reduction in replication cycles (i.e., the total number of times a cell can replicate before becoming postmitotic).

- Wolfram syndrome 2

Wolfram syndrome 2 is characterized by early onset of diabetes, optic atrophy, and a shortened lifespan. It is caused by a mutation in the CISD2 gene, which encodes a protein associated with the outer mitochondrial membrane. Cisd2-null mice develop a progressive mitochondriopathy associated with defective respiration and with mitochondrial breakdown. The mice demonstrate premature aging and early death [74]. [Glossary Mitochondria, Mitochondriopathy]

- Xeroderma pigmentosum

Xeroderma pigmentosum affects individuals who cannot efficiently repair DNA damage induced by the UV light. Skin cancers develop at an early age in sun-exposed skin. The mainstay of treatment is avoidance to daylight. Life expectancy is shortened, and the signs of premature aging are seen. Xeroderma pigmentosum is usually listed among the diseases of aging, but most of the changes are confined to the skin and are simply the result of excess skin damage; not of a constitutive process that accelerates the aging process.

Based on the observation that some of the premature aging diseases have defects in DNA repair, it was hypothesized that the longevity of animal species is determined by the species-specific rate of DNA repair. Species that had a high rate of DNA repair were expected to have a long life span. Species with low rates of DNA repair would be short-lived. Though some data supported this hypothesis, a reanalysis of the data found little evidence to favor earlier conclusions [75].

- Sui generis aging disorder

A sui generis condition (i.e., one of a kind) was observed in a 16-year-old girl who had the appearance and anthropometric traits of an 11-month infant [76]. External and internal organs were infantile, including brain structure. After fetal development and birth, she had failed to mature into early childhood or adolescence. In a sense, her condition is the opposite of the premature aging conditions. The extreme rarity of this condition (i.e., more rare than the very rare monogenic disorders that produce premature aging), suggests that a simple loss of function in a single gene is unlikely to be at fault. Most severe loss-of-function monogenic diseases occur as de novo conditions with frequency in the order of one in a million population. If we can believe that this is truly a sui generis condition that has never before occurred in any human, then it's incidence is 1 in 107 billion (the number of humans who have ever lived on earth). This would suggest that the disease may have arisen from two simultaneous and independent rare de novo mutations. This strange and sad case raises many questions about human development and aging, but, at this time, there are no answers.

When we review the diseases of premature aging, we find that the underlying mechanisms of these diseases are manifold: chromatin instability (Hutchinson-Gilford progeria); DNA instability (Werner syndrome); mitochondrial degeneration (Wolfram syndrome); and telomere shortening (dyskeratosis congenita). What do all these syndromes of diverse etiology have in common?

Without exception, every disease of premature aging creates a defect in the normal process of cellular renewal. If we understood how to control and maintain stem cell renewal, a feat that comes naturally to plants and nematodes, and to most of the animals phylogenetically preceding the bilaterians, then we might understand how to defeat the aging process.

Section 7.5 Why Good People Get Bad Diseases

The world appears to me to be put together in such a painful way that I prefer to believe that it was not created ... intentionally.

Stanislaw Lem

The late great Philip K. Dick wrote a short story, entitled "Expendable" that clarifies the distinction between individual organisms and species. As the story unfolds, the human protagonist finds himself trapped in a cabin that is under attack by billions of soldier ants. As the floor heaves and the walls bow to the pressure of the ant army, a telepathically endowed spider appears in the cabin. The spider explains that the ant kingdom has decided to annihilate the human species, but that the spider kingdom has launched a defensive counter-attack, which will certainly be successful. Paraphrasing the dialogue in the story, the spider reassures the man, saying "We will save you." Relieved, the man thanks the spider, and indicates how grateful he is that the spiders will save him from the attacking ants. The spider pauses a long moment, and says "I'm afraid there has been a misunderstanding. When I say that the spiders will save you, I am referring to the human race, as a species, not to you as an individual." At this moment, the walls of the cabin crash down, as the ants swarm over its two conversing inhabitants.

The story serves to remind us that evolution through natural selection serves the species; not the individual members of the species. The destiny of a species, made possible through natural selection, is to adapt and speciate. Genetic alterations that enhance the ability of a species to evolve and speciate, while reducing the survival of individuals of the species, are a natural consequence of evolution. When we survey the devastating effects of disease on animal populations, we should not be asking ourselves why we must endure such suffering. A more scientific question, and one that is likely to lead to a useful outcome, would be, "Does a species benefit from the diseases that occur among its individual members?" In this section, we shall learn why genetic diseases are just one more example of an evolutionary process that works to strengthen the species at the expense of its individual members. In a very real sense, genetic diseases come to us as naturally as breathing.

Let's look at life from the viewpoint of a species. The three primary functions of a species are survival, adaptation, and speciation. All three of these functions are achieved by expansions of the gene pool (i.e., the sum total of genes from its population of individuals). In the case of an existential threat to the species, which might include sudden, and intense changes in climate; the introduction of a new predator into the habitat of the species; new and unfamiliar toxins or pathogenic organisms; or a scarcity of a customary food source, the species will rely on its gene pool to survive. From within the gene pool, the species can often produce individuals who are endowed with the genetic wherewithal to survive any and all of these threats.

Sometimes, the genes that are best suited to protect the species happen to be genes that contribute to diseases [77]. When trying to imagine the beneficial role of disease genes, it is useful to contemplate the daily activities of the sporting bookmaker, affectionately known as the "bookie." Successful bookies never care which horse wins the race. They operate by assigning betting odds to all the horses, with highest returns going to the horses that have a history of losing their races, and lowest returns going to the historical winners, and by readjusting the payout odds based on the money being put on each horse. When all the bets are in, the payout odds have been gradually titrated so that the total payout, regardless of which horse wins, is less than the bookie's take. If the bookie has done his job well, the house always wins.

Every gene pool carries a contingent of "loser" genes that reduce the chances of survival among the individuals whose chromosomes contain such genes. Let's imagine that these "loser" genes are distributed among race horses, and that a bolt of lightning sends a scream of thunder just as the horses leave the starting gate. The horse with the gene for deafness, unperturbed by the noise, streaks ahead and wins the race. In this case, the bad gene beat all the good genes, and evolution's bookie was glad that he had handicapped the bets, backing every horse.

Let's look at another example where a large gene pool saved a species from extinction.

Myxomatosis is a fatal disease of rabbits caused by the myxomatosis virus. The disease is characterized by the rapid appearance of skin tumors (myxomas), followed by severe conjunctivitis, a variety of severe systemic reactions, and fulminant pneumonia. Death usually occurs 2–14 days following infection. In 1952, a French virologist, hoping to reduce the rabbit population on his private estate, inoculated a few rabbits with Myxoma virus. The results were much more than he had bargained for. The virus destroyed the rabbits living on his property and spread throughout France and environs. Within 2 years, 90% of the rabbit population of France had succumbed to myxomatosis [78]. This horrifying and inhumane example of germ warfare, launched against a helpless species of fellow mammals, was promptly seized upon as a simple solution to a man-made problem brewing on another continent.

European rabbits, introduced to Australia in the 19th century, became feral and multiplied. By 1950 the rabbit population of Australia was about three billion. Seizing upon the

Myxoma virus as a solution to rabbit overpopulation, the Australians launched a Myxoma virus inoculation program. In less than 10 years, the Australian rabbit population was reduced by 95% [79]. Nearly three billion rabbits died, a number very close to the number of humans living on the planet in the mid-1950s. This plague on rabbits was unleashed by a committee of humans who decided, one day, that it would be expeditious to use a lethal rabbit virus as a biological weapon [78].

The species did not die out. A population of rabbits emerged that had genetic resistance to the myxomatosis virus [80]. Eventually, the rabbit population in Australia was restored. The new generation of rabbits are mostly resistant to the Myxoma virus.

The genes conferring genetic resistance against Myxoma virus did not mutate into existence; those genes resided in the species' preexisting gene pool. How can we be sure that the genes did not arise as de novo mutations, appearing in response to the outbreak? Evolution is slow. It is unlikely that a new mutation would have occurred and spread through the population, in time to save a species that is killed in just a few days after exposure to a virus. It would be much more likely that the gene conferring viral resistance was already in the rabbit gene pool and emerged as an existing variant among a subpopulation of living rabbits.

How can we be certain that the genes that saved the rabbit species in Australia were bad genes? Might they not have been extraordinarily good genes that were present only in the strongest, most fit rabbits? Maybe, but probably not. Because nearly all the rabbit population promptly died, on exposure to the virus, we can infer that the genes conferring resistance to the Myxoma plague were rare in the population. Rare genes are never particularly good genes; elsewise natural selection would have increased their prevalence.

Healthy individuals carry putative disease-causing genes

We now know the root genetic cause of thousands of monogenic diseases. In the vast majority of these cases, we do not know the chain of biological events that lead from the root causal genetic mutation to the eventual expression of the disease. In many of these diseases, 100% of the individuals affected by the disease have a mutation in a specific gene, and based on these observations, we can infer that if the root genetic cause of the disease were eliminated, then the disease would not occur.

It has long been assumed that because many of the monogenic diseases only occur in individuals carrying a known disease gene, then any person who carries the disease gene is almost certain to develop disease. It is this assumption, that disease genes always cause disease, that has provided the impetus for the genetic screening industry.

You can imagine the consternation of geneticists upon learning that many of the genes known to be the root cause of rare diseases are found in apparently healthy individuals, in a frequency much higher than the frequency of occurrence of the diseases associated with the genes [81–83]. In one study, it was shown that healthy individuals carry on average, at least two disease-causing mutations (i.e., mutations that were formerly thought to be the necessary and sufficient causes of one or another diseases) [84]. As more and more presumptive disease-causing alleles are discovered, the estimate for the number of such mutations in the healthy population will almost certainly increase.

Many of the putative disease genes are now recognized as common polymorphisms [30]. Other disease genes uncovered by genome screening were confirmed as valid markers for genetic disease that happened to be present in healthy individuals. In all such cases, the presence of a disease gene, in the absence of any other confirmatory information, would not be sufficient to establish a diagnosis [30, 85].

Let's look at some specific examples:

- One of the first surprises came when the bcr/abl fusion gene, thought to be pathognomonic for chronic myelogenous leukemia, was discovered in many healthy individuals, a finding that was confirmed by a number of different laboratories [85–87]. [Glossary Pathognomonic]

- Milroy disease patients affected by the FLT4/VEGFR3 fusion gene had family members who were unaffected by disease, but who carried the same mutation [88].

- About 1% of individuals, from two large studies, were shown to carry variants of the mutations that cause the rare disease maturity-onset-diabetes of the young (MODY). The vast majority of these individuals were euglycemic (i.e., had normal glucose levels) through middle age [89].

- The t(14;18) translocation producing the IGH/BCL2 fusion product characteristic of some nonHodgkin lymphomas, is found in 16%–55% of the general population, with the frequency increasing with age. [Glossary Hodgkin lymphoma, Non-Hodgkin lymphoma]

- About 49% of healthy individuals have genetic rearrangements previously thought to be diagnostic for mixed lineage leukemia [83].

- BRAF mutations, of the kind associated with about 70% of malignant melanomas, are found in about 80% of benign nevi (i.e., moles) [90]. An individual with a benign nevus has only the very smallest likelihood of developing into malignant melanoma. This indicates that a putative genetic driver of the malignant phenotype in malignant melanomas has virtually no carcinogenic potential in nevi [91]. [Glossary Nevus]

- Somatic blood cells with JAK2 mutations are found in 10% of apparently healthy individuals [82]. As noted in Section 2.1, "Mutation Burden," mutations of the JAK2 gene are thought to be the root genetic cause of several myeloproliferative conditions, including polycythemia vera, myelofibrosis, and at least one form of hereditary thrombocythemia [92, 93].

We can draw two important conclusions from the aforementioned observations:

-1. Diseases are not caused by a single mutation. Diseases emerge following a sequence of events occurring over time. Simply having a mutation, even in those circumstances when every occurrence of the disease is associated with the mutation, does not guarantee that the individual carrying the mutation will progress through all the necessary subsequent cellular events that lead to the clinical expression of disease. [Glossary Penetrance]

−2.   Disease genes play a beneficial role for the species.

The second assertion requires some justification. We have seen that some of the disease genes that are the underlying causes of serious, even fatal, diseases are found in a large percentage of the healthy population. Because de novo mutations are always rare, the common occurrence of a particular genetic mutation in the general population implies that the gene is positively selected in the species gene pool. Species do not positively select genes that have a purely deleterious effect. Over time, natural selection always reduces the frequency of purely deleterious genes. Hence, we can infer that common disease genes occurring in a general population must have some benefit.

At this point, we are reminded that a gene that is purely detrimental to an individual may serve a beneficial, even a necessary role for the species. In point of fact, it is easy to find disease genes that have proven benefit to the human species. Oncogenes serve as a good example of genes we might think that we could live without. Their primary function seems to be to cause cancer when activated. We know, however, that all oncogenes are highly conserved within our genomes, and this should tell us that they serve some necessary function or functions. It should come as no surprise that oncogene knockout mutations are often embryo-lethal in mice [94–96].

The BRCA1 and BRCA2 genes are associated with a heightened risk of breast cancer and ovarian cancer in women. There is evidence that BRCA genes are under positive selective pressure, and hypotheses abound for the beneficial role of BRCA [97]. One such theory suggests that BRCA enhances the growth of neurons, possibly increasing the intelligence of BRCA gene carriers [98]. Similarly, it would appear that the disease gene conferring a heightened risk for Alzheimer's disease (i.e., the APOE4 gene), may also confer superior verbal skills in young carriers [99]. [Glossary BRCA]

In Section 8.2, "Specificities and Idiosyncrasies," we will see how a variety of disease genes that produce red cell fragility and hemolysis all serve to reduce the morbidity and mortality produced by malaria infections. The populations of humans who have the highest rates of these hemolysis-inducing disease genes correspond to the populations wherein malaria is endemic.

We should note that healthy individuals carry around pathogenic material other than genes. For any given infectious agent, no matter how virulent they may seem, we can find examples of infected individuals with no clinical consequences. Moreover, as a generalization, the majority of individuals who are infected with a pathogenic microorganism will never develop any clinically significant disease [100].

Organisms that were formerly thought to be purely pathogenic are now known to frequently live quietly within infected humans, without causing symptoms of disease. For example, parasites such as the agents that cause Chagas disease, leishmaniasis, and toxoplasmosis are commonly found living in apparently normal individuals [101]. Viruses, including the agents that cause herpes simplex infections and infections by hepatitis viruses B and C can be found in healthy individuals. *Mycobacterium tuberculosis* can infect an individual, produce a limited pathologic reaction in the lung, and remain in the body in a quiescent state for the life of the individual. In fact, it has been estimated that about one out of three individuals, worldwide, is infected with *M. tuberculosis*, but will never suffer any consequences. Luckily, asymptomatic carriers of tuberculosis in whom the there is no active pulmonary disease, are noninfective. *Staphylococcus aureus*, a bacterial pathogen that is known to produce abscesses, invade through tissues, and release toxins, is also known to circulate in the blood, without causing symptoms, in a sizeable portion of the human population [101]. [Glossary HACEK]

Neisseria meningitidis, a cause of bacterial meningitis, can be cultured from nasal swabs sampled from the general population. If *N. meningitidis* were a primary pathogen, then why does it not produce disease in the vast majority of infected individuals. If *N. meningitidis* were an opportunistic infection, then why does it typically cause disease in healthy college-age individuals (not immunocompromised individuals)? If this organism is neither a primary pathogen nor an opportunistic pathogen, then what kind of pathogen is it? More importantly, why is *N. meningitidis* a potentially fatal pathogen in some individuals and a harmless commensal in others [102]? [Glossary Opportunistic infection, Primary pathogen]

A question that should be asked is: "If we can identify populations of healthy individuals who carry pathogens (disease genes or infectious organisms), then should we be trying to learn the reason why such people are spared from illness?" Knowing how these people escape disease might lead to new ways to prevent and treat the diseases that would otherwise be caused. It happens that a new project, led by the Icahn Institute for Genomics at Mount Sinai, and dubbed "The Resilience Project" seeks out individuals unaffected by disease gene mutations, in hopes of solving the mystery of their good health [103].

Glossary

Aplastic anemia A profound reduction in circulating blood cells, resulting from the loss of bone marrow stem cells. Severe or prolonged cases of aplastic anemia has a high mortality rate.

Autosomal dominance Refers to a pattern of inheritance in which a one allelic variant in an autosomal location (i.e., not on the X or Y chromosome) is the root cause of a trait or a disease. For the most part, diseases that have an autosomal pattern of inheritance will involve a gene variant that codes for an abnormal structural protein; not an enzyme. Why is this? When an allelic mutation produces loss of function of an enzyme, its paired (normal) allele can often compensate for the deficiency. We see this commonly in recessive diseases that require both alleles to be affected by a mutation. Contrariwise, when a mutation produces an abnormal structural protein, then disease may result, even though one allele is producing a normal protein product. It is the presence of the abnormal protein that accounts for the disease. Hence, mutations that produce altered protein products may be expected to have a dominant pattern of inheritance.

There are numerous examples of autosomally dominant inherited disorders that are characterized by altered protein products: altered tau protein and autosomal dominant frontotemporal dementia and parkinsonism linked to chromosome 17 (FTDP-17) [104]; altered FGFR3 protein and autosomal dominant achondroplasia; altered neurofibromin and autosomal dominant neurofibromatosis type 1; altered elastin gene and autosomal dominant cutis laxa; altered huntingtin and Huntington disease; altered hamartin and tuberin in autosomal dominant tuberous sclerosis. There are exceptions. For example, sickle cell anemia is caused by a mutation that alters hemoglobin. We would expect sickle cell disease to have an autosomal dominant pattern of inheritance; but it does not. Apparently, one mutant allele is not sufficient to produce a full-blown case of sickle cell disease. Severely affected individuals must have the characteristic sickle cell mutation in both alleles, and for this reason, sickle cell disease is recessive. It can be mentioned, however, that individuals with only one allele affected by the sickle cell mutation do not have normal red blood cells. Their red blood cells have a tendency to deform (i.e., to sickle) in conditions of hypoxia; just not to the extent necessary to precipitate the devastating clinical symptoms that comprise sickle cell anemia.

BRCA BRCA1 and BRCA2 are Breast Cancer tumor suppressor genes that code for proteins that play a role in DNA repair. Inherited mutations within BRCA genes are present in a small percentage of women with breast cancer and ovarian cancer [105]. People born with a germline inactivating BRCA2 mutation have a variant type of Fanconi anemia. A germline mutation characterized by a large BRCA1 intragenic deletion has been associated with a variant type of Li-Fraumeni syndrome [106–108].

COPD An abbreviation for chronic obstructive pulmonary disease. COPD covers a range of lung disorders characterized by airway damage. COPD is often associated with chronic cigarette abuse.

Chromosomal disorder Disorders associated with physical abnormalities in the physical structure of the chromosome. An example is found in fragile X syndrome. In this disease, a leading cause of inherited mental retardation, chromosomes are studded by fragile sites appearing as poorly condensed regions of the chromosome. Under experimental conditions, these regions break easily. Fragile sites have been associated with CCG repeats. Other examples of chromosomal disorders include Pelger-Huet anomaly and Roberts syndrome [109–111].

Congenital anomaly A structural deformity observed at birth. A clinically significant abnormality is present in about 3% of human births, indicating that in the aggregate, congenital abnormalities represent a common disease. About 60% of congenital anomalies are considered to be sporadic (i.e., they occur without any recognized genetic or environmental or physical cause). About 12%–25% of congenital anomalies are associated with genetic alterations, and the majority of these are chromosomal disorders. Hence, only a small percentage of congenital anomalies (perhaps considerably less than 5%) are associated with an identified single gene mutation that is suspected of being the root cause of the deformity.

Contiguous gene deletion syndrome Also called "Contig disease." A syndrome caused by abnormalities in a stretch of chromosome covering at least two adjacent genes. When the abnormality is a deletion, a contiguous gene syndrome is equivalent to a microdeletion syndrome.

Dollo's law Evolutionary biologists have toyed with the notion of the reversibility of gains and losses of complex traits and have expressed a consensus opinion as Dollo's Law. Dollo's law asserts that complex traits, once lost, cannot re-evolve in the same form. This law, which is actually just a reasonable presumption that may or may not hold up to verification, is based on the observation that complex traits develop through many evolutionary steps, and involve modifications of many different genes and metabolic pathways. The odds of a species, having lost a complex trait, evolving the same trait, with all of its attendant genetic alterations, seems unlikely.

As we might expect, Dollo's law has been challenged. In limpets, coiled shells may have re-evolved from ancestors that had devolved to lose their coil. In phasmids (commonly known as stick insects), wings may have reevolved from ancestors that had devolved to lose their wings [112].

Does Dollo's law make any sense? Not if we take that position that evolved genetic machinery is rarely lost; it is merely repressed. One day, whales may reconsider their decision to move from land to water, finding a land-based life more to their liking. If so, they may evolve into a species with arms and legs, employing ancestral anlagen found in every member of Class Tetrapoda.

End-stage condition A set of pathologic features that represent the typical morphologic status of an organ that has suffered a series of irreversible events leading to organ failure. The term "end-stage" evokes the idea that many different diseases will converge to the same pathological outcome. Every organ has its own brand of "end-stage" condition. For example, dilated cardiomyopathy is the end-stage condition for many different types of heart disease (e.g., myocarditis, coronary artery disease, systemic diseases, and myocardial toxins). Dilated cardiomyopathy, characterized by cardiac dilation and reduced systolic function, may occur in children or in adults and is the primary indication for cardiac transplantation [113].

Genome Wide Association Study Abbreviation: GWAS. A method to find single nucleotide polymorphisms (SNPs) that are statistically associated with a polygenic disease. The methodology involves hybridizing DNA from individuals with disease, as well as individuals from a control group, against a DNA array of immobilized fragments of DNA known to contain commonly occurring SNPs (i.e., allele-specific oligonucleotides). The SNPs that hybridize against the DNA extracted from individuals with disease (i.e., the SNPs matching the case samples) are compared with the SNPs that hybridize against the controls. SNPs that show a statistical difference between case samples and control samples are said to be associated with the disease.

Of course, there are many weaknesses to this approach; one being that differences in SNPs do not necessarily imply any functional variance in the gene product [114]. In addition, differences in SNPs may lead to statistically valid results that nonetheless have no relevance to the pathogenesis of disease [115]. Aside from false positive GWAS associations, the methodology is virtually guaranteed to miss valid SNP associations, simply because SNP arrays are not exhaustive (i.e., do not contain all 50 million SNPs), and are limited to a selected set of commonly occurring polymorphisms. For example, a rare variant of the APOE gene has been shown to be strongly correlated with longevity [116]. This variant, because it is not included among the common APOE variants included in SNP arrays, would have been missed by a GWAS study. True associations are those that can be found repeatedly from laboratory to laboratory, and that can be shown to have pathogenetic relevance. To date, very few disease-associated SNPs found in GWAS studies have met these criteria. It has been suggested that the GWAS studies, in toto, have had little scientific merit and have been misleading [117].

A sympathetic evaluation of GWAS studies is that they help us to see recurrent sets of pathway genes involved in diseases. Knowing that a related set of genes seems to implicate a pathway in the development or expression of a common disease has great value [118]. By focusing attention on a pathway, scientists can start to dissect the important events in the pathogenesis of a disease [119]. In addition, we should keep in mind that a gene whose variant form plays only a very minor role in the expression of a polygenic disease, may actually serve as the target of a new drug that is highly effective for the disease. How so? Small variants in the enzyme (as observed in SNPs) may produce only a small change in the activity of the enzyme, and this may reveal itself as a very small disease association in a GWAS study. Nonetheless, we can imagine situations wherein a new drug may decrease the activity of a minor pathway enzyme by 95%, thus greatly reducing the overall output of the pathway. Such an effect might be crucial in key disease pathways. This seems to be the case observed in statins, a drug that targets one of the many enzymes involved in cholesterol synthesis (HMG-CoA reductase) but which produces profound alterations in total cholesterol levels. A small variation in a disease pathway gene (such as the gene coding for HMG-CoA reductase) might not produce a dramatic finding in a GWAS study, but even a small effect might serve as a significant finding, leading to the discovery of a new, major class of drugs [120].

HACEK A group of proteobacteria, found in elsewise healthy individuals, that are known to cause some cases of endocarditis, especially in children, and which do not grow well from cultured blood (due primarily to their slow growth rates). The term HACEK is created from the initials of the organisms of the group: *Haemophilus*, particularly *Haemophilus parainfluenzae*; *Aggregatibacter*, including *Aggregatibacter actinomycetemcomitans* and *Aggregatibacter aphrophilus*; *Cardiobacterium hominis*; *Eikenella corrodens*; and *Kingella*, particularly *Kingella kingae*.

For decades, every medical student was taught that the blood of a healthy individual is sterile (e.g., free from bacteria and parasites). Not so. The isolation of slow-growing HACEK organisms from the cultured blood of healthy individuals was an early indication that so-called pathogens can persist in our bodies for extended periods without any pathogenic consequences, in most cases. Their presence in blood caused us to reevaluate our long-held notion that all infections are pathogenic and must be treated. Instead, we may now wonder whether every infection is potentially nonpathogenic under some circumstances. Perhaps we should be trying to understand how we can achieve conditions wherein infectious organisms remain clinically dormant.

Hodgkin lymphoma A common type of lymphoma that departs, in many ways, from the behavior observed in all other lymphomas. Hodgkin lymphoma has a bimodal age distribution, peaking in young adults, and again in much older adults. Hodgkin lymphoma can be successfully treated in the majority of cases, with reported cure rates as high as 90%. Hodgkin lymphoma often has a gradual stepwise anatomic spread of tumor from its primary site of origin, whereas most other lymphomas are widely disseminated from an early phase of disease.

Hyperplasia An overgrowth of tissue. Hyperplasia can be diffuse or focal, and the two types of hyperplasia presumably have different pathogeneses.

When hyperplasia is diffuse, the entire involved tissue becomes larger. The basic pathogenesis of diffuse hyperplasia involves a physiologic growth stimulus affecting all the normal cells in a tissue. An example of diffuse hyperplasia is Menetrier disease, in which the folds of the stomach enlarge and the gastric mucosa thickens.

Focal hyperplasia involves a subset of growing cells within a tissue that yield clonal growths that stand out from the surrounding normal tissue. Focal hyperplasias, sparing the vast majority of cells in a tissue, target a small subset of scattered cells that are particularly sensitive to a growth stimulus. Focal hyperplasias, unlike diffuse hyperplasias, can be easily confused with benign neoplasms [121].

In situ Latin for "in its place." When referring to a cancer, in situ implies that the cancer has not invaded deeply (e.g., through the basement membrane in the case of mucosal and epidermal tumors) and has not metastasized to lymph nodes or to distant organs.

Incidence The number of new cases of a disease occurring in a time interval (e.g., 1 year), expressed as a fraction of a predetermined population size (e.g., 100,000 people). For example, if there were 11 new cases of a rare disease occurring in a period of 1 year, in a population of 50,000 people, then the incidence would be 22 cases per 100,000 persons per year.

Innate immunity An ancient and somewhat nonspecific immune and inflammatory response system found in plants, fungi, insects [122], and most multicellular organisms. This system recruits immune cells to sites of infection, using cytokines (chemical mediators). Innate immune mechanisms activate the complement system and the macrophage system (also called the reticuloendothelial system), both of which serve to clear dead cells, among numerous other functions.

Crohn's disease is the first of the common, genetically complex diseases that were shown to be associated with a germline polymorphism of a gene known to be a component of the innate immune system; specifically, the NF-kappa-B pathway [123, 124]. Drugs that inhibit TNF (tumor necrosis factor), a participant in the cytokine pathway recruited by the innate immune system, are active against Crohn's disease [125].

Examples of rare genetic disorders of the innate immune system include: Familial Mediterranean fever; TNF receptor-associated periodic syndrome; Hyperimmunoglobulin D syndrome; Familial cold autoinflammatory syndrome; Muckle-Wells syndrome; Neonatal-onset multisystem inflammatory disease, also known as chronic infantile neurologic, cutaneous, and arthritis syndrome; Pyogenic arthritis, pyoderma gangrenosum, and acne; Blau syndrome; Early-onset sarcoidosis, and Majeed syndrome [126, 127].

Mitochondria Self-replicating organelles wherein respiration and the production of cellular energy from oxygen occurs. As far as anyone knows, the very first eukaryote came fully equipped with a nucleus, one or more undulipodium, and one or more mitochondria. Similarities between mitochondria and bacteria of Class Rickettsia suggests that the eukaryotic mitochondrium was derived from an ancestor of a modern rickettsia. All existing eukaryotic organisms, even the so-called amitochondriate classes (i.e., organisms without mitochondria), contain vestigial forms of mitochondria (i.e., hydrogenosomes and mitosomes) [5, 13, 14, 128]. A single eukaryotic cell may contain thousands of mitochondria, as is the case for human liver cells, or no mitochondria, as is the case for human red blood cells. The control of mitochondrial number is determined within the nucleus, not within the mitochondrion itself. The mitochondria in a human body are descended from mitochondria contained in the cytoplasm of the maternal oocyte, with no contribution from the father's sperm. Hence, mitochondria have a purely maternal lineage.

Mitochondriopathy Diseases whose underlying cause is mitochondrial pathology (i.e., dysfunctional mitochondria, or an abnormal number of mitochondria). Mitochondriopathies can be genetic or acquired. Most of the genetic mitochondriopathies are caused by nuclear gene mutations. Though mitochondria live outside the nucleus and have their own genomes, mitochondrial DNA codes for only 13 proteins of the respiratory chain. All the other proteins and structural components of mitochondria are coded in the nucleus. As we might expect, mitochondriopathies affect the cells that are most dependent on their mitochondria for their functionality (e.g., skeletal muscle and heart cells). Mitochondriopathies may also involve: pigmentary retinopathy, ocular atrophy, deafness, gut motility disorder, and sideroblastic anemia, among many others. A mitochondriopathy should be in the differential workup for any unexplained multisystem disorder, especially those arising in childhood [129].

Myelodysplastic syndrome The myelodysplasias consist of several related disorders characterized by pancytopenia, disorders of myeloid maturation, the appearance of blast cells in the circulating blood, chromosomal abnormalities, and the frequent progression to leukemia. Listed in increasing grades hematologic severity, the myelodysplasias are: refractory anemia, refractory anemia with ringed sideroblasts, refractory anemia with excess blasts, refractory anemia with excess blasts in transformation, and (in some listings) chronic myelomonocytic leukemia. Progression to lower to higher grades is common. The myelodysplasias have a bimodal age distribution; most occurring in the elderly, some in the young. A secondary type of myelodysplasia occurs following a bout of aplastic anemia or following chemotherapy for some other neoplasm.

Nevus A common, benign growth of melanocytic cells, usually a mixture of epidermal melanocytes and dermal melanocytes, the latter often called nevocellular cells. Almost all nevi occur in the skin, the normal site of melanocytic growth. Nevi can be identified as small (usually several millimeters across), slightly raised brown spots. Because melanocytes have a neural crest origin, we can say that nevi are a type of benign neural crest neoplasm.

Non-Hodgkin lymphoma Lymphoid neoplasms are divided into two categories: Hodgkin lymphoma and non-Hodgkin lymphoma. Non-Hodgkin lymphoma comprises every lymphoid neoplasm other than Hodgkin lymphoma (and there are many). The diagnosis of Non-Hodgkin lymphoma, communicated to a patient, is uninformative and needlessly confusing. "Non-Hodgkin lymphoma" is an example of a diagnosis that informs patients what their diagnosis is not (i.e., it's not Hodgkin lymphoma), declining to specify what their diagnosis happens to be.

Opportunistic infection Opportunistic infections are diseases that do not typically occur in healthy individuals, but which can occur in individuals who have a physiologic status favoring the growth of the organisms (e.g., diabetes, malnutrition, and immune deficiency). For example, diabetics are more likely to contract systemic fungal diseases than are nondiabetic individuals. Most of the infectious diseases occurring in the setting of AIDS arise from the population of organisms that live within most humans, without causing disease under normal circumstances (i.e., commensals). Such organisms include *Pneumocystis jirovecii*, *Toxoplasma gondii*, *Cryptococcus neoformans*, and Cytomegalovirus, among many others.

The concept of an opportunistic organism is, at best, a gray area of medicine, as virtually all of the organisms that arise in immunocompromised patients will, occasionally, cause disease in seemingly immunocompetent patients (e.g., *Cryptococcus neoformans*). Moreover, the so-called primary infectious organisms that produce disease in healthy individuals, will tend to produce a more virulent version of the disease in immunosuppressed individuals (e.g., *Coccidioides immitis*). Recent observations would suggest that many seemingly immunocompetent individuals who contract opportunistic infections have identifiable genetic abnormalities that account, specifically, for their heightened susceptibilities [45, 78, 100, 102, 130].

Pathognomonic A feature that is present in every occurrence of a disease, and which is never found in individuals who are not affected by the disease.

Penetrance An individual may have inherited a disease gene without ever developing its associated disease. In medical genetics, the penetrance is the proportion of individuals with a disease-causing mutation who develop the disease (compared to those who do not).

There are several reasons why the penetrance of a disease-causing gene may be significantly lower than 100%. Some diseases, particularly the common diseases, are polygenic. It may take many genes to produce the disease phenotype. Epistasis may also modify penetrance (e.g., one gene may be influenced by a particular allele of another gene). Environmental and epigenetic factors can also influence gene function. An inherited mutation may require environmental triggers (e.g., excessive sunlight exposure in porphyria cutanea tarda) or conditional physiological conditions (e.g., fatigue or starvation precipitating the hyperbilirubinemia associated with Gilbert syndrome) to fully express the clinical trait. Such factors may influence the penetrance of the gene, or the age of the individual when the disease emerges, or the severity of the disease, or clinical phenotype of the disease (i.e., which clinical features will develop).

The concept of disease penetrance serves as a reminder that diseases develop over time, through a sequence of events. It is inaccurate to think of an inherited mutation as the "cause" of a disease.

Postmitotic Refers to fully differentiated cells that have lost the ability to divide. For example, the epidermis of the skin has a basal layer of cells that are capable of dividing to produce a postmitotic cell and another basal layer cell. The postmitotic cells sit atop the basal cells, where they gradually flatten out (i.e., become squamous shaped) and lose their nucleus as the cells rise through the epidermal layers. The top layer of the epidermis sloughs off into the air and is replaced by postmitotic epidermal cells in the next lower layer. Much of the small flakes of house dust that we see dancing in the beams of sunlight passing through our windows is composed of squamous cells, sloughed from our skin. This cycle of cell renewal from the bottom and cell sloughing from the top is typical of most epithelial surfaces of the body (e.g., epidermis of skin, gastrointestinal tract, and glandular organs). Aside from epithelial surfaces, postmitotic cells arise from populations of mitotic cells that have exhausted their regenerative potential. One theory of aging holds that certain cell types of the body (e.g., fibroblasts) have a limited number of mitotic cell cycles. When a determined number of cell cycles have elapsed, cells that cannot divide further become postmitotic.

Precision medicine Despite what you may have read in the popular press and in social media, precision medicine is not devoted to finding unique treatments for individuals, based on analyzing their DNA. To the contrary, the goal of precision medicine is to find general treatments that are highly effective for large numbers of individuals who fall into precisely diagnosed groups [18].

We now know that every disease develops over time, through a sequence of defined biological steps, and that these steps may differ among individuals, based on genetic and environmental conditions. We are using the tools and concepts of precision medicine to develop rational therapies and preventive measures, based on our new ability to study the steps leading to the clinical expression of diseases.

Primary pathogen An organism that causes disease in healthy individuals, without requiring comorbid conditions (e.g., young age, old age, immunosuppression, starvation, or concurrent diseases) for its virulence.

Progeria Any of several rare inherited conditions of premature aging, typically characterized by early onset atherosclerosis and diabetes. All of the progerias, despite differences in their root genetic causes, seem to involve defects in normal cellular renewal processes (e.g., stem cell regeneration, maintenance of stem cells, and DNA repair in dividing cell populations).

Somatic mosaicism If the new mutation occurs in an embryonic cell after the zygote has split to produce daughter cells, then the new mutation produces somatic mosaicism; meaning that it will only occur in those somatic cells that descended from the embryonic cells in which the mutation occurred. If the somatic cells that inherit the new mutation include germ cells (i.e., cells that will differentiate into ova or sperm), then the de novo mutation can be passed to the next generation as an inherited germline mutation. Proteus syndrome is an example of a disease that exhibits somatic mosaicism; never as a germline mutation [131]. Presumably, the gene causing Proteus syndrome, if present in the germline, would have been lethal to the embryo. Somatic mosaicism is a particular type of de novo mutation (a so-called postzygotic de novo mutation).

Sporadic Describes a disease or a specific case occurrence of a disease with no known cause, and without any discernible pattern of occurrence (e.g., genetic, environmental). Thus, diseases that have a familial pattern of inheritance are always considered nonsporadic, even when the root genetic cause is unknown. Likewise, diseases that occur as an epidemic or endemic pattern are always considered nonsporadic, even when the precise environmental cause is unknown. Rare diseases are seldom sporadic, as they typically exhibit some pattern of inheritance. Common diseases are often sporadic, but may contain subsets of disease occurrences that are nonsporadic. An example is schizophrenia. Schizophrenia is a common disease with a prevalence of about 1.1%. This translates to about 51 million individuals, worldwide, who suffer from this mental disorder. Many cases of schizophrenia occur in families and such cases are considered to be inherited and, thus, nonsporadic. Other cases seem to have no familial association and are considered sporadic. Are these sporadic cases caused by environmental factors, or are they caused by de novo mutations that arose in the affected individuals? Recent evidence would suggest that many of the so-called sporadic cases arise from new mutations in affected individuals [132]. When an association is made between a disease and some demographic factor, the distinction between sporadic and nonsporadic may be arbitrary. For example, if a disease occurs predominantly in women, can it be called sporadic? The cause may be completely unknown, but it has a definite pattern. It should be mentioned that the term "sporadic" is fraught with scientific ambiguity and should probably be abandoned altogether. To label a disease "sporadic" seems to legitimize and perpetuate the dubious notion that diseases can occur without cause (i.e., by chance). When you read an old textbook of medicine, and you see a disease listed as "sporadic," you would likely accept this as a fact that was substantiated sometime in the past. The term "sporadic" provides no compelling reason to check to see if the cause of the disease were subsequently discovered; if you've accepted the idea that the disease has no cause. Of course, this is a terrible way to think about diseases. Many of the diseases that were considered to be sporadic, decades ago, are now known to have specific causes. Would it not be more accurate to use the phrase "not as yet determined" in place of "sporadic," for occurrences of a disease whose cause is currently unknown?

Susceptibility Refers to a state of increased risk of harm. The frequently encountered term "susceptibility gene" would imply that individuals with the gene have an increased susceptibility to a particular disease. From the point of view of understanding pathogenesis, "susceptibility" is not a helpful concept, in that it doesn't signify a specific biological process. Is "susceptibility" an event, or is it a pathway, or is it a disease? Is it simply a constitutive condition of the individual that heightens risk of disease? We say that something increases our susceptibility to a disease if it increases the probability that the disease will occur. Shouldn't we be asking ourselves, "How does susceptibility influence a pathologic process?"

Confusion around the term "susceptibility" is demonstrated in the following example. Caucasians have a higher risk of developing skin cancers than do African-Americans. We say that being Caucasian increases susceptibility to skin cancer, but we never say that being Caucasian is the underlying cause of skin cancer. If a native of Nigeria is born with albinism, then that individual is highly susceptible to skin cancer, and we might say that albinism is the underlying cause of skin cancer in these cases. In these two instances, the word "susceptibility" does not tell us much about pathogenesis and does not help us resolve issues of causality with any consistency.

Telomere Chromosomes are built with a long padding sequence of repetitive DNA, at the chromosome tips, and this sequence is called the telomere. Animal cells lose a fragment of DNA from the tip of the chromosome with each cell division. This happens because one strand of DNA is replicated as sequential fragments, with each fragment requiring a template sequence beyond its end to initiate replication. The last fragment in the DNA strand has no template for itself and is not replicated. By furnishing a padding at the tips of chromosomes, the telomere sequence sacrifices fragments of itself for the sake of preserving the coding sequences of the chromosome. As all good things come to an end, the telomere padding exhausts itself, after about 50 rounds of mitosis. At this time, the cell ceases further replication and will eventually die.

Cells that continually renew throughout life, such as bone marrow stem cells, skin and hair follicle basal layer cells, and intestinal basal crypt cells, retain an enzyme, telomerase, that restores telomere length. When such cells contain a loss of function mutation in any of the genes encoding for components of the telomerase complex, their ability to divide throughout the lifetime of the organism is reduced.

A mutation in the telomerase gene is the root cause of dyskeratosis congenita, a rare inherited condition in which bone marrow failure frequently occurs [67]. Telomerase gene mutations have also been found in some cases of acquired bone marrow failure [69].

Cancer cells, like bone marrow cells, continuously divide, and have high concentrations of telomerase. The ability of some cancer cells to restore their telomeres contributes to their continuous capacity to divide without limit, a phenomenon sometimes called cancer cell immortality. Telomerase insufficiency has been suggested as an possible cause of spontaneous regression in tumors, a rarely observed phenomenon [133].

Thalassemia Alpha thalassaemia and beta thalassaemia are the most common inherited monogenic disorders worldwide. The disorder is characterized by ineffective red blood cell production due to a reduction in the synthesis of the alpha or beta chains of hemoglobin. As is the case with sickle cell disease, carriers of one copy of the thalassemia allele seems to confer some protection against malaria [134] and this beneficial effect may explain the conservation of the thalassaemia gene in those populations wherein malaria is endemic.

References

[1] Gould SJ, Lewontin R. The spandrels of San Marco and the panglossian paradigm: a critique of the adaptationist programme. Proc Roy Soc Lond B 1979;205:581–98.

[2] Reilly SM, White TD. Hypaxial motor patterns and the function of epipubic bones in primitive mammals. Science 2003;299:400–2.

[3] Simpson GG. The principles of classification and a classification of mammals. Bull Am Mus Nat Hist 1945;85:.

[4] Berthoz A, Weiss G. Simplexity: simplifying principles for a complex world. London: Editions Odile Jacob Book, Yale University Press; 2012.

[5] Burri L, Williams B, Bursac D, Lithgow T, Keeling P. Microsporidian mitosomes retain elements of the general mitochondrial targeting system. PNAS 2006;103:15916–20.

[6] Keeling PJ, Luker MA, Palmer JD. Evidence from beta-tubulin phylogeny that microsporidia evolved from within the fungi. Mol Biol Evol 2000;17:23–31.

[7] Sak B, Kvac M, Kucerova Z, Kvetonova D, Sakova K. Latent microsporidial infection in immunocompetent individuals: a longitudinal study. PLoS Negl Trop Dis 2011;5:e1162.

[8] Chang ES, Neuhof M, Rubinstein ND, Diamant A, Philippe H, Huchon D, et al. Genomic insights into the evolutionary origin of Myxozoa within Cnidaria. PNAS 2015;48:14912–7.

[9] Erb TJ, Zarzycki J. A short history of RubisCO: the rise and fall of nature's predominant CO_2 fixing enzyme. Curr Opin Biotechnol 2018;49:100–7.

[10] Koonin EV. How many genes can make a cell: the minimal-gene-set concept. Annu Rev Genomics Hum Genet 2000;1:99–116.

[11] Li K, Wahlqvist ML, Li D. Nutrition, one-carbon metabolism and neural tube defects: a review. Nutrients 2016;8:741.

[12] Knobloch J, Ruther U. Shedding light on an old mystery: thalidomide suppresses survival pathways to induce limb defects. Cell Cycle 2008;7:1121–7.

[13] Tovar J, Fischer A, Clark CG. The mitosome, a novel organelle related to mitochondria in the amitochondrial parasite *Entamoeba histolytica*. Mol Microbiol 1999;32:1013–21.

[14] Stechmann A, Hamblin K, Perez-Brocal V, Gaston D, Richmond GS, van der Giezen M, et al. Organelles in Blastocystis that blur the distinction between mitochondria and hydrogenosomes. Curr Biol 2008;18:580–5.

[15] Brinkmann H, Philippe H. The diversity of eukaryotes and the root of the eukaryotic tree. Adv Exp Med Biol 2007;607:20–37.

[16] Ramesh MA, Malik SB, Logsdon JM. A phylogenomic inventory of meiotic genes; evidence for sex in Giardia and an early eukaryotic origin of meiosis. Curr Biol 2005;15:185–91.

[17] Manolio TA, Collins FS, Cox NJ, Goldstein DB, Hindorff LA, Hunter DJ, et al. Finding the missing heritability of complex diseases. Nature 2009;461:747–53.

[18] Berman J. Precision medicine, and the reinvention of human disease. Cambridge, MA: Academic Press; 2018.

[19] Sperling EA, Frieder CA, Raman AV, Girguis PR, Levin LA, Knoll AH. Oxygen, ecology, and the Cambrian radiation of animals. Proc Natl Acad Sci U S A 2013;110:13446–51.

[20] Olsson M, Meadows JR, Truve K, Rosengren Pielberg G, Puppo F, Mauceli E, et al. A novel unstable duplication upstream of HAS2 predisposes to a breed-defining skin phenotype and a periodic fever syndrome in Chinese Shar-Pei dogs. PLoS Genet 2011;7:e1001332.

[21] Komazawa S, Sakai H, Itoh Y, Kawabe M, Murakami M, Mori T, et al. Canine tumor development and crude incidence of tumors by breed based on domestic dogs in Gifu prefecture. J Vet Med Sci 2016;78:1269–75.

[22] Kehrli ME, Ackermann MR, Shuster DE, van der Maaten MJ, Schmalstieg FC, Anderson DC, et al. Bovine leukocyte adhesion deficiency: beta(2) integrin deficiency in young Holstein cattle. Am J Path 1992;140:1489–92.

[23] Fisher RA. The correlation between relatives on the supposition of Mendelian inheritance. Trans R Soc Edinb 1918;52:399–433.

[24] Ward LD, Kellis M. Interpreting noncoding genetic variation in complex traits and human disease. Nat Biotechnol 2012;30:1095–106.

[25] Visscher PM, McEvoy B, Yang J. From Galton to GWAS: quantitative genetics of human height. Genet Res 2010;92:371–9.

[26] Sniekers S, Stringer S, Watanabe K, Jansen PR, Coleman JRI, Krapohl E, et al. Genome-wide association meta-analysis of 78,308 individuals identifies new loci and genes influencing human intelligence. Nat Genet 2017;49:1107–12.

[27] Zhang G, Karns R, Sun G, Indugula SR, Cheng H, Havas-Augustin D, et al. Finding missing heritability in less significant Loci and allelic heterogeneity: genetic variation in human height. PLoS ONE 2012;7:e51211.

[28] Billings LK, Florez JC. The genetics of type 2 diabetes: what have we learned from GWAS? Ann N Y Acad Sci 2010;1212:59–77.

[29] Mitchell JJ, Capua A, Clow C, Scriver CR. Twenty-year outcome analysis of genetic screening programs for Tay-Sachs and beta-thalassemia disease carriers in high schools. Am J Hum Genet 1996;59:793–8.

[30] Bell CJ, Dinwiddie DL, Miller NA, Hateley SL, Ganusova EE, Mudge J, et al. Carrier testing for severe childhood recessive diseases by next-generation sequencing. Sci Transl Med 2011;3:65ra4.

[31] Kaback MM. Population-based genetic screening for reproductive counseling: the Tay-Sachs disease model. Eur J Pediatr 2000;159:S192–5.

[32] Kronn D, Jansen V, Ostrer H. Carrier screening for cystic fibrosis, Gaucher disease, and Tay-Sachs disease in the Ashkenazi Jewish population: the first 1000 cases at New York University Medical Center, New York. NY Arch Intern Med 1998;158:777–81.

[33] Gersbach CA, Perez-Pinera P. Activating human genes with zinc finger proteins, transcription activator-like effectors and CRISPR/Cas9 for gene therapy and regenerative medicine. Expert Opin Ther Targets 2014;18:835–9.

[34] Wang D, Gao G. State-of-the-art human gene therapy: part II. Gene therapy strategies and applications. Discov Med 2014;18:151–61.

[35] Lv Q, Yuan L, Deng J, Chen M, Wang Y, Zeng J, et al. Efficient generation of myostatin gene mutated rabbit by CRISPR/Cas9. Sci Rep 2016;6:25029.

[36] Collins F. Statement on NIH funding of research using gene-editing technologies in human embryos, https://www.nih.gov/about-nih/who-we-are/nih-director/statements/statement-nih-funding-research-using-gene-editing-technologies-human-embryos; 2015.

[37] Rochman B. Five myths about gene editing. The Washington Post August 25, 2017.

[38] Shah RR, Cholewa-Waclaw J, Davies FCJ, Paton KM, Chaligne R, Heard E, et al. Efficient and versatile CRISPR engineering of human neurons in culture to model neurological disorders. Wellcome Open Res 2016;1:13.

[39] Mitton JB, Grant MC. Genetic variation and the natural history of Quaking Aspen. BioScience 1996;46:25–31.

[40] Schmich J, Kraus Y, De Vito D, Graziussi D, Boero F, Piraino S. Induction of reverse development in two marine Hydrozoans. Int J Dev Biol 2007;51:45–56.

[41] Martinez DE. Mortality patterns suggest lack of senescence in hydras. Exp Gerontol 1998;33:217–25.

[42] Rich N. Can a jellyfish unlock the secret of immortality? The New York Times November 28, 2012.

[43] Martinez DE, Bridge D. Hydra, the everlasting embryo, confronts aging. Int J Dev Biol 2012;56:479–87.

[44] Tan T, Rahman R, Jaber-Hijazi F, Felix DA, Chen C, Louis EJ, et al. Telomere maintenance and telomerase activity are differentially regulated in asexual and sexual worms. Proc Natl Acad Sci U S A 2012;109:4209–14.

[45] Berman JJ. Rare diseases and orphan drugs: keys to understanding and treating common diseases. Cambridge, MD: Academic Press; 2014.

[46] Fogli A, Rodriguez D, Eymard-Pierre E, Bouhour F, Labauge P, Meaney BF, et al. Ovarian failure related to eukaryotic initiation factor 2B mutations. Am J Hum Genet 2003;72:1544–50.

[47] Zhang X, Li J, Sejas DP, Pang Q. Hypoxia-reoxygenation induces premature senescence in FA bone marrow hematopoietic cells. Blood 2005;106:75–85.

[48] Ishii N, Fujii M, Hartman PS, Tsuda M, Yasuda K, Senoo-Matsuda N, et al. A mutation in succinate dehydrogenase cytochrome *b* causes oxidative stress and ageing in nematodes. Nature 1998;394:694–7.

[49] McLaughlin PJ, Bakall B, Choi J, Liu Z, Sasaki T, Davis EC, et al. Lack of fibulin-3 causes early aging and herniation, but not macular degeneration in mice. Hum Mol Genet 2007;16:3059–70.

[50] Vermulst M, Wanagat J, Kujoth GC, Bielas JH, Rabinovitch PS, Prolla TA, et al. DNA deletions and clonal mutations drive premature aging in mitochondrial mutator mice. Nat Genet 2008;40:392–4.

[51] Abbott MH, Murphy EA, Bolling DR, Abbey H. The familial component in longevity. A study of offspring of nonagenarians. II. Preliminary analysis of the completed study. Johns Hopkins Med J 1974;134:1–16.

[52] Jarvik LF, Falek A, Kallmann FJ, Lorge I. Survival trends in a senescent twin population. Am J Hum Genet 1960;12:170–9.

[53] Xue Q. The frailty syndrome: definition and natural history. Clin Geriatr Med 2011;27:1–15.

[54] Sayer AA, Robinson SM, Patel HP, Shavlakadze T, Cooper C, Grounds MD. New horizons in the pathogenesis, diagnosis and management of sarcopenia. Age Ageing 2013;42:145–50.

[55] NIH study offers insight into why cancer incidence increases with age. NIH news relases. Monday, February 3, 2014. https://www.nih.gov/news-events/news-releases/nih-study-offers-insight-into-why-cancer-incidence-increases-age [viewed 24.08.18].

[56] Partnership for Aging and Cancer Research (PAR-18-552). Department of Health and Human Services, National Institutes of Health. December 22, 2017. https://grants.nih.gov/grants/guide/pa-files/PAR-18-552.html [viewed 19.06.18].

[57] Tavassoli FA, Norris HJ. A comparison of the results of long-term follow-up for atypical intraductal hyperplasia and intraductal hyperplasia of the breast. Cancer 1990;65:518–29.

[58] Westbrook KC, Gallagher HS. Intraductal carcinoma of the breast. A comparative study. Am J Surg 1975;130:667–70.

[59] Seer Cancer Stat Fact Sheets. Cancer of the breast. Available from: http://seer.cancer.gov/statfacts/html/breast.html.

[60] Mortality, total U.S. (1969–2005). Surveillance, Epidemiology, and End Results (SEER) Program (www.seer.cancer.gov). National Cancer Institute, DCCPS, Surveillance Research Program, Cancer Statistics Branch, released April 2008. Underlying mortality data provided by NCHS (www.cdc.gov/nchs).

[61] Lindor NM, McMaster ML, Lindor CJ, Greene MH. Concise handbook of familial cancer susceptibility syndromes—second edition. J Natl Cancer Inst Monogr 2008;38:1–93.

[62] Sinnhuber RO, Hendricks JD, Wales JH, Putnam GE. Neoplasms in rainbow trout, a sensitive animal model for environmental carcinogenesis. Ann N Y Acad Sci 1977;298:389–408.

[63] Kimura M, Fukuda T, Sato K. Effect of aging on the development of gastric cancer in rats induced by N-methyl-N′-nitro-N-nitrosoguanidine. Gan 1979;70:521–5.

[64] Berman JJ. Neoplasms: principles of development and diversity. Sudbury: Jones & Bartlett; 2009.

[65] Mohaghegh P, Hickson ID. DNA helicase deficiencies associated with cancer predisposition and premature ageing disorders. Hum Mol Genet 2001;10:741–6.

[66] Calado R, Neal Young N. Telomeres in disease. F1000 Medicine Reports 2012;4:8.

[67] Vulliamy T, Beswick R, Kirwan M, Marrone A, Digweed M, Walne A, et al. Mutations in the telomerase component NHP2 cause the premature ageing syndrome dyskeratosis congenita. Proc Natl Acad Sci U S A 2008;105:8073–8.

[68] Shtessel L, Ahmed S. Telomere dysfunction in human bone marrow failure syndromes. Nucleus 2011;2:24–9.

[69] Yamaguchi H. Mutations of telomerase complex genes linked to bone marrow failures. J Nippon Med Sch 2007;74:202–9.

[70] Mallampalli MP, Huyer G, Bendale P, Gelb MH, Michaelis S. Inhibiting farnesylation reverses the nuclear morphology defect in a HeLa cell model for Hutchinson-Gilford progeria syndrome. PNAS 2005;102:14416–21.

[71] Scaffidi P, Misteli T. Lamin A-dependent misregulation of adult stem cells associated with accelerated ageing. Nat Cell Biol 2008;10:452–9.

[72] Shumaker DK, Dechat T, Kohlmaier A, Adam SA, Bozovsky MR, Erdos MR, et al. Mutant nuclear lamin A leads to progressive alterations of epigenetic control in premature aging. Proc Natl Acad Sci U S A 2006;103:8703–8.

[73] Liu GH, Barkho BZ, Ruiz S, Diep D, Qu J, Yang SL, et al. Recapitulation of premature ageing with iPSCs from Hutchinson-Gilford progeria syndrome. Nature 2011;472(7342):221–5.

[74] Chen YF, Kao CH, Chen YT, Wang CH, Wu CY, Tsai CY, et al. Cisd2 deficiency drives premature aging and causes mitochondria-mediated defects in mice. Genes Dev 2009;23:1183–94.

[75] Promislow DE. DNA repair and the evolution of longevity: a critical analysis. J Theor Biol 1994;170:291–300.

[76] Walker RF, Pakula LC, Sutcliffe MJ, Kruk PA, Graakjaer J, Shay JW. A case study of "disorganized development" and its possible relevance to genetic determinants of aging. Mech Ageing Dev 2009;130:350–6.

[77] Crespi BJ. The emergence of human-evolutionary medical genomics. Evol Appl 2011;4:292–314.

[78] Berman JJ. Taxonomic guide to infectious diseases: understanding the biologic classes of pathogenic organisms. Cambridge, MA: Academic Press; 2012.

[79] Spiesschaert B, McFadden G, Hermans K, Nauwynck H, Van de Walle GR. The current status and future directions of myxoma virus, a master in immune evasion. Vet Res 2011;42:76.

[80] Ross J, Sanders MF. The development of genetic resistance to myxomatosis in wild rabbits in Britain. J Hyg (Lond) 1984;92:255–61.

[81] Basecke J, Griesinger F, Trumper L, Brittinger G. Leukemia- and lymphoma-associated genetic aberrations in healthy individuals. Ann Hematol 2002;81:64–75.

[82] Sidon P, El Housni H, Dessars B, Heimann P. The JAK2V617F mutation is detectable at very low level in peripheral blood of healthy donors. Leukemia 2006;20:1622.

[83] Brassesco MS. Leukemia/lymphoma-associated gene fusions in normal individuals. Genet Mol Res 2008;7:782–90.

[84] Xue Y, Chen Y, Ayub Q, Huang N, Ball EV, Mort M, et al. Deleterious- and disease-allele prevalence in healthy individuals: insights from current predictions, mutation databases, and population-scale resequencing. Am J Hum Genet 2012;91:1022–32.

[85] Bose S, Deininger M, Gora-Tybor J, Goldman JM, Melo JV. The presence of typical and atypical BCR-ABL fusion genes in leukocytes of normal individuals: biologic significance and implications for the assessment of minimal residual disease. Blood 1998;92:3362 7.

[86] Biernaux C, Loos M, Sels A, Huez G, Stryckmans P. Detection of major bcr-abl gene expression at a very low level in blood cells of some healthy individuals. Blood 1995;86:3118–22.

[87] Bayraktar S, Goodman M. Detection of BCR-ABL positive cells in an asymptomatic patient: a case report and literature review, Case Rep Med 2010;939706. Available from: https://www.hindawi.com/journals/crim/2010/939706/ [viewed 22.09.16].

[88] Gordon K, Spiden SL, Connell FC, Brice G, Cottrell S, Short J, et al. Hum Mutat FLT4/VEGFR3 and Milroy disease: novel mutations, a review of published variants and database update. Hum Mutat 2013;34:23–31.

[89] Flannick J, Beer NL, Bick AG, Agarwala V, Molnes J, Gupta N, et al. Assessing the phenotypic effects in the general population of rare variants in genes for a dominant Mendelian form of diabetes. Nat Genet 2013;45:1380–5.

[90] Pollock PM, Harper UL, Hansen KS, Yudt LM, Stark M, Robbins CM, et al. High frequency of BRAF mutations in nevi. Nat Genet 2003;33:19–20.

[91] Kato S, Lippman SM, Flaherty KT, Kurzrock R. The conundrum of genetic "drivers" in benign conditions. J Natl Cancer Inst 2016;108:.

[92] Mead AJ, Rugless MJ, Jacobsen SEW, Schuh A. Germline JAK2 mutation in a family with hereditary thrombocytosis. N Engl J Med 2012;366:967–9.

[93] Barosi G, Bergamaschi G, Marchetti M, Vannucchi AM, Guglielmelli P, Antonioli E, et al. JAK2 V617F mutational status predicts progression to large splenomegaly and leukemic transformation in primary myelofibrosis. Blood 2007;110:4030–6.

[94] Saleque S, Cameron S, Orkin SH. The zinc-finger proto-oncogene Gfi-1b is essential for development of the erythroid and megakaryocytic lineages. Genes Dev 2002;16:301–6.

[95] Jacks T, Fazeli A, Schmitt EM, Bronson RT, Goodell MA, Weinberg RA. Effects of an Rb mutation on the mouse. Nature 1992;359:259–300.

[96] Mercer K, Giblett S, Green S, Lloyd D, DaRocha DS, Plumb M, et al. Expression of endogenous oncogenic V600E B-raf induces proliferation and developmental defects in mice and transformation of primary fibroblasts. Cancer Res 2005;65:11493–500.

[97] Huttley GA, Easteal S, Southey MC, Tesoriero A, Giles GG, McCredie MR, et al. Adaptive evolution of the tumour suppressor BRCA1 in humans and chimpanzees. Australian Breast Cancer Family Study. Nat Genet 2000;25:410–3.

[98] Cochran G, Hardy J, Harpending H. Natural history of Ashkenazi intelligence. J Biosoc Sci 2006;38:659–93.

[99] Alexander DM, Williams LM, Gatt JM, Dobson-Stone C, Kuan SA, et al. The contribution of apolipoprotein E alleles on cognitive performance and dynamic neural activity over six decades. Biol Psychol 2007;75:229–38.

[100] Casanova J-L. Human genetic basis of interindividual variability in the course of infection. Proc Natl Acad Sci U S A 2015;112:e7118–27.

[101] Banuls A, Thomas F, Renaud F. Of parasites and men. Infect Genet Evol 2013;20:61–70.

[102] Casanova J. Severe infectious diseases of childhood as monogenic inborn errors of immunity. Proc Natl Acad Sci 2015;112:e7128–37.

[103] Giller G. Genetic heroes may be key to treating debilitating diseases: the resilience project seeks to find people who are unaffected by genetic mutations that would normally cause severe and fatal disorders. Scientific American, May 30, 2014. Available from: https://www.scientificamerican.com/article/genetic-heroes-may-be-key-to-treating-debilitating-diseases/ [viewed 12.03.17].

[104] Wszolek ZK, Tsuboi Y, Ghetti B, Pickering-Brown S, Baba Y, Cheshire WP. Frontotemporal dementia and parkinsonism linked to chromosome 17 (FTDP-17). Orphanet J Rare Dis 2006;1:30.

[105] Walsh T, Casadei S, Coats KH, Swisher E, Stray SM, Higgins J, et al. Spectrum of mutations in BRCA1, BRCA2, CHEK2, and TP53 in families at high risk of breast cancer. JAMA 2006;295:1379–88.

[106] D'Andrea AD. Susceptibility pathways in Fanconi's anemia and breast cancer. N Engl J Med 2010;362 (20):1909–19.

[107] Stecklein SR, Jensen RA. Identifying and exploiting defects in the Fanconi anemia/BRCA pathway in oncology. Transl Res 2012;160:178–97.

[108] Silva AG, Ewald IP, Sapienza M, Pinheiro M, Peixoto A, de Nóbrega AF, et al. Li-Fraumeni-like syndrome associated with a large BRCA1 intragenic deletion. BMC Cancer 2012;12:237.

[109] Wang E, Boswell E, Siddiqi I, Lu CM, Sebastian S, Rehder C, et al. Pseudo-Pelger-Huet anomaly induced by medications: a clinicopathologic study in comparison with myelodysplastic syndrome-related pseudo-Pelger-Hu t anomaly. Am J Clin Pathol 2011;135:291–303.

[110] Juneja SK, Matthews JP, Luzinat R, Fan Y, Michael M, Rischin D, et al. Association of acquired Pelger-Huet anomaly with taxoid therapy. Br J Haematol 1996;93:139–41.

[111] Schule B, Oviedo A, Johnston K, Pai S, Francke U. Inactivating mutations in ESCO2 cause SC phocomelia and Roberts syndrome: no phenotype-genotype correlation. Am J Hum Genet 2005;77:1117–28.

[112] Cruickshank RH, Paterson AM. The great escape: do parasites break Dollo's law? Trends Parasitol 2006;22:509–14.

[113] Olson TM, Keating MT. Mapping a cardiomyopathy locus to chromosome 3p22-p25. J Clin Invest 1996;97:528–32.

[114] Ikegawa S. A short history of the genome-wide association study: where we were and where we are going. Genom Inf 2012;10:220–5.

[115] Platt A, Vilhjalmsson BJ, Nordborg M. Conditions under which genome-wide association studies will be positively misleading. Genetics 2010;186:1045–52.

[116] Beekman M, Blanch H, Perola M, Hervonen A, Bezrukov V, Sikora E, et al. Genome-wide linkage analysis for human longevity: genetics of healthy aging study. Aging Cell 2013;12:184–93.

[117] Couzin-Frankel J. Major heart disease genes prove elusive. Science 2010;328:1220–1.

[118] Field MJ, Boat T. Rare diseases and orphan products: accelerating research and development. Institute of Medicine (US) Committee on Accelerating Rare Diseases Research and Orphan Product Development, Washington, DC: The National Academics Press; 2010. Available from: http://www.ncbi.nlm.nih.gov/books/NBK56189/.

[119] Cantor RM, Lange K, Sinsheimer JS. Prioritizing GWAS results: a review of statistical methods and recommendations for their application. Am J Hum Genet 2010;86(1):6–22.

[120] Panagiotou OA, Evangelou E, Ioannidis JP. Genome-wide significant associations for variants with minor allele frequency of 5% or less—an overview: a HuGE review. Am J Epidemiol 2010;172:869–89.

[121] Solt D, Medline A, Farber E. Rapid emergence of carcinogen-induced initiated hepatocytes in liver carcinogenesis. Am J Pathol 1977;88:595–618.

[122] Vilmos P, Kurucz E. Insect immunity: evolutionary roots of the mammalian innate immune system. Immunol Lett 1998;62:59–66.

[123] Hugot JP, Chamaillard M, Zouali H, Lesage S, Lesage S, Cezard JP, et al. Association of NOD2 leucine-rich repeat variants with susceptibility to Crohn's disease. Nature 2001;411:599–603.

[124] Ogura Y, Bonen DK, Inohara N, Nicolae DL, Chen FF, Ramos R, et al. A frameshift mutation in NOD2 associated with susceptibility to Crohn's disease. Nature 2001;411:603–6.

[125] McGonagle D, McDermott MF. A proposed classification of the immunological diseases. PLoS Med 2006;3:e297.

[126] Glaser RL, Goldbach-Mansky R. The spectrum of monogenic autoinflammatory syndromes: understanding disease mechanisms and use of targeted therapies. Curr Allergy Asthma Rep 2008;8:288–98.

[127] Masters SL, Simon A, Aksentijevich I, Kastner DL. Horror autoinflammaticus: the molecular pathophysiology of autoinflammatory disease. Annu Rev Immunol 2009;27:621–68.

[128] Tovar J, Leon-Avila G, Sanchez LB, Sutak R, Tachezy J, van der Giezen M, et al. Mitochondrial remnant organelles of Giardia function in iron-sulphur protein maturation. Nature 2003;426:172–6.

[129] Finsterer J. Mitochondriopathies. Eur J Neurol 2004;11:163–86.

[130] Ku CL, Picard C, Erd s M, Jeurissen A, Bustamante J, Puel A, et al.. IRAK4 and NEMO mutations in otherwise healthy children with recurrent invasive pneumococcal disease. J Med Genet 2007;44:16–23.

[131] Lindhurst MJ, Sapp JC, Teer JK, Johnston JJ, Finn EM, Peters K, et al. A mosaic activating mutation in AKT1 associated with the Proteus syndrome. N Engl J Med 2011;365:611–9.

[132] Xu B, Roos JL, Dexheimer P, Boone B, Plummer B, Levy S, et al. Exome sequencing supports a de novo mutational paradigm for schizophrenia. Nat Genet 2011;43:864–8.

[133] Bodey B. Spontaneous regression of neoplasms: new possibilities for immunotherapy. Expert Opin Biol Ther 2002;2:459–76.

[134] Wambua S, Mwangi TW, Kortok M, Uyoga SM, Macharia AW, Mwacharo JK, et al. The effect of alpha +-thalassaemia on the incidence of malaria and other diseases in children living on the coast of Kenya. PLoS Med 2006;3:e158.

8

Animal Models of Human Disease: Opportunities and Limitations

Section 8.1 The Animal Model Problem, in a Nutshell

There are no good animal models, but some are useful.

Herbert Slade.

Back in Chapter 6.4, "Eutherians to Humans," we read a highly provocative assertion: "Basically, all the eutherians are interchangeable" and "By the time that eutherians evolved, our cellular operating system and our anatomic arrangements were all finished." If there were no fundamental differences between mice and men, then shouldn't we be able to choose any eutherian species (e.g., mice, rats, rabbits) to screen for carcinogens, to test for toxicity, or to determine the effectiveness and safety of new drugs for humans? As it happens, we are finding that rodents and other eutherian animal models have a dismal record when it comes to helping us discover new, effective, and safe drugs [1–11].

The purpose of this section is to explain why it is that nonhuman eutherians are never ideal and are often inappropriate for testing the safety and effectiveness of drugs. Furthermore, we see that noneutherians, including fish, insects, and nematodes are often preferable to eutherian species as models for the metabolic pathways drive human pathogenesis and serve as the targets of new and effective treatments for human diseases.

So as not to leave you in suspense, here is the chain of reasoning that we will be exploring that explains why mice and other eutherians are disappointing models for testing potential new drugs.

    –1.   Although all eutherian species are fundamentally equivalent, with the same basic embryological development, the same cell types, a strictly homologous skeletal anatomy, and equivalent organs, the different eutherian species have diverged from one another in terms of habitat and diet, and have evolved with very different collections of endogenous microorganisms.

−2. The differences in our diets and our habitats have modified the manner in which exogenous chemicals are metabolized in our livers and in other tissues. The differences in the organisms that live within our bodies has greatly modified the manner in which we launch inflammatory responses. [Glossary Inflammasome]

−3. Differences in physiological responses in respect to metabolism and inflammation account for most of the differences between humans and nonhuman eutherians in experimental trials of new drugs.

−4. Rodents are difficult and expensive to employ in laboratory systems. In general, nonmammalian species have short generation times, and can be cultivated in large numbers, at low cost. Flies, worms, and fish have been proven to be excellent models for many monogenic diseases of humans and for examining the metabolic pathways that drive complex, polygenic diseases.

Now that the premise of this chapter has been laid out, we can determine in the next few sections of this chapter, whether we actually have the scientific evidence that supports these four assertions.

Section 8.2 Specificities and Idiosyncrasies

Because all of biology is connected, one can often make a breakthrough with an organism that exaggerates a particular phenomenon, and later explore the generality.

Thomas R. Cech

If man and man's best friend (his dog) are both eutherians, and if all eutherian organisms are just about alike, then why is it that a chunk of chocolate, or a grape, or an onion, will poison the dog, while the man may enjoy these foods with relative impunity?

While we're on the subject, how can we explain the difference in toxicity produced by Gram-negative organisms in mice and in men? We know that Gram-negative bacteria produce shock in animals via a particular lipopolysaccharide molecule found in their cell walls. Mice, unlike humans, have high resistance to the shock-inducing effect of lipopolysaccharide. The weight-adjusted dose of lipopolysaccharide causing death in mice happens to be one million times the dose that causes fever in humans and about 1000–10,000 times more than the dose that causes shock in humans [12].

Here's yet another mystery. In 2006, eight paid healthy volunteers were assembled. These subjects would be the first humans to receive a test drug, in an early phase clinical trial. In a single session, six of the volunteers were infused with TGN1412, and two volunteers were infused with a placebo. In about an hour, all six of the treated subjects developed cytokine storm, a life-threatening condition in which an immune-response precipitates shock, and a wide range of extreme system-wide responses, including multi-organ failure. Prompt treatment saved all their lives, but two of the six had prolonged hospital courses. All six patients must now deal with the long-term medical consequences of the event. Again, in 2016, in France, an early clinical trial for a new drug led to serious adverse effects in five subjects, including the death of one [13]. In both the 2006 trial and the 2016 trial, preclinical animal studies failed to predict human toxicity. Why not? [Glossary Clinical trial]

All eutherians share the same basic methods of drug metabolism (e.g., the cytochrome p450 system in hepatocytes), and they have the same physiologic responses to the bioactive molecules that are produced in all eutherians (e.g., acetylcholine, epinephrine, and cortisone). This being the case, why don't all eutherians have the same responses to administered drugs?

We think that the answer to these questions lies in our phylogenetic history of exposures to various exogenous toxins and invasive organisms. To survive such exposures, our ancestors evolved responses that carried to the genomes of their descendant species. There is every reason to expect that any given eutherian species will have inherited different response genes than those found in their cousin species (i.e., species descending from a different line of descent). Our responses to infections stimulate the expression of many genes, with estimates ranging from the hundreds to in excess of 1000 [14–17]. Regardless of the precise number of genes involved, we have ample reason to believe that coordinated inflammatory response are complex and involve multiple pathways. Furthermore, there is evidence to suggest that every species has its own orchestrated responses to a variety of pathologic stimuli [18].

In the case of the individuals who nearly died after administration of a drug (TGN1412) that was deemed safe based on animal trials, the affected subjects experienced an exaggerated inflammation pathway reaction known as a cytokine storm [19, 20]. It wasn't the drug that exerted a toxic effect; it was the inflammatory response to the drug that caused all the damage. Nonhuman species will not have the same responses to exogenous agents, because each species and its recent ancestors will have had a different history of exposures, with a different set of evolved responses. We should not be surprised to learn that in a review of human clinical trials based on research data collected from mouse models, all 150 clinical trials failed to predict inflammatory responses in humans [18].

Historically, the drug development process employs rodent models to identify candidate drugs for clinical trials in humans, but only a small fraction of mouse-inspired trials have shown success [8–11]. The National Academy of Sciences recently convened a workshop entitled, "Therapeutic development in the absence of predictive animal models of nervous system disorders [21]." Mouse models for common neurological disorders, such as Alzheimer's disease and Parkinson's disease have been largely unsuccessful [21, 22]. In fact, multiple mouse models for Parkinson's disease have been developed, including the "parkin knockout mouse" and LRRK2 knockouts, but none have shown dopamine degeneration, motor dysfunction, or even the synuclein bodies that are characteristic of human Parkinson's disease [23]. Likewise, mouse models of Alzheimer's disease fail to develop the neurofibrillary tangles and neuronal losses that are hallmarks for the human disease. Even if refinements in the mouse model may someday produce mice with a disease having the same pathological changes that are observed in the human disease, the treatments developed for mice may not carry over to clinical trials in humans [10, 24]. One of the repeated themes emerging from the National Academy of Sciences workshop is that animal models cannot predict how humans will respond to drugs that perturb complex pathways [21]. The situation in the neurosciences, wherein clinical responses of mice and men are not strictly comparable, is sufficiently discouraging that there are signs of a withdrawal from animal research by the pharmaceutical industry [25].

All microorganisms are potential pathogens

For the sake of discussion, let us accept that there are 50 million species of organisms on earth (a gross underestimate by some accounts). There have been about 1400 pathogenic organisms reported in the medical literature. This means that if you should stumble randomly upon a member of one of the species of life on earth, the likelihood that it is a known infectious pathogen is considerably less than 0.000028. Still, we must not drop our guard, insofar as organisms once thought to be harmless can sometimes surprise us.

In 1950, the US navy conducted an ill-advised experiment on the unsuspecting citizens of San Francisco [26]. Large hoses sprayed out a fog of supposedly harmless strains of *Serratia marcescens* and *Bacillus globigii*, to determine whether this kind of dispersal mechanism might be an effective way of exposing a large population to a biological warfare agent. As hoped, the bacteria were distributed widely over the Bay area, proving that their delivery system was effective.

The US navy declared victory in its undeclared war on San Francisco. At the time, the generals did not dream that there might be collateral damage. Very soon thereafter, a small epidemic of *Serratia marcescens* infections were reported among the exposed population. Eleven individuals required hospitalization, and one individual died. No cases of *Serratia marcescens* infections had been previously reported in the hospital where the death occurred, and no clusters of *Serratia marcescens* infections had occurred in the years preceding or following the navy's experiment. It seems that *Serratia marcescens*, though harmless to most individuals, was pathogenic to a tiny subpopulation of the population. For several decades, civilian epidemiologists could not explain why this short-lived mini-epidemic had hit San Francisco. It was not until 1976, when the navy experiment was declassified, that the truth came to light.

We tend to assume that rare individuals who succumb to "nonpathogenic" microorganisms must suffer from an immune deficiency disorder. Not so. If we want to understand why certain individuals are susceptible to infections and other individuals are not, we must understand that infectious disease, like all disease, develops in steps. It stands to reason that there must be many different pathways through which those steps can be enhanced or blocked. In point of fact, we are constantly being surprised when organisms, formerly believed to be harmless, produce devastating illness in otherwise healthy individuals, with no history of heightened susceptibility to other infectious diseases, and with absolutely no evidence of immune deficiency [27–30].

What does it take to be a pathogen? Apparently, not very much. Even organisms that don't grow well in human blood and tissues can be pathogenic in a select group of individuals. The proteobacteria *Eikenella corrodens* is a normal inhabitant of the mouth, that is harmless under most, but not all, conditions. When the organism is mechanically forced into the blood stream (e.g., by accidentally biting through the oral mucosa while eating), it can produce a cellulitis or a bacteremia with subsequent endocarditis. *Eikenella corrodens* can also produce disease in diabetics and immunocompromised individuals, apparently without inadvertent biting. Genus *Prevotella* contains oral inhabitants that can produce plaque, halitosis, and periodontal disease. *Prevotella dentalis*, like *Eikenella corrodens*, produces the so-called bite infections, wherein oral bacteria are inoculated, by a bite or abrasion, into adjacent tissues, producing abscesses, wound infections, or bacteremia.

As our ability to detect and diagnose infectious organisms improves, we encounter instances wherein once-obscure organisms have risen to the level of common pathogens. *Blastocystis hominis* is a eukaryotic organism that was observed as an incidental finding, of no known significance, on stool examinations. For a long time, the proper taxonomic classification of this organism was undetermined, and it has been variously referred to as a yeast, a fungus, an amoeba, a flagellated protozoa, and a sporozoan [31]. Today, *Blastocystis* is considered a genus belonging to the eukaryotic heterokonts and is the only heterokont known to produce a human infection. Infection follows ingestion of the cyst, through the fecal-oral route. Most infections do not result in any clinical symptoms, but sometimes, a syndrome mimicking irritable bowel syndrome may occur. Among individuals who have their stool specimens examined microscopically, up to 25% of specimens contain *Blastocystis* [32]. Because *Blastocystis* is found in the stools of healthy individuals, the finding of the organism in the stool of a symptomatic patient does not necessarily establish a causal relationship. Treatment with metronidazole, an antibiotic effective against eukaryotes and prokaryotes, has its advocates [33, 34], but can we do better? At this point, we know almost nothing about the heterokonts that would help us design a drug that would be effective against *Blastocystis*. In hindsight, we should have paid closer attention to this large class of eukaryotes. [Glossary Heterokonts]

Naegleria fowleri is an example of a eukaryotic pathogen that affects some individuals, but not others, for undetermined reasons. *Naegleria fowleri* is often found in warm fresh water. Swimmers in contaminated waters may develop an infection that spreads from the nasal sinuses to the central nervous system to produce an encephalitis that is fatal in 97% of cases [35]. Despite the hazard posed by *Naegleria*, health authorities do not generally test freshwater sources to determine the presence of the organism. Do not expect to find warning signs posted at swimming holes announcing that the water is contaminated by an organism that produces a disease that has a nearly 100% fatality rate. It is simply assumed that anyone who spends any time around freshwater will eventually be exposed to *Naegleria*. As it happens, although many thousands of individuals are exposed each year to *Naegleria* in the United States, only a few cases of Naegleria encephalitis occur in this country. In fact, since *Naegleria* was recognized as a cause of encephalitis, in 1965, less than 150 cases have been reported [36]. Most of the reported cases have occurred in children and adolescents and are associated with recreational water activities [37, 38].

The children who develop Naeglerian encephalitis, though exhibiting no signs of immune deficiency, are nonetheless susceptible to infection. We do not know why some exposed children develop Naeglerian encephalitis, while others are unaffected.

It should be noted that *Naegleria* is a member of Class Percolozoa, a single-celled eukaryote that occupies a subclass of Class Excavata. Why is this significant? *Naegleria* happens to be the only pathogenic species in Class Percolozoa, and we know almost nothing about the Percolozoan pathways that might render *Naegleria* susceptible to treatment. At present, Naegleria encephalitis is treated just as though it were an amoebic encephalitis, which it most certainly is not. Like the amoebic encephalitides, naeglerian infections are treated with amphotericin B. With or without amphotericin B treatment, nearly all cases of naeglerian (i.e., percolozoan) encephalitis are fatal [39]. Clearly, we need to learn a lot more about the biological pathways of members of Class Percolozoa, so that we can design a class-based strategy to prevent and treat Percolozoan encephalitis. Most importantly, we need to stop pretending that *Naegleria* is a genus in Class Amoebozoa, simply because both classes of eukaryotic organisms may infect the brain.

While the list of common infections is growing slowly, the list of rare infectious diseases is exploding. A source of new, rare infections are invasive instruments and catheters, particularly those that dwell inside the body for prolonged periods, such as: bladder catheters, ventilator tubes and pulmonary assistive devices, shunts, venous and arterial lines, and indwelling drains and tubes. These devices provide a path of entry for a wide variety of organisms that would otherwise be halted by normal anatomic barriers. Of the different organisms that invade via indwelling devices, most are bacteria. Fungal disease has occurred in adults who receive intravenous parental nutrition; the fungi growing in the lipid-rich alimentation fluids [40]. The bacterial organisms that invade via indwelling devices include species of *Pseudomonadales*, *Bacillales*, *Bacteroidetes*, *Fusobacteria*, and *Legionellales*. Despite their taxonomic diversity, all these organisms seem to share an ability to secrete biofilms over surfaces, and to glide through the biofilms they create. Biofilms are invisible, slimy coatings, composed of polysaccharides and cellular debris that provide sanctuary from the antibacterial sprays and solutions used in hospitals. Bacterial species that can glide through a biofilm can track a catheter into the body. For example, *Staphylococcus epidermidis* is a commensal organism that lives on human skin. Some of the organisms now known to cause catheter-associated hospital infections were previously considered obscure contaminants, of no pathogenic potential [41].

Because the list of common infectious organisms is short, while the list of potential rare pathogenic organisms is very, very long. It seems reasonable to assume that we, as a species, are continually fine-tuning our gene pool to best cope with the thousands of potential infectious organisms in our environment. Is there any evidence that would support this assumption?

The history of infections are written in our genes

The Taino Indians were the indigenous people of the island of Hispaniola; today recognized as the island containing the nations of the Dominican Republic and Haiti.

When Columbus first landed on Hispaniola, in 1492, hundreds of thousands of Taino lived on the island. The Taino population was soon thereafter decimated by plagues of the flu (1493) and smallpox (1518), and various other diseases introduced by European colonialists. These infections included hepatitis, measles, tuberculosis, diphtheria, cholera, and typhus. By 1550, there were only about 500 Tainos in Hispaniola [42].

Why did the Tainos die in large number, while the Europeans, who brought these disease to the new world, managed to survive? The pat answer to this question is that the Taino were new to these infections, and were thus immunologically unprepared to launch an effective biological defense. Hence, many of the Taino died from infections that produced relatively minor illness in Europeans. Does this answer really make much sense? Every individual, whether Taino or European, is immunologically naive to an organism prior to their first infectious encounter. If the Taino died because they have never before been infected by organisms of European origin, then we would expect Europeans to die, with equal frequency, when they are exposed to indigenous organisms, for the first time.

In retrospect, the Taino were doomed from the moment that Columbus set foot on Hispaniola. Their problem was not that they had never encountered the pathogens carried by the Europeans. Their problem was that their ancestors had never encountered those pathogens. Hepatitis, measles, tuberculosis, diphtheria, cholera, and typhus had not been written into the chapter of the human genome titled "The History of Taino Infections."

If subpopulations of humans have widely different responses to the same infectious pathogen based on the infectious history of their ancestors, then the same should apply to other species of eutherians. We would expect each unique species to adjust their gene pools according to the infectious organisms that they, and their ancestors, encountered. This seems to be the case, insofar as we can observe wide differences among different species of eutherians, in the types of organisms that cause infection, and in the response of the species to infection.

We now know that environmental pathogens, largely microorganisms, set the conditions for natural selection (i.e., who dies, and who survives) among humans; more so than climate or exposure to naturally occurring toxins, or famine [16]. Differences in molecular signatures among populations can be correlated with endogenous exposure to pathogenic parasites. No such differences can be correlated with climate or other factors related to demographic ecosystems [17]. We shouldn't be surprised. Numerous epidemics, several wiping out the majority of living humans in a large geographic area, have been thoroughly described in recorded history. In the last century alone, the 1917–18 influenza pandemic caused somewhere between 50 and 100 million deaths, in just a few months.

The 16th century poet, John Donne, wrote that humans are "a volume of diseases bound together." He wasn't far wrong. Let's take a look at some of the most prevalent infections, worldwide, in descending order of incidence.

- Demodex is a tiny mite that lives in facial skin. Demodex mites can be found in the majority of humans.

- The BK polyomavirus rarely causes disease in infected patients, and the majority of humans carry the latent virus.

- The JC polyomavirus persistently infects the majority of humans, but it is not associated with disease in otherwise healthy individuals.

- About one third of the human population has been infected (i.e., about 2.3 billion people) by the only species of Genus Toxoplasma that produces human toxoplasmosis: *Toxoplasma gondii*.

- About two billion people (of the world's 7 billion population) have been infected with *Mycobacterium tuberculosis*.

- Ascaris lumbricoides, the cause of ascariasis, infects about 1.5 billion people worldwide, making it the most common helminth (worm) infection of humans [43].

- Various estimates would suggest that worldwide, more than half a billion people are infected with one or another subtypes of *Chlamydia trachomatis*. This would include the various *Chlamydia* organisms and serotypes that account for trachoma eye infections [44] and chlamydial urethritis.

- Hookworms infect about 600 million people worldwide. Two species are responsible for nearly all cases of hookworm disease in humans: *Ancylostoma duodenale* and *Necator americanus*.

- Genus *Plasmodium* is responsible for human and animal malaria. About 300–500 million people are infected with malaria worldwide, causing 2 million deaths each year [45, 46].

- Scabies is an exceedingly common global disease, with about 300 million new cases occurring annually.

- About 200 million people are infected by schistosomes (i.e., have some form of schistosomiasis).

- Hepatitis B infects more than 200 million people worldwide, causing two million deaths each year.

- Bubonic plague is credited with killing one third of the population of Europe in the mid-1300s. Altogether, bubonic plague is estimated to have caused about 200 million deaths. In modern times, plague is rare, but not extinct. Each year, several thousand cases of plague occur worldwide, resulting in several hundred deaths. Virtually all of the contemporary cases occur in Africa.

- About 150 million people are infected by the filarial nematodes (genera Brugia, Loa, Onchocerca, Mansonella, and Wuchereria) [47]. *Wuchereria bancroft* and *Brugia malayi* together infect about 120 million individuals [47]. Most cases occur in Africa and Asia.

- Smallpox is reputed to have killed about 300 million people in the 20th century, prior to the widespread availability of an effective vaccine. Smallpox, now extinct, has been referred to as the greatest killer in human history.

- Worldwide, about 100 million cases of acute diarrhea are caused by rotavirus. In 2004, rotavirus infections accounted for about a half million deaths in young children, from severe diarrhea [48].

Evolution, by playing with our gene pools, has yielded a variety of protective mechanisms against organisms. Consider malaria, an infection of humans and other animals caused by a single-celled eukaryotes of various *Plasmodium* species (Class Apicomplexa). Malaria infects nearly 500 million people and causes 2 million deaths each year worldwide [45, 46]. To defend against malaria, humans have preserved various disease mutations that render red cells unsuitable hosts for malarial guests. Heightened red cell fragility and a tendency to hemolysis (i.e., red cell rupture) would normally reduce human fitness. In red blood cells infected with malaria, increased hemolysis renders the parasite less able to survive in its host. Hence, a heightened rate of red cell lysis can be beneficial in malaria-stricken populations. The eternal threat of malaria has been the driving force for the preservation of recessive hemolysis-causing disease genes in the human gene pool [49]. As a result, we now have, in the human gene pool a host of mutant genes that combat malaria. These include the following:

- Variants of beta-globin (HbS, HbC, HbE).

- Defective regulation of alpha and beta globin syntheses, causing alpha and beta thalassemias.

- Variation in the structural protein SLC4A1, which causes ovalocytosis.

- Variation in the chemokine receptor FY, which causes the Duffy-negative blood group. [Glossary Chemokine]

- Polymorphisms of the red cell enzyme gene G6PD, which causes glucose-6-phosphate dehydrogenase deficiency.

- Common erythrocyte variants that affect resistance to malaria: FY Duffy antigen, G6PD glucose-6-phosphatase dehydogenase, glycophorin A, B, and C sialoglycoproteins, alpha globin, beta-globin, and SLC4A1 (erythrocyte band 3 protein chloride/bicarbonate exchanger) [49].

- Variants of at least one nonerythrocytic protein: haptoglobin (a hemoglobin binding protein in plasma) [49].

- Proteins that influence red cell cytoadhesion with *Plasmodium falciparum* that have been associated with enhanced resistance or susceptibility to malaria: CD36 antigen (thrombospondin receptor), Complement receptor 1, intercellular adhesion molecule-1, platelet-endothelial cell adhesion molecule [49].

- Immune genes suspected of influencing the course of malaria infections that include the low affinity receptor for Fc fragment of IgG, HLA-B53, Interferon-a receptor component, Interferon-g Cytokine, Interferon-g receptor component, Interleukin-1a and -1b pro-inflammatory cytokines, Interleukin-10 antiinflammatory cytokine, Interleukin-12b subunit, IL4 Interleukin-4, Mannose-binding protein, Inducible NO synthase, Tumor necrosis factor, and TNFSF5 CD40 ligand [49].

- A mutation of the TIRAP gene that is associated with protection against invasive pneumococcal disease, bacteremia, malaria, and tuberculosis [50].

In summary, natural selection does not provide any single protective mechanism against environmental pathogens. In the case of malaria, an assortment of gene variants are preserved in the human gene pool. It is easy to see that a member of a species may carry hundreds, perhaps thousands, of gene variants that provide some level of protection against infectious organisms. Furthermore, the preserved gene variants will vary greatly from species to species, depending on the species-specific history of past infections. Hence, there is little reason to expect that any nonhuman species will serve as an adequate predictor for the human response to organisms and other exogenous agents.

Section 8.3 New Animal Options

Medical researchers are currently faced with the following dilemma: We are discovering the root mutations responsible for genetic diseases at a rate that far exceeds our ability to understand how the mutation produces its clinical phenotype. This is a big, big problem for medical researchers who need to understand how a gene leads to the clinical expression of disease, if they have any hope of finding effective preventions or treatments for human diseases.

Perhaps, a solution to the dilemma faced by medical researchers will come in the form of orthodisease research. Orthodiseases are conditions observed in nonhuman species that result from alterations in genes that are homologous to the genes known to cause diseases in humans. For example, if a loss of function mutation in a particular gene in humans were associated with an inherited blood cell disorder; and if a loss of function mutation in the homologous gene in a zebrafish resulted in lymphocytosis (proliferation of lymphocytes), then we would consider the condition in zebrafish to be an orthodisease of the human genetic counterpart. [Glossary Lymphoproliferative disease]

Organisms such as the nematode *Caenorhabditis elegans*, the fruit fly *Drosophila melanogaster*, the zebrafish *Danio rerio* or the yeast *Saccharomyces cerevisiae*, can be propagated, manipulated genetically, and studied in the laboratory. Experiments on these nonmammalian species are conducted at much lower cost, and in much less time, than comparable experiments on rodents and other mammals [51].

Why would an orthodisease in the yeast *Saccharomyces cerevisiae* have any relevance to a disease occurring in humans? Surely, the two diseases will have completely different clinical phenotypes. A yeast cannot develop heart disease or gout or an ovarian neoplasm!

The justification for orthodisease models of human pathologic processes is as follows:

-1. The mutations that account for genetic diseases in humans occur in conserved genes.

The reason being that conserved genes are essential (otherwise, they would not be conserved). If a malfunction occurs in an essential gene, it is likely to produce some pathological consequences.

-2. Conserved genes nearly always have homologs that can be found throughout the eukaryotic lineage.

If a gene is essential for us, it is probably essential for other organisms.

-3. The homologous conserved genes of humans that are found in other animals are likely to participate in metabolic pathways that are at least similar to the pathways found in humans.

The reason being that the protein products of conserved genes tend to have a similar function (i.e., similar substrates and similar products) in every organism. Hence, we might expect that the proteins encoded by conserved genes will participate in conserved metabolic pathways. As it happens, nearly 75% of human disease-causing genes are believed to have a functional homolog in the fly [52].

-4. The metabolic pathways affected by gene mutations are likely to be involved in disease pathogenesis.

-5. It is much easier and faster to study genetic mutations, and their subsequent effects on metabolic pathways and on the development of disease, in model organisms (such as worms, flies, fish, and yeast) than in mammalian systems.

-6. Drugs that are effective in modifying metabolic pathways in model organisms are likely to have similar biological effects in humans.

There is abundant observational and experimental evidence that seems to support all of these logical assertions, and the study of human disease pathogenesis using nonmammalian orthodisease models is flourishing [51–56]. Where traditional animal models are failing, biologists are finding success with single-cell eukaryotes and insects. Though we can expect disease phenotypes to diverge among species affected by orthologous genes, we might be able to study specific pathways that have been conserved through most of the history of eukaryotic evolution. For example, the 2013 Nobel Prize in Physiology or Chemistry was awarded for work on vesicular transport disorders. Progress in this area came from studies of human inherited transport disorders [57]. However, the vesicular transport pathway was dissected by studying orthologous genes in yeast [55].

Though yeast has many homologous genes with humans, there has been concern that the homologs may not participate in the same pathways in yeast as they do in humans. In a large study of proteins involved in human spinocerebellar ataxias, it was found that the human genes coding for such proteins have yeast homologs that code for proteins that participate in pathways that are similar to the pathways used by the human encoded proteins [58]. This tells us that for yeast models of the spinocerebellar ataxias, homologous pathways exist for homologous genes, supporting the relevance of the yeast model for human disease (Fig. 8.1).

The nematode *Caenorhabditis elegans* is a well-studied organism. Despite its different phylogenetic lineage (i.e., Class Protostomia, not Class Deuterostomia, containing humans) more than 65% of human disease-causing genes currently identified have a counterpart in the worm [59]. As a hermaphroditic organism, *C. elegans* has a particularly useful property that enhances its value in disease research. When organisms are exposed to mutagens, the first generation progeny self-fertilize producing some second generation worms that are homozygous for the mutation. This has allowed researchers to study how specific mutations disrupt development and cause disease.

When using nematodes (roundworms) to study human disease processes, we must be very careful to remember that biological systems are always complex, and the final phenotype resulting from a gene mutation is an emergent property of the total system, developing over time. For example, the root cause of human retinoblastoma (a cancer of retinal

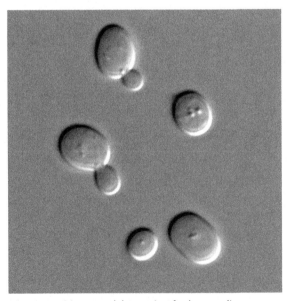

FIG. 8.1 The yeast *Saccharomyces cerevisiae*, a model organism for human disease research. *Source: Wikipedia, entered into the public domain by its copyright holder, Masur.*

stem cells) is a mutation in the RB1 gene. Mutating the homologous gene in the nematode results in ectopic vulvae, a condition that is unmistakably different from retinoblastoma [56]. Nonetheless, pathways involved in causing retinoblastoma (in humans) and in causing ectopic vulvae (in *C. elegans*) may share many important commonalities, including sensitivity to gene-targeted drugs. [Glossary Trilateral retinoblastoma] (Fig. 8.2)

Another example of a useful orthologous organisms is the fruit fly, *Drosophila melanogaster*. *Drosophila* contains homologs of the genes that cause tuberous sclerosis, a hamartoma-cancer syndrome in humans. The brain tubers (hamartomas of the neuroectoderm, also called phakomas), for which tuberous sclerosis takes its name, contain large multinucleate neurons. Loss of function of the same genes in *Drosophila* produce enlarged cells with many times the normal amount of DNA [60]. The tuberous sclerosis orthodisease in *Drosophila* is being studied to help us understand cell growth control mechanisms in humans. [Glossary Ortholog, Tuberous sclerosis]

The zebrafish (*Danio rerio*) is a species of small freshwater minnows. Their common name comes from the distinctive horizontal stripes on the sides of their bodies; a somewhat inaccurate appellation insofar as most of the stripes on zebras (*Equus quagga*) are nearly vertical. Zebrafish eggs are fertilized outside the mother's body, allowing scientists to inject DNA or RNA into one-cell-stage embryos, to produce transgenic or knock-out strains of zebrafish strains. Zebrafish are easy to grow, in large or small schools, and their development can be closely monitored, from egg onwards(Fig. 8.3).

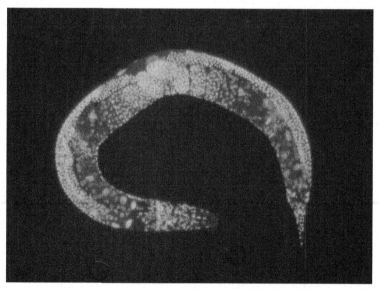

FIG. 8.2 *Caenorhabditis elegans*, a nematode (roundworm) about 1 mm in length. *C. elegans* serves as a model organism for human disease research. *Source: Wikipedia, from a public domain work from the US National Institutes of Health.*

FIG. 8.3 Zebrafish, a model organism for the study of human diseases. *Source: Wikipedia, and entered into the public domain by its author, Azul.*

The zebrafish share with humans the same evolutionary descent, to the level of Class Euteleostomi. At this point, the ancestors of the zebrafish branched to Class Actinopterygii (i.e., ray-finned fish accounting for at least 30,000 species of extant fish), while the ancestors of humans branched to Class Sarcopterygii (i.e., the lobe-finned fish). En route to Class Euteleostomi, the ancestors of humans and of ray-finned fish descended through Class Metazoa, Class Bilateria, Class Craniata, and Class Vertebrata and Gnathostomata. In doing so, zebrafish acquired nearly all the cell types that are present in humans today. Consequently, zebrafish and humans are not all that distant from one another, developmentally, and these two species have similar immunologic systems, central nervous systems and peripheral nervous systems. This being the case, it should not be surprising to learn that 70% of human genes are found in zebrafish [61].

In the past few decades, the zebrafish has become a very useful model for studying developmental and disease-related pathways in humans [62]. For example, zebrafish, as members of Class Chordata (like humans) develop chordomas (as do humans). As a fellow member of Class Craniata, the zebrafish has been an excellent model for the neurocristopathies, including Waardenburg-Shah syndrome and Hirschsprung's disease [62–65].

As was the case with *C. elegans*, it is important not to overinterpret experiments using zebrafish. In one large study, a gene in the zebrafish was shown to modulate its susceptibility to mycobacterial infection [66]. Naturally, there was hope that the orthologous gene in humans would be associated with human susceptibility to tuberculosis. Despite a large study involving 9115 subjects, no such association was found [67]. Again we learned that the genetic root cause of a disease does not account for pathogenesis, which is an emergent property of the system in which the mutation is expressed. [Glossary Multi-event process, Multistep process]

Non-vertebrate models for cancer research

Oncogenes and tumor suppressor genes are core constituents of metazoans, and we can often find nonvertebrate organisms that recapitulate some of the steps in human carcinogenesis. As it happens, all multicellular organisms of the plant and animal kingdoms share a susceptibility to cancer. Crown gall is a tumorous disease that infects many different kinds of woody and herbaceous plants. It is caused by a bacteria, *Agrobacterium tumefaciens*, which is transmitted from plant to plant. The tumors grow near the soil line. Bacteria from the periphery of the tumor infect the soil and spread to nearby plants. The bacteria produce a hyperplastic reaction in adjacent plant cells. Tumor cell proliferation continues in response to extracellular bacterial colonies. Mature galls consist of mitotically active cells that can be multinucleate and that are less differentiated than the cells in the adjacent normal plant tissue. Vessels can grow within the gall, but tumor vessels are disorganized and not well-connected to the normal vasculature of the plant. Consequently, galls tend to die, sloughing off the plant. Aside from bacterial-induced gall tumors, plants are known to develop neoplasms much like humans, as inherited genetic traits [68].

Neoplasms also occur in insects, but rarely. Insects are highly resistant to neoplastic development because most cells in an adult insects are postmitotic; they cannot divide and hence cannot propagate as a cancer. However, insects have an Achilles heel; imaginal discs. Imaginal discs are foci of stem cells that live in insect larvae and that grow to become the various parts of adult insects. Flying insects have an imaginal disc to create a wing and another imaginal disc to create a leg and another for an antenna, and so forth. Imaginal discs have been the fundamental tool that developmental biologists have used, for decades, to study the control of organogenesis [69]. In insects, tumors may arise from imaginal disc cells, and these insect neoplasms are transplantable, invasive, and metastatic [70]. *Drosophila* is a genus of fly, and *Drosophila melanogaster* is the fly species most studied by biologists. Loss-of-function mutations in *Drosophila* genes, that are homologous to human tumor suppressor genes, have resulted in neoplasms growing from imaginary disc cells. These findings support the hypothesis that carcinogenesis involves conserved genes that have been evolving through the insect and the human evolutionary lineages [71].

Koch's postulates and their reliance on animal models are obsolete

Koch's postulates are a set of observations and experimental requirements proposed by Heinrich Hermann Robert Koch in the late 1800s, intended to prove that a particular organism causes a particular infectious disease. Koch's postulates require that the suspected causal organism be extracted from the infected lesion (i.e., from diseased, infected tissue); then isolated; then cultured in a laboratory; and that the lesion be reproduced in animals after inoculation with the cultured organism.

Whipple disease was first described in 1907 [72], but its cause was unknown until 1992, when researchers isolated and amplified, from Whipple disease tissues, a 16s ribosomal RNA sequence that could only have a bacterial origin [73]. Based on molecular features of the ribosomal RNA molecule, the researchers assigned it to Class Cellulomonadacea, and named the species *Tropheryma whipplei*, after the man who first described the disease, George Hoyt Whipple.

Particularly noteworthy, in the case of Whipple disease, is that Koch's postulates never came close to being satisfied. The putative pathogen has never been isolated and grown in culture. There is no animal model for Whipple disease. The discovery of the organism causing Whipple disease was an achievement of molecular biologists who had the audacity to ignore Koch's postulates [73, 74].

Side-stepping Koch's postulates, including its reliance on animal models, has become de rigueur in modern medicine. For example, the United States has experienced a recent increase in cases of acute flaccid myelitis, a rare disease of children [75]. Diagnosis is based on a metagenomic analysis (i.e., culture-independent sequence searches conducted on an assemblage of microbial gene sequences in a biologic sample) of DNA obtained from nasopharyngeal swabs. The organism that is present in most of the examined cases is enterovirus-D68, and this virus is the presumed causal organism of acute flaccid myelitis, until proven otherwise. Like so much of modern medicine, this research waived Koch's postulates and its reliance on animal models of human disease.

Section 8.4 The Proper Study of Mankind

The proper study of mankind is man.

From "An Essay on Man" by Alexander Pope

Our rapidly growing understanding of animal evolution has provided us with fresh insights that have changed the way we think about animal experimentation. It is now possible to design experiments with a deep understanding of the phylogenetic relevance of non-vertebrate models, and an awareness of species-specific traits that cannot be modeled in any nonhuman organism. For many human diseases that have an inflammatory component and for most of the common, chronic diseases of humans, rodent models have little chance of answering basic questions of pathogenesis and cannot reliably indicate the treatments that are safe and effective. We should stop wasting time and money on animal models that will almost certainly fail to give us the information we need.

In Section 7.5, "Why Good People Get Bad Diseases," we described how the myxoma virus was unleashed on Australian rabbits. About 3 billion rabbits died in the aftermath. Looking back, we see that humans exercise enormous power over the welfare of nonhuman animals [76]. There is ample evidence to suggest that all eutherians are roughly equivalent in terms of neurologic organization, cognition, and intellectual responses. Any animal lover will tell you that the emotional needs of a pet are much like those of a human, if not greater. We can be quite certain that experimentation on animals can result in pain and emotional distress, much like the pain and distress that would occur in humans subjected to the same kind of treatment. If we can reduce the pain and suffering of experimental animals, by using our new-found understanding of the relationships linking evolution, genes, and pathogenesis, then should we not do so?

Glossary

Chemokine An inflammatory cytokine that stimulates white blood cells to move to a tissue target.

Clinical trial Before a drug can be approved for use, it must undergo and pass three phased trials, using human subjects. Phase 1 is the safety phase; the drug must be safe for humans. Phase 2 is the effectiveness phase; the drug must have some desired biological effect. Phase 3 is the large, expensive trial wherein individuals are tested against a control group treated with a placebo or with the standard-of-care medication. Phase 3 trials are very expensive to conduct, and many trials are negative (i.e., fail to indicate that the drug is effective in a phase 3 trial) or demonstrate only incremental success. Of the successful phase 3 trials, a significant number of drugs will eventually be withdrawn, because their effectiveness in clinical practice could not meet the earlier expectations observed in the phase 3 trial results [77].

Heterokonts Also known as stramenopiles. A class of single-celled eukaryote and a subclass of Class Chromista. There are over 25,000 species of heterokonts, including some types of algae, unicellular diatoms, and single-celled components of plankton. Also includes organisms that are pathogenic in plants (oomycetes) and in humans (blastocystis).

Inflammasome A protein complex expressed by white blood cells that activates inflammatory cytokines which, in turn, attract inflammatory cells. Examples of inflammasome proteins are caspase 1 and 5, PYCARD, and NALP. The inflammasome is part of the innate immune system.

Lymphoproliferative disease Sustained hyperplasias of lymphoid cells, as distinguished from a transient or reactive lymphocytosis (increase in circulating lymphocytes). Though there is no universal agreement over which conditions should be included in this category, a list of lymphoproliferative lesions might include: angioimmunoblastic lymphadenopathy with or without dysproteinemia, hairy cell leukemia, posttransplant lymphoproliferative disease, Helicobacter-responsive maltomas, X-linked lymphoproliferative disease, Epstein-Barr virus associated lymphoproliferative disease, and the histiocytoses of childhood.

Multievent process Same as multistep process.

Multistep process All of life can be described as a multistep process, wherein each cellular events is directly preceded by some other event. As every biological event has a preceding event, it can be inferred that every cellular event that occurs in any organism can be iteratively traced backwards through history to the first cellular event that occurred on the planet some 4 billion years ago. For practical reasons, determining the root cause of a disease requires us to choose an arbitrary cut point where we say that pathogenesis begins and we call this cut point the root cause.

Ortholog Refers to a gene found in different organisms that evolved from a common ancestor's gene through speciation. As an empiric observation, orthologs in different species often have the same or similar functionality. Orthology is a type of homology.

Trilateral retinoblastoma Hereditary bilateral retinoblastomas are sometimes accompanied by the occurrence of a pineoblastoma, and this three-tumor combination is referred to as trilateral retinoblastoma. The pineal gland has an evolutionary anlage identical to that of the eye. The development of a pineal equivalent of a retinoblastoma in the same individual with bilateral retinoblastomas, suggests that, in the case of these hereditary tumors, pathogenesis follows a sequence of steps that recapitulates an embryonic structure that is homologous to the retina [78].

Tuberous sclerosis Also known as tuberous sclerosis complex (TSC) and formerly known as Bourneville disease. Tuberous sclerosis can be caused by one of at least two different genetic lesions at two different cytogenetic loci; TSC1 on chromosome 9q34 encoding hamartin, and TSC2 on chromosome 16p13.3 encoding tuberin. Hamartin and tuberin lock together in a protein complex; hence, a defect in either gene disrupts the same pathway [79].

The types of hamartomas associated with this disease are: ash-leaf spots, shagreen patches, periungual fibromas, so-called adenoma sebaceum (actually facial angiofibromas), retinal hamartomas ("mulberry lesions"), cardiac rhabdomyomas, renal angiomyolipomas, and subependymal giant cell astrocytomas [80].

References

[1] Baker D, Lidster K, Sottomayor A, Amor S. Two years later: journals are not yet enforcing the arrive guidelines on reporting standards for pre-clinical animal studies. PLoS Biol 2014;12(1)e1001756.

[2] Vesterinen HM, Sena ES, Ffrench-Constant C, Williams A, Chandran S, et al. Improving the translational hit of experimental treatments in multiple sclerosis. Mult Scler 2010;16:1044–55.

[3] Cumberland Consensus Working Group, Cheeran B, Cohen L, Dobkin B, Ford G, et al. The future of restorative neurosciences in stroke: driving the translational research pipeline from basic science to rehabilitation of people after stroke. Neurorehabil Neural Repair 2009;23:97–107.

[4] Billiau A, Heremans H, Vandekerckhove F, Dijkmans R, Sobis H, et al. Enhancement of experimental allergic encephalomyelitis in mice by antibodies against IFN-gamma. J Immunol 1988;140:1506–10.

[5] Baker D, Butler D, Scallon BJ, O'Neill JK, Turk JL, et al. Control of established experimental allergic encephalomyelitis by inhibition of tumor necrosis factor (TNF) activity within the central nervous system using monoclonal antibodies and TNF receptor-immunoglobulin fusion proteins. Eur J Immunol 1994;24:2040–8.

[6] Panitch HS, Hirsch RL, Haley AS, Johnson KP. Exacerbations of multiple sclerosis in patients treated with gamma interferon. Lancet 1987;1:893–5.

[7] The Lenercept Multiple Sclerosis Study Group. TNF neutralization in MS: results of a randomized, placebo-controlled multicenter study. The Lenercept Multiple Sclerosis Study Group and The University of British Columbia MS/MRI Analysis Group. Neurology 1999;53:457–65.

[8] Pound P, Ebrahim S, Sandercock P, Bracken MB, Roberts I, Reviewing Animal Trials Systematically (Rats) Group. Where is the evidence that animal research benefits humans? BMJ 2004;328:514–7.

[9] Hackam DG, Redelmeier DA. Translation of research evidence from animals to humans. JAMA 2006;296:1731–2.

[10] Van der Worp HB, Howells DW, Sena ES, Porritt MJ, Rewell S, O'Collins V, et al. Can animal models of disease reliably inform human studies. PLoS Med 2010;7:e1000245.

[11] Rice J. Animal models: not close enough. Nature 2012;484:S9.

[12] Warren HS, Fitting C, Hoff E, Adib-Conquy M, Beasley-Topliffe L, Tesini B, et al. Resilience to bacterial infection: difference between species could be due to proteins in serum. J Infect Dis 2010;201:223–32.

[13] Blamont M. French drug trial disaster leaves one brain dead, five injured. Reuters; 2016.

[14] Zheng SL, Liu W, Wiklund F, Dimitrov L, Balter K, Sun J, et al. A comprehensive association study for genes in inflammation pathway provides support for their roles in prostate cancer risk in the CAPS study. Prostate 2006;66:1556–64.

[15] Natoli G, Ghisletti S, Barozzi I. The genomic landscapes of inflammation. Genes Dev 2011;25:101–6.

[16] Fumagalli M, Sironi M, Pozzoli U, Ferrer-Admettla A, Pattini L, Nielsen R. Signatures of environmental genetic adaptation pinpoint pathogens as the main selective pressure through human evolution. PLoS Genet 2011;7:e1002355.

[17] Banuls A, Thomas F, Renaud F. Of parasites and men. Infect Genet Evol 2013;20:61–70.

[18] Seok J, Warren HS, Cuenca AG, Mindrinos MN, Baker HV, Xu W, et al. Genomic responses in mouse models poorly mimic human inflammatory diseases. Proc Natl Acad Sci U S A 2013;110:3507–12.

[19] D'Elia RV, Harrison K, Oyston PC, Lukaszewski RA, Clark GC. Targeting the cytokine storm for therapeutic benefit. Clin Vaccine Immunol 2013;20:319–27.

[20] Lee DW, Gardner R, Porter DL, Louis CU, Ahmed N, Jensen M, et al. Current concepts in the diagnosis and management of cytokine release syndrome. Blood 2014;124:188–95.

[21] National Academies of Sciences, Engineering, and medicine. Therapeutic development in the absence of predictive animal models of nervous system disorders: proceedings of a workshop. Washington, DC: The National Academies Press; 2017.

[22] Watts JC, Prusiner S. Mouse models for studying the formation and propagation of prions. J Biol Chem 2014;289:19841–9.

[23] Johns Hopkins. The mouse model: less than perfect, still invaluable. Johns Hopkins Medicine; 2010. Available from: http://www.hopkinsmedicine.org/institute_basic_biomedical_sciences/news_events/articles_and_stories/model_organisms/201010_mouse_model.html [viewed 15.03.19].

[24] Dawson TM, Ko HS, Dawson VL. Genetic animal models of Parkinson's disease. Neuron 2010;66 (5):646–61.

[25] Choi DW, Armitage R, Brady LS, Coetzee T, Fisher W, Hyman S, et al. Medicines for the mind: policy-based "pull" incentives for creating breakthrough CNS drugs. Neuron 2014;84:554–63.

[26] Loria K. One of the largest human experiments in history was conducted on unsuspecting residents of San Francisco. Business Insider; 2015. July 9.

[27] Sak B, Kvac M, Kucerova Z, Kvetonova D, Sakova K. Latent microsporidial infection in immunocompetent individuals: a longitudinal study. PLoS Negl Trop Dis 2011;5:e1162.

[28] Tehmeena W, Hussain W, Zargar HR, Sheikh AR, Iqbal S. Primary cutaneous mucormycosis in an immunocompetent host. Mycopathologia 2007;164:197–9.

[29] Jiang Y, Huang A, Fang Q. *Disseminated nocardiosis* caused by *Nocardia otitidiscaviarum* in an immunocompetent host: a case report and literature review. Exp Ther Med 2016;12:3339–46.

[30] Permi HS, Sunil KY, Karnaker VK, Kishan PHL, Teerthanath S, Bhandary SK. A rare case of fungal maxillary sinusitis due to *Paecilomyces lilacinus* in an immunocompetent host, presenting as a subcutaneous swelling. J Lab Physicians 2011;3:46–8.

[31] Silberman JD, Sogin ML, Leipe DD, Clark CG. Human parasite finds taxonomic home. Nature 1996;380:398.

[32] Amin OM. Seasonal prevalence of intestinal parasites in the United States during 2000. Am J Trop Med Hyg 2002;66:799–803.

[33] Sekas G, Hutson WR. Misrepresentation of academic accomplishments by applicants for gastroenterology fellowships. Ann Intern Med 1995;123:38–41.

[34] Samuelson J. Why metronidazole is active against both bacteria and parasites. Antimicrob Agents Chemother 1999;3:1533–41.

[35] Centers for Disease Control and Prevention. *Naegleria fowleri*—primary amebic meningoencephalitis—amebic encephalitis, http://www.cdc.gov/parasites/naegleria/general.html; 2016 [viewed 18.04.17].

[36] Budge PJ, Lazensky B, Van Zile KW, Elliott KE, Dooyema CA, Visvesvara GS, et al. Primary amebic meningoencephalitis in Florida: a case report and epidemiological review of Florida cases. J Environ Health 2013;75:26–31.

[37] Grace E, Asbill S, Virga K. *Naegleria fowleri*: pathogenesis, diagnosis, and treatment options. Antimicrob Agents Chemother 2015;59:6677–81.

[38] Hebbar S, Bairy I, Bhaskaranand N, Upadhyaya S, Sarma MS, Shetty AK. Fatal case of *Naegleria fowleri* meningo-encephalitis in an infant: case report. Ann Trop Paediatr 2005;25:223–6.

[39] Garrison FH. History of medicine. Philadelphia: WB Saunders; 1921.

[40] Inamadar AC, Palit A. The genus *Malassezia* and human disease. Indian J Dermatol Venereol Leprol 2003;69:265–70.

[41] de Mauri A, Chiarinotti D, Andreoni S, Molinari GL, Conti N, De Leo M. *Leclercia adecarboxylata* and catheter-related bacteremia: review of the literature and outcome of catheters and patients. J Med Microbiol 2013;62:1620–3.

[42] Mann CC. 1493: Uncovering the new world Columbus created. New York: Knopf; 2011.

[43] Crompton DW. How much human helminthiasis is there in the world? J Parasitol 1999;85:397–403.

[44] Resnikoff S, Pascolini D, Etyaale D, Kocur I, Pararajasegaram R, Pokharel GP, et al. Global data on visual impairment in the year 2002. Bull World Health Organ 2004;82:844–51 2004.

[45] World Health Organization. The state of world health. Chapter 1 in world health report 1996, World Health Organization; 1996. Available from: http://www./whr/1996/en/index.html.

[46] Lemon SM, Sparling PF, Hamburg MA, Relman DA, Choffnes ER, Mack A. Vector-borne diseases: understanding the environmental, human health, and ecological connections, workshop summary. Institute of medicine (US) forum on Microbial threats. Washington (DC): National Academies Press (US); 2008.

[47] Foster J, Ganatra M, Kamal I, Ware J, Makarova K, Ivanova N, et al. The Wolbachia genome of *Brugia malayi*: endosymbiont evolution within a human pathogenic nematode. PLoS Biol 2005;3:e121.

[48] World Health Organization. Weekly epidemiological record. 32:World Health Organization; 2007285–96.

[49] Kwiatkowski DP. How malaria has affected the human genome and what human genetics can teach us about malaria. Am J Hum Genet 2005;77:171–92.

[50] Khor CC, Chapman SJ, Vannberg FO, Dunne A, Murphy C, Ling EY, et al. A Mal functional variant is associated with protection against invasive pneumococcal disease, bacteremia, malaria and tuberculosis. Nat Genet 2007;39:523–8.

[51] Strange K. Drug discovery in fish, flies, and worms. ILAR J 2016;57:133–43.

[52] Pandey UB, Nichols CD. Human disease models in *Drosophila melanogaster* and the role of the fly in therapeutic drug discovery. Pharmacol Rev 2011;63:411–36.

[53] Washington NL, Haendel MA, Mungall CJ, Ashburner M, Westerfield M, Lewis SE. Linking human diseases to animal models using ontology-based phenotype annotation. PLoS Biol 2009;7:e1000247.

[54] Chow CY, Reiter LT. Etiology of human genetic disease on the fly. Trends Genet 2017;33:391–8.

[55] Novick P, Field C, Schekman R. Identification of 23 complementation groups required for post-translational events in the yeast secretory pathway. Cell 1980;21:205–15.

[56] McGary KL, Park TJ, Woods JO, Cha HJ, Wallingford JB, Marcotte EM. Systematic discovery of non-obvious human disease models through orthologous phenotypes. Proc Natl Acad Sci U S A 2010;107:6544–9.

[57] Gissen P, Maher ER. Cargos and genes: insights into vesicular transport from inherited human disease. J Med Genet 2007;44:545–55.

[58] Rubinsztein DC. Protein-protein interaction networks in the spinocerebellar ataxias. Genome Biol 2006;7:229.

[59] Palikaras K, Tavernarakis N. *Caenorhabditis elegans* (Nematode). In: Brenner's encyclopedia of genetics. 2nd ed. Philadelphia: Elsevier; 2013. p. 404–8.

[60] No attributed author. Tuberous sclerosis complex in flies too? a fly homolog to TSC2, called gigas, plays a role in cell cycle regulation, Available from: http://www.ncbi.nlm.nih.gov/books/bv.fcgi?rid=coffeebrk.chapter.25; 2000.

[61] Howe K, Clark MD, Torroja CF, Torrance J, Berthelot C, Muffato M, et al. The zebrafish reference genome sequence and its relationship to the human genome. Nature 2013;496:498–503.

[62] Spitsbergen JM, Kent ML. The state of the art of the zebrafish model for toxicology and toxicologic pathology research—advantages and current limitations. Toxicol Pathol 2003;31(Suppl):62–87.

[63] Kelsh RN, Eisen JS. The zebrafish colourless gene regulates development of non-ectomesenchymal neural crest derivatives. Development 2000;127:515–25.

[64] Smolowitz R, Hanley J, Richmond H. A three-year retrospective study of abdominal tumors in zebrafish maintained in an aquatic laboratory animal facility. Biol Bull 2002;203:265–6.

[65] Wojciechowska S, van Rooijen E, Ceol C, Patton EE, White RM. Generation and analysis of zebrafish melanoma models. Methods Cell Biol 2016;134:531–49.

[66] Tobin DM, Vary Jr. JC, Ray JP, Walsh GS, Dunstan SJ, Bang ND, et al. The lta4h locus modulates susceptibility to mycobacterial infection in zebrafish and humans. Cell 2010;140:717–30.

[67] Curtis J, Kopanitsa L, Stebbings E, Speirs A, Ignatyeva O, Balabanova Y, et al. Association analysis of the LTA4H gene polymorphisms and pulmonary tuberculosis in 9115 subjects. Tuberculosis (Edinb) 2011;91:22–5.

[68] Jina Y, Heo K, Han W, Lim H, Wang M. Morphological and genetic characteristics of *Nicotiana langsdorffii*, *N. glauca* and its hybrid. EXCLI J 2005;4:25–33.

[69] Baker NE. Patterning signals and proliferation in *Drosophila* imaginal discs. Curr Opin Genet Dev 2007;17:287–93.

[70] Woodhouse E, Hersperger E, Shearn A. Growth, metastasis, and invasiveness of *Drosophila* tumors caused by mutations in specific tumor suppressor genes. Dev Genes Evol 1998;207:542–50.

[71] Hariharan IK, Bilder D. Regulation of imaginal disc growth by tumor-suppressor genes in *Drosophila*. Annu Rev Genet 2006;40:335–61.

[72] Whipple GH. A hitherto undescribed disease characterized anatomically by deposits of fat and fatty acids in the intestinal and mesenteric lymphatic tissues. Bull Johns Hopkins Hosp 1907;18:382–93.

[73] Relman DA, Schmidt TM, MacDermott RP, Falkow S. Identification of the uncultured bacillus of Whipple's disease. N Engl J Med 1992;327:293–301.

[74] Marth T, Roux M, von Herbay A, Meuer SC, Feurle GE. Persistent reduction of complement receptor 3 alpha-chain expressing mononuclear blood cells and transient inhibitory serum factors in Whipple's disease. Clin Immunol Immunopathol 1994;72:217–26.

[75] Iverson SA, Ostdiek S, Prasai S, Engelthaler DM, Kretschmer M, Fowle N, et al. Notes from the field: cluster of acute flaccid myelitis in five pediatric patients—Maricopa County, Arizona, 2016. Morb Mortal Wkly Rep 2017;66:758–60.

[76] Russell WMS, Burch RL. The principles of humane experimental technique. London: Methuen; 1959.

[77] Leaf C. Do clinical trials work? The New York Times; 2013 July 13.

[78] Kivela T. Trilateral retinoblastoma: a meta-analysis of hereditary retinoblastoma associated with primary ectopic intracranial retinoblastoma. J Clin Oncol 1999;17:1829–37.

[79] Van Slegtenhorst M, Nellist M, Nagelkerken B, Cheadle J, Snell R, van den Ouweland A, et al. Interaction between hamartin and tuberin, the TSC1 and TSC2 gene products. Hum Mol Genet 1998;7:1053–7.

[80] Kufe D, Pollock R, Weichselbaum R, Bast R, Gansler T, Holland J, Frei E, editors. Holland-Frei cancer medicine. ON, Canada: BC Decker; 2003.

9

Medical Proof of Evolution

Section 9.1 What Does Proof Mean, in the Biological Sciences?

America is the only country where a significant proportion of the population believes that professional wrestling is real but the moon landing was faked.

David Letterman

Those of us living in the United States have come to accept that a significant portion of our population does not believe that men have walked on the moon, or that climate change is real, or that the earth is billions of years old, or that vaccinations have medical value, or that evolution through natural selection accounts for all the millions of living species inhabiting planet earth. We often assume, without much evidence, that physicians, biologists, and scientists all believe in evolution; but this assumption is wrong. Some scientists are evolution skeptics. Some are outright evolution deniers. Others consider the theory of evolution to be a branch of philosophy, with no particular relevance to their own professional pursuits. Based on a personal sampling, conducted over the past decade, it seems that there are a considerable number of scientists who have no opinion on the subject. Whether evolution is right or wrong, fact or theory, the whole subject is of no concern to them, in any case.

For myself, the theory of evolution was proven to be true long ago, through the efforts of specialists in chronometry, who established the temporal succession of organisms in rock strata; thus confirming the work of zoologists, such as Charles Darwin, who studied the process of speciation through natural selection.

In the preface of this book, there appeared a quotation from the evolutionary biologist Theodosius Dobzhansky, who asserted that "Nothing in biology makes sense except in the light of evolution" [1]. The previous chapters of this book may have convinced readers that the stunning advances in molecular biology in the past 50 years, have fully validated Dobzhansky's claim. For readers who remain skeptical, or who simply enjoy a vigorous argument, this chapter is written to address three important issues.

Evolution's Clinical Guidebook. https://doi.org/10.1016/B978-0-12-817126-4.00009-6

- To discuss the meaning of biological proof.
- To prove the theory of evolution, drawing arguments from the field of medicine (and setting aside earlier proofs from the fields of chronometry, paleontology, geology, and zoology).
- To describe how our world would suffer, if the theory of evolution were wrong.

Let's begin by addressing the seemingly trivial question, "What is proof?" Strictly speaking, proof is something that a mathematician does. It is only in the heady world of mathematics that real proof is achieved. Most impressively, once a mathematical law is proven, it stays proven forever and must be obeyed by everyone, regardless of stature. If you believe in an all-powerful god, and if you believe in mathematical proof, then you must also believe that the all-powerful god must conform to the mathematical law, just like you and me. Such is the power of mathematics.

In the nonmathematical sciences, nothing is really proven, but the burden of trying to prove a theory is, nonetheless, much greater than anything faced by mathematicians. Here is a list of the various requirements of a credible biological theory.

–1.   The theory must conform to all known facts.

You cannot pick and choose the observations that ought to fit the theory. If there is one exception to the rule, then the rule is no longer valid.

–2.   The theory must make logical sense.

It is not sufficient for a theory to fit the facts. The theory must be plausible and must make sense.

–3.   All future observations are consistent with the theory.

The theory must be true forever.

–4.   The theory is testable, and all tests of the theory uphold the theory.

Unless there is a way to test the theory, by experiment or by observation, then the theory cannot be accepted. Untestable assertions can be dismissed as pseudoscience.

–5.   Predictions can be inferred from the theory.

A good theory allows us to predetermine the behavior of biological systems. If not, the theory has no practical purpose. When a theory fails to correctly predict a biological response, then the theory must be either modified or abandoned.

–6.   A theory must generate new theories.

New hypotheses, based on a proven theory, can be tested and proven, inspiring another generation of testable hypotheses. This is how science advances.

Although biological proof never achieves mathematical perfection, the theory of evolution achieves five of the six criteria for biological proof. Criteria number three ("All future observations are consistent with the theory") is a long reach. So far, the theory of evolution has withstood the test of time, and many scientists consider the theory of evolution to be as proven as any nonmathematical assertion may be.

Section 9.2 The Differences Between Designed Organisms and Evolved Organisms

Omnis cellula e cellula (all cells come from cells).

Commonly attributed to Rudolph Virchow, pathologist (1821–1902)

Cellular pathology was able to develop when it was demonstrated that the organism of every higher animal is a federation of elementary organisms, endowed with fundamental biological activities; it was the demonstration of the proliferation of the cells in the normal and pathological tissues (omnis cellula e cellula), which has given the greatest impulse to this doctrine.

Achilli Monti, 1900 [2]

Adherents of intelligent design, who do not believe in evolution, pose a fascinating argument, known as the "watchmaker analogy." If you look at an intricate device, such as a fine watch, you can pretty much assume that the watch was designed by a watchmaker. The watch could not have evolved its way into existence. The mechanisms and the craftsmanship and the subtle intricacies of a fine watch are beyond anything that evolution could achieve.

Looking at a watch, or a jet plane, or a television set, we can concede that these items could not have come into existence without a designer. Adherents of the theory of evolution would make the following two counterpoints:

–1.   There are fundamental differences between a designed object and an evolved object. In most cases, we can infer whether an object has been designed or whether it has been evolved, by looking for these differences.
–2.   It is possible for a designed object to be created by an evolved object. In a sense, even designed objects, such as watches, are the result of evolution.

In this section, we provide an argument that evolved objects (such as humans) are fundamentally different from designed systems (such as watches) and are burdened by a collection of diseases and biological conditions that simply could not occur in designed organisms (even if those organisms were purposely designed to be imperfect).

Designed organisms would not have malfunctioning subsystems inherited from ancestral species

Several years ago, a 24-year-old woman, complaining of nausea and vomiting, was admitted to a hospital in Shandong Province, China. Her medical history indicated that she had had life-long problems walking steadily. Her mother indicated that she could not walk until the age of seven, and that she had trouble articulating her speech, prior to the age of 6. Routine imaging of the head, at the time of admission, revealed that the patient lacked a cerebellum [3]. This case tells us that the cerebellum might have its benefits, but that a human can survive fairly well in its absence.

Be assured that the cerebellum is not a minor appendage to the brain. Though the cerebellum accounts for only 10% of the brain volume, it happens to contain more than three times the number of neurons than the cortex. The higher number of neurons in the cerebellum is accounted for by the increased density of neurons in cerebellar tissue. The neurons of the cerebellum are apparently quite busy, as judged by their high levels of metabolic activity. Indeed, more than half of the ATP consumed by the brain is utilized by the cerebellum [4] (Fig. 9.1).

A not-strictly-essential and oversized cerebellum is found in mammals other than humans. The Australian kelpie is a dog breed that is prone to a genetic disease known as cerebellar abiotrophy, wherein the cerebellum deteriorates some time after birth. Affected dogs may display loss of proper balance and a change in gait. Tremors and seizures may also develop. Some dogs having the same disease may suffer only mild defects and little disability.

No reasonable brain scientist would have guessed, based on our understanding of the cerebellum as an organ whose chief purpose is to fine-tune muscle movements, that the cerebellum needs to be as large as it is [5]. Why is it that an organ, that humans and dogs can survive fairly well without, contains more neurons than all the other parts of the brain, combined? Why would anyone design our brains in such a manner?

Of course, the reason for the cerebellum's disproportionate presence within our skulls has nothing to do with design. This is because we humans were never designed; we evolved. If we want to understand anything about the cerebellum, we need to study evolution.

Our ancient ancestors relied on their brains for motor control, navigation, and the various skills related to pursuing, seizing, and killing their prey. Hence, ancient species evolved a large cerebellum, to hold most of their neurons. *Homo sapiens*, the thinking man, devotes much of the day to sitting on the couch and watching sports on television. It seems that we have inherited a super-sized cerebellum that we don't strictly need.

Diseases that result from persistent traits, found throughout an ancestral lineage of animals, would not be observed in a designed animal.

In point of fact, many of the diseases that occur in humans are not the result of design flaws in our anatomy; they result from our inheritance of ancestral anatomic traits. We presume that these preserved traits served some useful purpose in an ancestor; otherwise the trait would not have evolved. As an example, let's look at our preposterous left recurrent laryngeal nerve.

The right and left recurrent nerves are branches of the vagus nerve that supply most of the laryngeal musculature. Without these nerves, we could not talk. Most small nerves

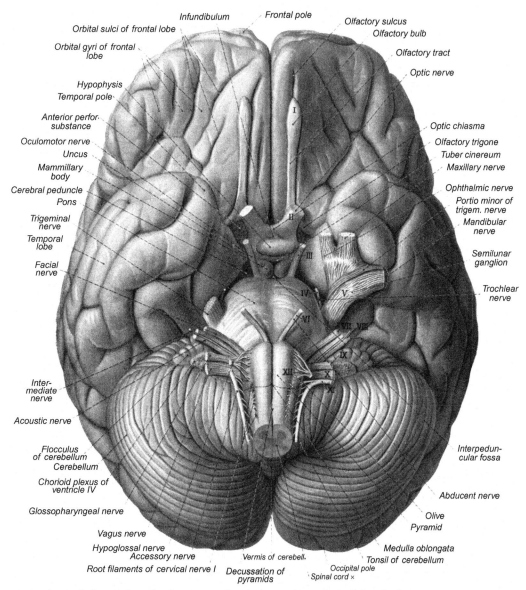

Infundibulum

Frontal pole

Olfactory sulcus

Orbital sulci of frontal lobe

Olfactory bulb

Orbital gyri of frontal lobe

Olfactory tract

Optic nerve

Hypophysis

Temporal pole

I

Anterior perfor. substance

Optic chiasma

Oculomotor nerve

Olfactory trigone

Uncus

Tuber cinereum

Mammillary body

Maxillary nerve

Cerebral peduncle

Ophthalmic nerve

Pons

Portio minor of trigem. nerve

II

Trigeminal nerve

Mandibular nerve

Temporal lobe

Semilunar ganglion

III

Facial nerve

Trochlear nerve

IV

V

VI

VII VIII

IX

Inter- mediate nerve

XII

X

X

Acoustic nerve

Flocculus of cerebellum

Interpedun- cular fossa

Cerebellum

Chorioid plexus of ventricle IV

Abducent nerve

Glossopharyngeal nerve

Olive

Pyramid

Vagus nerve

Hypoglossal nerve

Medulla oblongata

Accessory nerve

Vermis of cerebell.

Tonsil of cerebellum

Root filaments of cervical nerve I

Decussation of pyramids

Occipital pole

Spinal cord ×

FIG. 9.1 The cerebellum is the rather large appendage sitting underneath the left and right cerebri, in the lower third of the diagram, It is recognized by its long, thin, parallel, foliations. The cerebellum accounts for 10% of the human brain, by volume, but it doesn't rate as a vital organ. *Source: Wikipedia, from an illustration in Johannes Sobotta's "Human Anatomy," 1908.*

branch off their major nerve trunk and proceed directly for their target tissue. The recurrent nerves, due to the circumstances of their embryologic development, must swing below the major branches of the aortic vasculature. The right recurrent laryngeal nerve dives down the neck and loops below the right subclavian artery. The left recurrent laryngeal nerve dives down the neck, into the thorax, and under the aortic arch, near the heart. This nerve, due to its long and treacherous path, is vulnerable to cuts, torsions, compressions, and all manner of mayhem. Inadvertent bisection of the right recurrent laryngeal nerve is a leading cause of legal action taken against surgeons who specialize in neck and chest surgery. The same hazardous anatomic route, but many feet longer, applies to the giraffe, whose recurrent nerve must travel the full length of the neck, and down into the chest. The left recurrent laryngeal nerve is one, among many, reminders that animals are the product of evolution; not design [6].

The list of evolved flaws that living organisms must endure is quite long. Here are just a few [6, 7]:

- Whales have vestigial, and useless arms and legs, a consequence of their membership in, and descent from, Class Tetrapoda. Likewise, the vestigial femur and pelvis serves no useful purpose for the whale. They do, however, serve to inform us that whales evolved as a species of Class Mammalia that lived on land.
- The enzyme nitrogenase preferentially binds with acetylene over di-nitrogen, its intended substrate, despite its being the key enzyme used in nitrogen fixation in many bacteria and archaea.
- Birds evolved light, hollow bones for flight. These bones are of little value for flightless birds, such as penguins and ostriches, but a bird is a bird, whether it flies or runs, or waddles. Contrariwise, bats are flying mammals, descendants of Class Laurasiatheria, and are much more closely related to whales and giraffes than to birds. Consequently, flying bats have heavy bones, when they would certainly prefer a lighter-weight skeleton.
- All birds have wings, even the flightless ones. Given a design option, wouldn't ostriches prefer arms to wings?
- In Section 7.4, "The Evolution of Aging, and the Diseases Thereof," we learned that aging is an acquired trait. Along the line of evolution, an ancestral species found it advantageous to grow old and die. The descendants have been cursed by aging, ever since, while animals that evolved in a lineage that branched prior to the evolution of aging enjoy virtually unlimited life span.
- Humans, unlike our prognathous ancestors, have rather flat faces. As a consequence, our evolved teeth are crowded and we have poor sinus drainage. If this weren't bad enough, we inherited wisdom teeth, which we do not need and have no room for. To counteract our ancestral history, evolved humans designed dentists, who deal with our evolutionary shortcomings on a daily basis.
- The appendix, an evolutionary relic from ancestors that fermented their intestinal contents in alimentary holding sacs, serves no important purpose in humans.

The appendix contains some lymphoid tissue, but so does the every other part of the alimentary tract. In 2013, appendicitis accounted for 72,000 deaths worldwide. Aside from deaths due to appendiceal inflammation and rupture, the appendix is the site of a particularly aggressive, and sometimes fatal, variant of carcinoid tumor; the goblet cell carcinoid. Despite claims that the appendix is a useful part of the human anatomy, it is notable that persons from whom the appendix is removed never seem to miss it.

The types of errors occurring in designed systems are fundamentally different from the errors found in evolved systems.

Evolved systems have errors that served some useful purpose, in ancestral organisms. Errors in designed organisms are never ancestral (because designed organisms do not have ancestral species).

Evolved objects cannot just scrap individual subsystems and replace them with something new. Designed systems do this all the time.

Generations of individuals relied on wristwatches with a stempiece screw, mounted just above the "12" position. Every few days, we would wind the watch by turning the stem piece between our thumb and forefinger. It only took a few seconds of our day, but if we forgot, the clock would slow, then stop altogether. In a matter of a decade or two, stem-wound watches disappeared, replaced by newly designed watches with internal batteries.

In an evolved system, new traits do not appear in a single generation. They evolve in steps, over time, by modifying genes involved in component pathways. For example, there is simply no mechanism to evolve a feather, in one step, on an animal that had nothing previously growing from its skin. Presumably, feathers evolved from some proto-feather skin appendage. We can infer backwards that there was some precursor to the precursor, and so on. That is how evolution works. Similarly, all metabolic pathways developed from metabolic pathways that served some different purpose, before they were modified to serve the purpose that we can observe today. As a consequence, we must settle for a final product composed of physiological processes that were available to our ancestors.

Designed organisms would not have diseases affecting homologous parts of males and females

If humans were designed, we would expect to find parts of the male that were not included in females; and parts of females that were not included in males. Gender-specific anatomic parts would be determined by gender-specific functions. For example, a male might have a prostate and a penis and testicles, and these organs would be absent from females. A woman might have breasts and a uterus and a vagina and ovaries, and these parts would be absent in males. On a very superficial level, gender-specific anatomy would seem to be what we have, consistent with the assertion that humans were designed.

As we described in Section 3.1, "The Tight Relationship Between Evolution and Embryology," we find that male and female organs arise from the same embryonic structures. Hence, every male organ has either a homolog or a vestigial remnant in the female (e.g., the clitoris as the homolog of the penis). The same is true for female genitalia and adnexal organs, which have homologs in the male, or vestigial remnants in the male (e.g., nipples and breasts in males). The size and the function of homologous organs in males and females are accounted nearly entirely by gender-specific hormonal influences. A designed organism would not have a repertoire of vestigial organs, in males and females, that serve no function.

Designed species would not contain junk DNA

As described in Section 2.5, "Rummaging Through the DNA Junkyard," the human genome is full of DNA that serves no current function. The primary purpose of nonfunctional DNA is to serve as the fuel for evolution, by providing sequences that can be mutated to become functioning genes. A designed organism does not evolve and would not have a genome that is 90% nonfunctional.

Designed organisms do not speciate

A species is an evolving gene pool. As such, it's primary destiny is to speciate, and we see plenty of evidence of speciation in the many millions of living species that inhabit the earth. Because speciation is a product of evolution, we can infer that designed organisms do not speciate; they simply live the lives that they were designed to live.

Designed organisms would not conserve some genes, and not others

We see abundant evidence that some genes are conserved (i.e., their sequences are maintained with high fidelity over many generations), while other genes are allowed to degenerate over time. For example, by studying phylogenetic lineages, it has been shown that alpha-A crystallin, a protein that confers flexibility on the eye lens, is conserved in vertebrates, including most rodents. The rodent Spalax ehrenbergi has a subterranean habitat and is completely blind. In Spalax ehrenbergi, the gene coding for alpha-A crystallin is poorly conserved, mutating four times as fast as the same gene in sighted rodents [8].

In a designed species, we would expect every gene to perform its designed function. We would not expect to find the alpha-A crystallin gene in Spalax ehrenbergi, but if there were some reason to include such a gene in the design of a blind rodent, we would certainly expect the gene to be conserved.

In a designed organism, the genome is not a document of the organism's history

For today's evolutionary biologists and bioinformaticists, a cell's genome is a living historical document; a diary for the species [9]. By observing sequences and matching homologies, we can determine when a species split from an ancestor, when a species acquired a retrovirus, when a species acquired genes that cope with environmental pathogens. We can see how genes have adapted, over the ages, to perform new functions, and we can determine when those new functions were acquired.

It is not easy to read and comprehend the genome as if it were a document, and scientists are continually running up against obstacles, and misinterpreting their observations. Error is a prerequisite for scientific advancement. Nonetheless, as more data is obtained and as past errors are uncovered, the process clarifies and validated conclusions are drawn. Today, we have learned a great deal about biology and medicine by decrypting the genomic histories of many different organisms.

If organisms were designed and did not evolve from ancestral species, then the genome of an organism could not contain its evolutionary history and the knowledge and medical advancements that we have made, thereupon, would vanish.

Designed organisms cannot be positioned within our existing taxonomy

We did not always classify organisms based on evolution. Modern classification is based on finding relationships among species. In the classification of living organisms, these relationships are class traits that are passed from ancestral species to their descendants. As we have shown in our two phylogeny chapters (Chapters 5 and 6), we can trace the lineage of humans back to the first eukaryotes, and we can determine much of what we now know about the human genome by studying the genes, pathways, and diseases that evolved throughout our ancestry.

If humans and other species had not evolved from ancestral species, then the classification of organisms would no longer apply, and all the medical advances that issued from the classification would be lost. Instead, we would need to come up with some new classification, built on the relationships among the designed organisms that inhabit earth. To do so, we would need to know something about how the organisms were designed, what decisions were made regarding the parts to include and exclude, the purposes of the organisms, and so forth. Because living organisms are highly complex and because we know nothing about how or why such organisms may have been designed, it would be difficult to contrive a classification for intelligently designed life forms. Of course, we could group organisms by some feature of interest (e.g., size, function, type of habitat, and geographic location), but doing so would not produce a classification that generates new hypotheses, or that can be tested. Basically, we would have nothing more than a simple index of organisms.

Intelligently designed organisms would not have subsystems whose function is to correct design deficiencies in other subsystems

Evolved systems handle flaws differently than designed systems. Evolved systems are always adapting to new circumstances and such adaptations almost always involve a trade-off wherein one set of deficiencies are favored over some other set of deficiencies to enhance the survival of the species. Evolution neither seeks nor attains perfection. If perfection were attainable, then evolution would simply cease; and there is simply nothing to indicate that the process of evolution has ever taken a moment's respite.

Rather than eliminating every known flaw or limitation, as is the goal for designed systems, evolution rigs devices to cope with errors as they occur. For example, there are three different operational immune systems intended to protect humans from invading organisms and at least seven different DNA repair systems intended to preserve the integrity of the genome. Our cells have an elaborate self-destructive mechanism in place to commit cellular suicide when genomic damage is irreversible. **When we get sick and stand on the**

brink of death, we need not ask ourselves why we were designed to be so fragile. We are fragile because we were never designed; we simply evolved.

Section 9.3 What if Evolution Were Just a Foolish Fantasy

Whatever Nature has in store for mankind, unpleasant as it may be, men must accept, for ignorance is never better than knowledge.

Enrico Fermi

From careful observations about a few animals or plants we can make generaliza-tions about other species we have not yet encountered, and this ability to predict the unknown is what makes taxonomy a science rather than simply a technical pro-cedure. Recognizing that plants related to nightshades are likely to be poisonous is clearly valuable, whereas wrongly classifying bats as a kind of bird might entail a long and fruitless wait for them to lay eggs!

Peter C. Barnard [10]

Maybe we scientists got it all wrong. Maybe, despite all the evidence to the contrary, a flaw in our thinking led us astray. Perhaps it is true that natural selection cannot account for the existence of the human soul. Is it possible that we are missing the bigger picture? Scientists can only hope to understand what we can measure and verify. Our conclusions are always limited by what we can observe through our senses or through the use of instruments. Everything else is anybody's guess.

Nonetheless, much of what we have accomplished in the field of medicine has stemmed from the assumption that evolution by means of natural selection is a real thing. We can imagine the consequences if we were flat-out wrong and evolution were just a fan-tasy. Let's take a look and see how the world of medicine would fare in an evolution-less planet.

We could no longer infer gene function by identifying and studying homologous genes

The process of isolating a human gene and determining its nucleotide sequence sel-dom tells us much of anything about the gene's function in human cells. In many cases, the function of a gene is determined by finding the gene's homolog in another species (i.e., finding a gene that, like the human gene, was inherited from a common ancestral species). We presume that homologous genes will have similar function, and if we know the func-tion of the homolog in some other species, we can guess what the gene's function may be in humans.

As it happens, homologs are abundant throughout the many millions of species inha-biting the earth for the simple reason that all eukaryotic species have similar gene sets, inherited from a common eukaryotic ancestor (see Section 1.3, "Our Genes, for the Most Part, Come from Ancestral Species").

If evolution did not exist, then homologous genes would not exist, and everything that we have learned by studying the function of homologous genes would simply vanish.

Every newly encountered bacterial pathogen would be inscrutable

Let's imagine that evolution is just nonsense, phylogenetic relationships do not exist and that the taxonomy of living organisms is simply wrong. If you are a thoroughly modern expert in bioinformatics, you might think that you can reconstruct an alternative taxonomy based on examining a database of genomic sequences for each organism and then clustering organisms based on sequence similarities.

Imagine an experiment wherein you take DNA samples from every organism you encounter: bacterial colonies cultured from a river, unicellular nonbacterial organisms found in a pond, small multicellular organisms found in soil, crawling creatures dwelling under rocks, and so on. You own a powerful sequencing machine that produces the full-length sequence for each sampled organism and you have a powerful computer that sorts and clusters every sequence. At the end, the computer prints out a huge graph, wherein groups of organisms with the greatest sequence similarities are clustered together. You may think you've created a useful classification, but you haven't really, because you don't know anything about the properties of your clusters. You don't know whether each cluster represents a species, or a class (a collection of related species), or whether a cluster may be contaminated by organisms that share some of the same gene sequences, but are otherwise unrelated (i.e., the sequence similarities result from chance or from convergence, but not by descent from a common ancestor). The sequences do not tell you very much about the biological properties of specific organisms, and you cannot infer which biological properties characterize the classes of clustered organisms. You have no certain knowledge whether the members of any given cluster of organisms can be characterized by any particular gene sequence (i.e., you do not know a characterizing gene sequences that applies to every member of a class, and to no members of other classes). You do not know the genus or species names of the organisms included in the clusters, because you began your experiment without a presumptive taxonomy. Basically, you simply know what you knew before you started; that organisms have unique gene sequences that can be grouped by sequence similarity. Such groupings are biologically uninformative and do not establish a useful classification. [Glossary Phenetics, Non-phylogenetic property]

You may have begun your project with a small database of a few hundred organisms, but there are trillions of organisms on earth and as you add additional samples, you'll find that the clusters computed by your similarity-based computer analysis change every time you run the program. You quickly learn that similarities among sequences do not establish the identify of species. To your horror, you find that you have lost the concept of a species (i.e., an evolving gene pool) when you determined that evolution was a hoax.

Don't despair. Pre-Darwinian zoologists prepared a taxonomy of animals that looks much like our modern taxonomy, and they managed to accomplish their task without knowing anything about evolutionary theory. They did, however, have a keen, intuitive understanding of the essential features that define a species and that relate one species with another.

They sought and found anatomic and embryologic features that characterized classes of organisms, and they grouped their species in a class hierarchy that closely matches an ancestral phylogenetic hierarchy. All biologically valid classifications must be equivalent because they all encapsulate the same set of biological properties that define their members [11]. The observation that the zoological/paleontological classification of animals looks just like the phylogenetic classification of animals serves to validate both approaches.

Let's return to where we stood in the mid-twentieth century, when bacterial taxonomy was in a miserable state. Infectious organisms were classified by their growth in different culture media, by the staining properties (e.g., gram positive, gram negative, and acid fast), by the speed of their growth, by the temperature at which they grew, and by whether they produced pigment. Along with morphology, all these diagnostic procedures were helpful in identifying pathogenic bacteria, but they did not help the taxonomists understand the biological relationships among different organisms. [Glossary Classification system versus identification system]

When molecular biologists developed methods for sequencing the genomes of various organisms, there was hope that bacterial classification could be based on genomic sequencing. This hope was somewhat deflated when it was soon determined that horizontal gene transfer (i.e., the exchange of genetic material among different species) was rampant in the bacterial kingdom. The fundamental concept of a species is that of an evolving gene pool. If the gene pool is constantly being flooded with the genes of other species, then there is no stable gene pool. At the time, it was feared that bacteria were just a mish-mash of genes popping in and out of organisms, with no stable genetic composition and no stable classification. Bacterial taxonomy seemed hopeless.

In 1977, the field of bacterial taxonomy changed for the better when Carl Woese and George E. Fox announced that there existed a class of bacteria that contained species that were demonstrably different from all other species of prokaryotes. They named these bacteria Archaebacteria, later known as Archaea (from the Greek meaning original or first in time); the name indicating the Archaea predated all other classes of bacteria. When you compare species of Class Eubacteria with species of Class Archaea, you're not likely to notice any big differences. The eubacteria have the same shapes and sizes as the archaea. All species of eubacteria and all species of archaea are single-celled organisms, and they all have a typical prokaryotic structure (i.e., lacking a membrane-bound nucleus to compartmentalize their genetic material). As it happens, Class Eubacteria contains all of the bacterial organisms that are known to be pathogenic in humans. The archaeans, so far as we can currently tell, are nonpathogenic. Many archaeans are extremophiles, capable of living in hostile environments (e.g., hot springs, salt lakes), but some Archaean species occupy less demanding biological niches (e.g., marshland, soil, and human colon). Class Archaea does not hold a monopoly on extremophilic prokaryotes; some eubacterial species live in extreme environments (e.g. the alkaliphile *Bacillus halodurans*).

Woese and Fox showed that despite ongoing horizontal gene transfer in ancient and modern prokaryotes, there are fundamental differences among prokaryotes that establish excellent criteria for assigning biological classes. Differences in the sequence and structure of ribosomal RNA (the 16sRNA component in particular) distinguish archaean species from all other bacteria [12]. Furthermore, these differences between classes can be exploited to establish the different subclasses of prokaryotic organisms, in the chronologic order in which they evolved. It would seem that prokaryotes never swap their ribosomal RNA and that prokaryotes exist as valid species (i.e., evolving gene pools). Later studies indicate that genes other than ribosomal RNA serve as keys to the phylogenetic organization of prokaryotes, if we just look for them [13].

There are many millions of bacterial species and we will never be able to identify all of them, but now that we have a fairly accurate way of assigning species to phylogenetic classes, we can do a reasonable job determining any organism's genus (i.e., parent class of the species) by sequencing its 16sRNA [14–17]. Of course, there are limitations to this technique, but when we combine our analysis of 16sRNA with our accumulated knowledge of the morphology and growth characteristics of the class, we can often arrive at a helpful diagnosis [15, 18, 19]. [Glossary Long branch attraction]

If evolution were discredited, then we would lose our ability to classify microorganisms; we would lose the concept of "genus"; and sequence data obtained from pathogenic organisms would be inscrutable and medically unhelpful.

We would have no rational way of selecting model organisms for new drug development and testing

As discussed in Section 8.3, "New Animal Options," medical research has advanced to a stage where many of the most important questions that we ask are best studied in organisms other than rodents. The proverbial question: "What caused this disease," has been supplanted by the far more intellectually challenging question, "What is the sequence of cellular events and pathways that have produced this disease, and how might we intervene in the process and stop this disease in an early stage of its development?" [20].

When we begin to dissect diseases into their developmental steps and pathways, we enter a new realm, wherein studies involving individual cells (e.g., organelles and metabolic pathways observed in cell cultures), inherited molecular pathways found in simple organisms (e.g., single-cell eukaryotes, flies, worms, yeast), and evolved embryonic tissues (e.g., the neural crest in all craniates) are more easily controlled, repeated, and validated than studies conducted in rodents.

If all living organisms were designed and did not evolve from ancestral species, then there would be no rational way of choosing organisms for the study of human diseases. It would be anybody's guess which organism, if any, mimicked a human metabolic pathway.

We would no longer be able to find treatments for taxonomic classes of organisms

There are about 1400 fairly well studied microorganisms that are known to cause infectious diseases in humans, but this number is quite deceptive, insofar as every year brings us a fresh new crop of pathogens. As a case in point, let's look at the infectious fungi.

Approximately 54 fungi account for the vast majority of fungal infections, but the true number of fungi that are pathogenic in humans is much higher. Fungi happen to be a ubiquitous presence in air and in water. It is estimated that, on average, humans inhale about 40 conidia (spores from Class Ascomycota) each hour. Most of these organisms are nonpathogenic under normal circumstances. However, in the case of immunocompromised patients, or in the case of patients who provide a specific opportunity for ambient fungi to attach and grow within a body (e.g., an indwelling vascular line), an otherwise harmless fungus may produce a life-threatening illness. It is estimated that there are about 20 new fungal diseases reported each year [21]. If the number of diseases caused by other types of organisms (i.e., bacteria, single-cell eukaryotes, animals, viruses, and prions) were to remain steady, then it will not be long before the number of different fungal diseases exceeds the total number of described diseases produced by all other organisms.

As it happens, the variety of infections of all types is increasing as our diagnostic techniques improve. It is now possible to identify heretofore undiagnosed cases of pathogenic species [22]. In the past, when clinical laboratories lacked the sophisticated tests available today, many pathogens were simply missed. For example, *Aspergillus fumigatus* was the presumptive cause of aspergillosis arising as severe pulmonary infections in immunocompromised patients. With advanced typing techniques, an additional 34 species of *Aspergillus* have been isolated from the clinical specimens [21].

The number of known infectious organisms is climbing at a rate that vastly outpaces our ability to find new organism-specific treatments. Luckily for us, we have options. Because each species of infectious organism is phylogenetically classified, we can infer that every species within a class will have much the same pathways and physiologic processes and will be amenable to the same treatments. This generalization is not always true, most notably so when a strain within a species develops antibiotic resistance, but it has worked well enough to save countless lives.

As just one example of the value of taxonomy, let's consider a recent manuscript reporting the efficacy of proteosome inhibitors in the treatment of three common diseases that are together responsible for a good portion of the morbidity and mortality of infections diseases, worldwide [23]. These three are Chagas disease, sleeping sickness, and leishmaniasis. *Trypanosoma cruzi* is the cause of Chagas disease, also known as American trypanosomiasis. Chagas disease affects about eight million people [24]. *Trypanosoma brucei* is the cause of African trypanosomiasis (sleeping sickness). It has been reported that *Trypanosoma brucei* accounts for about 50,000 deaths each year. *Leishmania* species cause leishmaniasis, a disease that infects about 12 million people worldwide. Each year, about 60,000 people die from the visceral form of the disease. How did researchers suspect that these three organisms would be susceptible to the same drug? Taxonomy holds the answer. All three organisms descend from the same grandparent class, Trypanosomatida.

```
Trypanosoma cruzi, the root cause organism of Chagas disease
Order: Trypanosomatida
Genus: Trypanosoma
Species: T. cruzi

Trypanosoma brucei, the root cause organism of sleeping sickness
Order: Trypanosomatida
Genus: Trypanosoma
Species: T. brucei
```

```
Leishmania donovani, the root cause organism of leishmaniasis
Order: Trypanosomatida
Genus: Leishmania
Species: L. donovani
```

Notice that L. *donovani* has a different parent class (i.e., *Leishmania*) than *T. cruzi* and *T. brucei* (i.e., *Trypanosoma*). Nonetheless, all three organisms have the same grandparent class (Trypanosomatida), and all three organisms inherit the class properties associated with Trypanosomatida (Fig. 9.2).

Without a phylogenetic taxonomy, we would have no rational method for finding drugs that treat classes of organisms, and their descendants. When we work through the phylogenetic ancestry of organisms, we can most easily find the closely related protein families, molecular pathways, and cellular functions for known genes [13] and can use this knowledge to prepare treatments targeted to classes of organisms.

If the species of microorganisms were individually designed and did not arise through evolution-driven speciation, then we would have no rational way of developing treatments that might apply to multiple species. Every new cure, assuming we could find any cures, would be tailored to a single organism. We would have no way to predict which additional organisms might be susceptible to treatment, if any.

Diseases would be diagnosed by symptoms; not by biological principles

Suppose you are a statistician and are magically ported through time and space to Southern Italy, in the year 1640, where people are dying in great number of malaria, and you are a doctor trying to cope with the situation. You're not a microbiologist, but you know something about designing clinical trials and one of the local cognoscenti

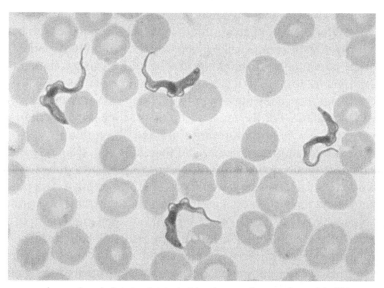

FIG. 9.2 Trypanosomes, observed as dark wavy flakes, in blood smear from patient with African trypanosomiasis. *Source: The US Centers for Disease Control and Prevention public domain image.*

has just given you an herb. "Take this drug today, and your fever will be gone by the next morning," he tells you. As it happens, the herb is an extract of bark from the Cinchona tree, recently imported from Brazil. It is a surefire cure for malaria, a disease endemic to the region. But you don't know any of this. Before you start treating your patients, you'll want to conduct a clinical trial.

At this time, physicians knew nothing about the pathogenesis of malaria. Current thinking was that it was a disease caused by breathing in insalubrious swamp vapors; hence the word roots "mal" meaning bad, and "aria" meaning air. You have just been handed a substance derived from the Cinchona tree, but you do not trust the herbalist. Insisting on a rational approach to the practice of medicine you design a clinical trial using 100 patients, all of whom have the same symptoms (delirium and fever) and all of whom carry the diagnosis of malaria. You administer the cinchona powder, also known as quinine, to all the patients. A few improve, but most don't. You recall that the symptoms of malaria wax and wane, with the fever subsiding on its own every few days. It is not uncommon for malaria victims to recover, without any treatment. Everything considered, you call the trial a washout. You decide not to administer quinine to your patients.

What happened? We know that quinine was a miracle cure for malaria. It should have been effective in a population of 100 patients. The problem with this hypothetical clinical trial is that the patients under study were assembled based on their mutual symptoms: fever and delirium. These same symptoms could have been accounted for by any of hundreds of other diseases that were prevalent in England at the time. The criteria employed at the time to render a diagnosis of malaria was imprecise, and the trial population was diluted with non-malarial patients who were guaranteed to be nonresponders. Consequently, the trial failed and you missed a golden opportunity to treat your malaria patients with quinine.

It isn't hard to imagine present-day dilemmas not unlike our fictitious quinine trial. If you are testing the effectiveness of an antibiotic on a class of people with bacterial pneumonia, the accuracy of your results will be jeopardized if your study population includes subjects with viral pneumonia, or smoking-related lung damage. The consequences of class blending are forever with us. In regions where malaria is endemic, it is commonplace to initiate antimalarial treatment for any individual with fever, without direct evidence that the patient has malaria. For the most part, this practice prevails wherever resources and expertise are scarce. When diseases are classified by symptoms, it is impossible to conduct rational trials for appropriate targeted therapies [25]. [Glossary Blended class]

Class noise refers to inaccuracies (e.g., misleading results) introduced in the analysis of classified data due to errors in class assignments (e.g., assigning a data object to class A when the object should have been assigned to class B). Many new and promising therapeutic technologies have yielded little or no scientific advancements simply because their practitioners have paid insufficient attention to the issue of class noise.

The past half century has witnessed remarkable advances in the field of brain imaging, including the introduction of computed tomography and nuclear magnetic resonance imaging. Scientists can now determine the brain areas that are selectively activated for specific physiologic functions. These imaging techniques include: positron emission tomography, functional magnetic resonance imaging, multichannel electroencephalography, magnetoencephalography, near infrared spectroscopic imaging, and single photon emission computed tomography. With all of these available technologies, you would naturally expect that neuroscientists would be able to correlate psychiatric conditions with abnormalities in anatomy or function, mapped to specific areas of the brain. Indeed, the brain research literature has seen hundreds, if not thousands of early studies purporting to find associations that link brain anatomy to psychiatric diseases. Alas, none of these early findings have been validated. Excluding degenerative brain conditions (e.g., Alzheimer's disease, Parkinson's disease), there is, at the present time, no psychiatric condition that can be consistently associated with a specific functional brain deficit or anatomic abnormality [26]. The reasons for the complete lack of validation for what seemed to be a highly promising field of research, pursued by an army of top scientists, is a deep and disturbing mystery.

In 2013, a new version of the diagnostic and statistical manual of mental disorders (DSM) was issued. The DSM is the standard classification of psychiatric disorders and is used by psychiatrists and other healthcare professionals worldwide. The new version was long in coming, following its previous version by 20 years. Spoiling the fanfare for the much-anticipated update was a chorus of loud detractors, who included among their ranks a host of influential and respected neuroscientists. Their complaint was that the DSM classifies diagnostic entities based on collections of symptoms; not on biological principles. For every diagnostic entity in the DSM, all persons who share the same collection of symptoms will, in most cases, be assigned the same diagnosis; even when the biological cause of the symptoms are unknown or unrelated.

When individuals with unrelated diseases, are studied together, simply because they have some symptoms in common, the results of the study are unlikely to have any validity [25]. A former Director of the National Institute of Mental Health, was quoted as saying, "As long as the research community takes the DSM to be a bible, we'll never make progress" [27].

We can not have class-related discoveries, class-based clinical trials, and class-targeted medical treatments until we had true and valid biological classes of diseases. The fact that we are now achieving all these important goals is due to our classification of living organisms, which is based on evolutionary relationships and which provides us with a rational basis to classify all biological systems, including molecular pathways [28], embryologic development, cellular physiology, and every type of disease.

If evolution did not exist, and organisms could not be classified, then we would have no rational basis for classifying anything in the biological sciences, including diseases. All medical science would struggle in the same quagmire as we currently see in the field of psychiatry, wherein diseases are grouped by symptom, not biological relationships.

There would be no ongoing evolution (which we see all around us)

Returning one last time to the Starship Enterprise, in the episode entitled "Transfigurations," the crew played host to the birth of a new species. An alien life form, made of flesh and bone, endured a painful metamorphosis, as a novel mutation transformed him into the first member of a new energy-only species. As discussed in Section 4.2, "The Biological Process of Speciation," this is simply not how the speciation process unfolds here on earth.

The first members of any new species are indistinguishable from the members of the parent species. The defining feature of any new species is its evolving gene pool. Just as long as a subpopulation of a species maintains the integrity of its own gene pool (i.e., does not mate with member of the parent species), the new species will eventually develop a characteristic set of physical traits.

Because we cannot observe the creation of the first organisms of a new species (there really is nothing to observe), advocates of intelligent design have concluded that the biological process of speciation is a fiction and that all species of animals came into existence at one moment in time, by design.

Despite objections to the contrary, speciation is a proven phenomenon. If we look at fossils in rock strata that accumulated in chronological order, over time, we see that species appear at a particular moment, and not before; and that the species persist from that moment on until such time as they are no longer found. After a species disappears from the rock strata, we see that species resembling the extinct species emerge; the presumed descendant species. We never ever see fossils whose position in the rock strata is out of sequence. For example, we never see examples of modern birds in the same layer as we find trilobites. We never see trilobite fossils mixed into the strata where dinosaurs dwell. Species appear and disappear, in accordance with a strict chronological schedule. These observations, taken alone, are ample proof of speciation.

While evidence of new species materializing before our very eyes is a phenomenon known only to science fiction aficionados, we see evidence of evolution everywhere around us, every day. For examples, rattlesnakes in the western United States had been suffering a population decline resulting from their enthusiastic use of the rattle. Evolution provided rattlesnakes with a mechanism intended to instill fear in their enemies and their prey. Modern-day humans, on hearing the tell-tale rattle, reach for their guns and shoot the clueless snake. Consequently, a new breed of rattle-free rattlesnake is repopulating the land. Under modern circumstances, the snakes most likely to survive a chance encounter with an armed human have evolved to keep quiet.

Likewise, there are many examples of species evolving to cope with climate change and other environmental stresses. It is useful to remember that a species is an evolving gene pool. Wherever you have a species, you can expect to find evolution.

Suppose this were not the case and that evolution were a fantasy. In that case, every species would be stuck in place; unable to benefit from any of the genomic variation in its population and unable to produce variant organisms within its species (i.e., within its gene pool).

Medical research would be in the hands of those who have promoted the countervailing theory of intelligent design

All the living organisms on earth must have arrived here somehow. If evolution was scientifically discredited, along with this book, then our fallback paradigm would be intelligent design. Scientists who support intelligent design believe that some powerful being (such as God) or beings (such as an advanced alien race) used their intelligence to design all the species that we see on earth today. If evolution were false, and intelligent design were true, then the same scientists who have championed intelligent design will take the lead roles in medical research.

Until now, intelligent design scientists have made no contribution to medical science. There are no published manuscripts introducing new and effective medicinal treatments based on the principles of intelligent design. In a world of the internet, where anyone can publish anything they like, the intelligent design community has conducted no laboratory experiments based on the theory of intelligent design. Nor have they conducted clinical trials, or outcome analyses, based on the theory of intelligent design. The theory of intelligent design has not inspired any new medical advances.

If it happens that evolution is just a foolish fantasy, then all future medical research will be in the hands of intelligent design proponents, a group with an unblemished record of nonproductivity.

Glossary

Blended class Also known as class noise. Blended class refers to inaccuracies (e.g., misleading results) introduced in the analysis of data due to errors in class assignments (e.g., inaccurate diagnosis). If you are testing the effectiveness of an antibiotic on a class of people with bacterial pneumonia, the accuracy of your results will be forfeit when your study population includes subjects with viral pneumonia, or smoking-related lung damage. Errors induced by blending classes are often overlooked by data analysts who erroneously assume that the experiment was designed to ensure that each tested population group is composed of a single, valid class of subjects [29].

Classification system versus identification system It is important to distinguish classification systems from an identification systems. An identification system matches an individual organism with its assigned species name. Identification is based on finding several features that, taken together, can help determine the name of an organism. For example, if you have a list of characteristic features: large, hairy, strong, African, jungle-dwelling, knuckle-walking; you might correctly identify the organisms as a gorilla. These identifiers are different from the phylogenetic features that were used to classify gorillas within the hierarchy of organisms (Animalia: Chordata: Mammalia: Primates: Hominidae: Homininae: Gorillini: Gorilla). Specifically, you can identify an animal as a gorilla without knowing that a gorilla is a type of mammal. Conversely, you can classify a gorilla as a member of Class Gorillini without knowing that a gorilla happens to be large.

One of the most common mistakes in the biological sciences is to confuse an identification system with a classification system. The former simply provides a handy way to associate an object with a name; the latter is a system of relationships among objects.

Long branch attraction When gene sequence data is analyzed and two organisms share a similar sequence in a stretch of DNA, it can be very tempting to infer that the two organisms have a common ancestor. This inference is not necessarily correct. Because DNA mutations arise stochastically over time, two species with different ancestors may achieve the same sequence in a chosen stretch of DNA, by chance alone (i.e., not through inheritance). When mathematical phylogeneticists began modeling inferences for gene data sets, they assumed that most such errors would occur when the branches between compared classes were long (i.e., when a long time elapsed between evolutionary divergences). They called this phenomenon, wherein non-sister taxa were mistakenly assigned the same ancient ancestor class, based on sequence similarities, "long branch attraction." In practice, errors of this type can occur whether the branches are long, or short, or in-between. Over the years, the accepted usage of the term "long branch attraction" has been extended to just about any error in phylogenetic grouping due to gene similarities acquired through any mechanism other than inheritance from a shared ancestor. This would include random mutational and adaptive convergences [30].

Non-phylogenetic property Properties that cannot be used by taxonomists to build a class structure. For example, we do not classify animals by height, or weight because animals of greatly different heights and weights may occupy the same biological class. Similarly, animals within a class may have widely ranging geographic habitats; hence, we cannot classify animals by locality. Case in point: penguins can be found virtually anywhere in the southern hemisphere, including hot and cold climates. Hence, we cannot classify penguins as animals that live in Antarctica or that prefer a cold climate. Scientists commonly encounter properties, once thought to be class-specific that prove to be uninformative for classification purposes. For many decades, all bacteria were assumed to be small; much smaller than animal cells. However, the bacterium *Epulopiscium fishelsoni* grows to about 600 μm by 80 μm, much larger than the typical animal epithelial cell (about 35 μm in diameter) [31]. *Thiomargarita namibiensis*, an ocean-dwelling bacterium, can reach a size of 0.75 mm, visible to the unaided eye. What do these admittedly obscure facts teach us about the art of classification? Superficial properties, such as size seldom inform us how to classify objects. The ontologist must think very deeply to find the essential phylogenetic properties that serve as defining features of classes.

Phenetics The classification of organisms by feature similarity, rather than through relationships. Taxonomists, who have long held that a species is a natural unit of biological life and that the nature of a species is revealed through the intellectual process of building a consistent taxonomy, are generally opposed to the process of phenetics-based classification [32].

References

[1] Dobzhansky T. Nothing in biology makes sense except in the light of evolution. Am Biol Teach 1973;35:125–9.

[2] Monti A. The fundamental data of modern pathology. London: The New Sydenham Society; 1900.

[3] Thomson H. Woman of 24 found to have no cerebellum in her brain. New Scientist; 2014. September 10.

[4] Howarth C, Gleeson P, David Attwell D. Updated energy budgets for neural computation in the neocortex and cerebellum. J Cereb Blood Flow Metab 2012;32:1222–32.

[5] Marcus G. The trouble with brain science. The New York Times; 2014. July 11.

[6] Selim J, Aguilera-Hellweg M. Useless body parts: what do we need sinuses for, anyway? Discover Magazine; 2004. June 26.

[7] Siebert C. Unintelligent design. Discover Magazine; 2006. March 15.

[8] Land MF, Fernald RD. The evolution of eyes. Annu Rev Neurosci 1992;15:1–29.

[9] Zuckerkandl E, Pauling L. Molecules as documents of evolutionary history. J Theor Biol 1965;8:357–66.

[10] Barnard PC. The royal entomological society book of British insects. Sussex, England: Wiley-Blackwell; 2011.

[11] Mayr E. The growth of biological thought: diversity, evolution and inheritance. Cambridge: Belknap Press; 1982.

[12] Woese CR, Fox GE. Phylogenetic structure of the prokaryotic domain: the primary kingdoms. PNAS 1977;74:5088–90.

[13] Wu D, Hugenholtz P, Mavromatis K, Pukall R, Dalin E, Ivanova NN, et al. A phylogeny-driven genomic encyclopaedia of Bacteria and Archaea. Nature 2009;462:1056–60.

[14] Schoenborn L, Abdollahi H, Tee W, Dyall-Smith M, Janssen PH. A member of the delta subgroup of proteobacteria from a pyogenic liver abscess is a typical sulfate reducer of the genus *Desulfovibrio*. J Clin Microbiol 2001;39:787–90.

[15] Janda JM, Abbott SL. 16S rRNA gene sequencing for bacterial identification in the diagnostic laboratory: pluses, perils, and pitfalls. J Clin Microbiol 2007;45:2761–4.

[16] Woo PC, Lau SK, Teng JL, Tse H, Yuen KY. Then and now: use of 16S rDNA gene sequencing for bacterial identification and discovery of novel bacteria in clinical microbiology laboratories. Clin Microbiol Infect 2008;14:908–34.

[17] Barlett DL, Steele JB. Monsanto's harvest of fear. Vanity Fair; 2008. May.

[18] Srinivasan R, Karaoz U, Volegova M, MacKichan J, Kato-Maeda M, Miller S, et al. Use of 16S rRNA gene for identification of a broad range of clinically relevant bacterial pathogens. PLoS ONE 2015;10:e0117617.

[19] Wang Q, Garrity GM, Tiedje JM, Cole JR. Naive Bayesian classifier for rapid assignment of rRNA sequences into the new bacterial taxonomy. Appl Environ Microbiol 2007;73:526–5267.

[20] Berman JJ. Precancer: the beginning and the end of cancer. Sudbury: Jones and Bartlett; 2010.

[21] Guarro J, Gene J, Stchigel AM. Developments in fungal taxonomy. Clin Microbiol Rev 1999;12:454–500.

[22] Pounder JI, Simmon KE, Barton CA, Hohmann SL, Brandt ME, Petti CA. Discovering potential pathogens among fungi identified as nonsporulating molds. J Clin Microbiol 2007;45:568–71.

[23] Khare S, Nagle AS, Biggart A, Lai YH, Liang F, Davis LC, et al. Proteasome inhibition for treatment of leishmaniasis, Chagas disease and sleeping sickness. Nature 2016;537:229–33.

[24] Rassi Jr. A, Rassi A, Marin-Neto JA. Chagas disease. Lancet 2010;375:1388–402.

[25] Committee on a Framework for Developing a New Taxonomy of Disease, Board on Life Sciences, Division on Earth and Life Studies, National Research Council of the National Academies. Toward precision medicine: building a knowledge network for biomedical research and a new taxonomy of disease. Washington, DC: The National Academies Press; 2011.

[26] Borgwardt S, Radua J, Mechelli A, Fusar-Poli P. Why are psychiatric imaging methods clinically unreliable? Conclusions and practical guidelines for authors, editors and reviewers. Behav Brain Funct 2012;8:46.

[27] Belluck P, Carey B. Psychiatry's guide is out of touch with science, experts say. The New York Times; 2013. May 6.

[28] National Academy of Sciences, Engineering, and Medicine. Enabling precision medicine. The role of genetics in clinical drug development: proceedings of a workshop. Washington, DC: The National Academies Press; 2017.

[29] Berman JJ. Data simplification: taming information with open source tools. Waltham, MA: Morgan Kaufmann; 2016.

[30] Bergsten J. A review of long-branch attraction. Cladistics 2005;21:163–93.

[31] Angert ER, Clements KD, Pace NR. The largest bacterium. Nature 1993;362:239–41.

[32] DeQueiroz K. Ernst Mayr and the modern concept of species. PNAS 2005;102(Suppl. 1):6600–7.

Index

Note: Page numbers followed by *f* indicate figures.